MACHAUT'S WORLD:

Science and Art in the Fourteenth Century

t se tu nies ceuls tresbien congnoissans
onme sont seurs Retorique et musique
as seurs auras ton engin enforme
e tout ce que tu voudras conforme
etorique nauva Rien enferme
ne ne tenroit en guerre ne en Riuez
t musique te douua chans
aur que voudras auers et de mains
nsi tu fais seront faigne
ar faure ne peus estre faillaye
ar tu as seus Retorique et musique
y fait seront plus quatre Ronime
ml mauua Rene qui fuct a blasmer
t si seront de toutes Riens ame
oudez soyant toles qui sans amer
ono ce beul que sores enerans
enfaire asses petis moyens a grans
et fay tost si ty applique
n nen neu Dois pas estre resisant
m te bail seus Retorique et musyque

omment guillaume de machaut Respont

E natire
ieue ne me Dit esprises ne Refuse
ue ne face le bon commandement
e Bous Dame se ie vous fay entendu
ar qui las trop bie a entendement
out Drois est quant bons cueur deurengs
faire Des amoureux ordenec
na ce faire ie me soubnie
t ie Bueil bien estre a ce fait Donnee
ant quen ce monde done plaire se me
ais sceusant faut nose croue entreprendre
e ie nauoie diles mes presentement
es trois enfans pour mon Duy vo aprendre
om Dit manes roy presentement
t Dist que ainsi me Bounoumet
nees de moy puis de me tous nates
ne auus ne autre chose sontune
Je ne naury y ne men pouruees
ant quen ce monde Bous plaurd festine
y me Bueil Dame Dntone mettre en tas Dre
us Dittes faure amonuissement
et de plusieurs luy quant e sand mielDre

omme natire voullant orendroit plus que on
ques mais honeller et faire essaucer les biens et
honneurs qui sont en amours. Bient a guillaume
De machant et ly ordonne et encharge a faire sur
ce nounianx Dis amoureux. Et luy baille
pour luy aidier et conseillier a ce faire trois
De ses enfans. Cest assauour. Sens. Reto
rique et musique. et luy Dit par ceste ma
mere.

E natire par qui tout est fourme
Enhaudie et en hue et en terre denpar
Bieus qui a tou puille qui sontune
Lay apparte pour faire par toy fourmer
Nouuianlx Dis Bons amoureux et plasa
Douuce bail ct auis Dames enfans
Qu ty teuDrvont la ce y la pratique

ANNALS OF THE NEW YORK ACADEMY OF SCIENCES

VOLUME 314

MACHAUT'S WORLD:
Science and Art in the Fourteenth Century

Edited by MADELEINE PELNER COSMAN AND BRUCE CHANDLER

THE NEW YORK ACADEMY OF SCIENCES

NEW YORK, NEW YORK
1978

LIBRARY OF CONGRESS CATALOGING IN PUBLICATION DATA

Conference on Machaut's World: Science and Art in the Fourteenth Century, New York Academy of Sciences, 1977.
    Machaut's world.

    (Annals of the New York Academy of Sciences; v. 314 ISSN 0077-8923)
    Held on Dec. 2-3, 1977; cosponsored by the Institute for Medieval and Renaissance Studies, City College of the City University of New York and others.
    Bibliography: p.
    1. Guillaume de Machaut, d. 1377. 2. Science, Medieval—Congresses. 3. Art, Medieval—Congresses. I. Cosman, Madeleine Pelner. II. Chandler, Bruce, 1931-III. New York Academy of Sciences. IV. New York (City). City College. Institute for Medieval and Renaissance Studies. V. Title. VI. Series: New York Academy of Sciences. Annal; v. 314.
Q11.N5 vol. 314   508'.ls [509'.023]
ISBN 0-89072-072-X   78-16447

CCP
Printed in the United States of America

ISBN 0-89072-072-X
ISSN 0077-8923

ANNALS OF THE NEW YORK ACADEMY OF SCIENCES

VOLUME 314

OCTOBER 20, 1978

MACHAUT'S WORLD:
SCIENCE AND ART IN THE FOURTEENTH CENTURY*

Editors
MADELEINE PELNER COSMAN AND BRUCE CHANDLER

Conference Organizers
BRUCE CHANDLER, MADELEINE PELNER COSMAN, AND LEO MUCHA MLADEN

✍ ℰ

CONTENTS

* This volume is the result of a conference entitled *Conference on Machaut's World: Science and Art in the Fourteenth Century*, held on 2-3 December 1977 and cosponsored by The Institute for Medieval and Renaissance Studies of The City College of The City University of New York and The New York Academy of Sciences, with the cooperation of the Maison Française of Columbia University and various divisions of The City University of New York and Columbia University. Activities of the Institute for Medieval and Renaissance Studies, CCNY of CUNY, not directly supported by CCNY are funded by generous benefactions, notably an $800,000 grant from the National Endowment for the Humanities, whose funds, courage and collaborations in New York City are gratefully acknowledged.

Foreword

MADELEINE PELNER COSMAN

⋖⋗

This book especially honors
MARION GIRARD FEGAN
Lady of Quality and Gracious Gentlewoman
who has shared administrative talents,
keen tastes for art and humor,
and pleasure in living

⋖⋗

Machaut's World: Science and Art in the Fourteenth Century was the title of a conference held in New York City, cosponsored by The Institute for Medieval and Renaissance Studies and The New York Academy of Sciences, that united scholars and students from numerous disciplines to share great cultural achievements of the Middle Ages. The papers presented were original contributions to scholarly understanding of the period.

Since the listeners and readers attending the conference were an intelligent popular audience consisting not only of scholars in each speaker's field of competence but also of biologists, literary historians, physicians, astronomers, chemists, linguists, art historians, and similarly learned persons, a style of presentation was sought that would be at once simple and elegant, of interest to the cognoscenti yet accessible to the intelligent neophyte.

The music and poetry of Guillaume de Machaut (1300-1377) serve as the central point from which intellectual excursions are made into literature, technology, astronomy, manuscript illumination, paper book production, natural science, theatrical presentations, politics of feasting, and other elements of the cultural context.

Machaut's world variously determined the qualities of his genius. He and his contemporaries experimenting with the great new musical art, the *Ars Nova*, shared interests, libraries, and experiences with mathematicians, physicians, and clockmakers. Singers and shawm-players probably practiced musical theory consistent with ideas of academic technologists.

Machaut's World: Science and Art in the Fourteenth Century has been an interdisciplinary adventure with these objectives: to promote a wider diffusion of knowledge and a clarity of understanding while celebrating the glories of an age and its art. Festivities were interspersed amongst learned papers: concerts, feasts, and ceremonies graced the days.

Much music of Machaut was performed during the conference at mid-morning, at the medieval feasts, and at night. *The Western Wind's* musicale in the nave of the Cathedral of Saint John the Divine preceded an architectural tour led by Canon Edward West. *Guido's Other Hand* (directed by Louise Schulman), *The Elizabethan Enterprise* (directed by Lucy Cross), and *The Ensemble for Early Music* (directed by Frederick Renz) provided glorious sounds of Machaut's music, the excuse for our convivial, intellectually exuberant adventure into *Machaut's World*.

✥ ✥

An academic statesman dare not remain an active scholar without a spectacularly devoted staff. I am particularly blessed by the presence of numerous intelligent, gracious, nimble-fingered and strong-backed human beings who have in various ways contributed to the editing of *Machaut's World*. Generous guardian of my office's order is Marion Fegan, who always balances efficiency with dignity despite trials. Dr. Henry Grinberg, Assistant to the Director of the Institute, made numerous suggestions tempered by learning and humor. Two graduate student staff members typed, thought, and commented: Miriam Mandlebaum and Laura Ann Johnston. Prefatory paragraphs to individual papers represent collaboration between myself and some students who worked with intelligent aplomb. From aspects of design through research checking to fetchings and carryings, several Institute staff members and students graciously performed the necessaries: Claudine Nemejanski Fischer, Barbara Verdi, Drew Kovach, Gardy St. Joy, Gurnel Rose, Ceil Michelle Lindner, Frank Urbancic, and Dr. Barbara Hanning.

Three friends labored on my contribution to this volume. Lore Strich Schirokauer, photographer, produced glossies with her usual skill and gentle humor. Clifford Cosman, a magazine editor, graciously read proof with his accustomed strictness and taste. Marin Cosman, of Yale University, caught several stylistic infelicities.

✥ ✥

As the Director of the Institute for Medieval and Renaissance Studies and the principal investigator of substantial grants, I am often asked to translate grand academic notions from idea into action. Leo Mucha Mladen presented one such notion, the impetus for *Machaut's World*. Describing himself as "a poor priest who married a rich couple, the Academy and the Institute," he persistently pursued people and institutions likely to join our venture. Encouraging the conference, they indirectly influenced this book.

Three university presidents blessed the enterprise: Robert Marshak of CCNY, CUNY; William McGill of Columbia; and Edmond Volpe of Staten Island, CUNY. At Columbia, several colleagues worked mightily, particularly Jacqueline Hellermann, director of the Maison Française. Donald Puchala, director of the Institute for Western Europe, and Elizabeth Minnich, Dean of Faculty at Barnard College, performed gracious favors. Deutsches Haus' Alan Castleman, the Music Department's Howard Shanet, the Renaissance collegium's Eugene Rice, and the University Seminars' Aaron Warner gave welcome support. Professors James Beck, Paul Oskar Kristeller, Meyer Schapiro, and Craig Trimberlake suggested speakers. Robert Myers and the *Collegium Musicum* presented a fine rendition of Machaut's *Messe de Nostre Dame*.

I am grateful to the Editorial Staff of The New York Academy of Sciences, Bill Boland, Executive Editor, and Joyce Hitchcock, Associate Editor, for their patient and skillful editorial work.

A BRIEF BIOGRAPHY OF GUILLAUME DE MACHAUT
(1300-1377)

BORN IN 1300 at Machaut in the Ardennes region of eastern France, Guillaume de Machaut became secretary and chaplain to John of Luxembourg, King of Bohemia (who was later killed at the Battle of Crécy, 1346); Guillaume then worked for Bonne of Luxembourg, wife of the future Jean II of France, and still later, King Charles II of Navarre. Settling at Reims, he remained there until his death in 1377, enjoying favors, patronage, and fame in France, Italy, and the European world. Celebrated as a lyric poet and composer of such long works as the *Dit dou Vergier*, the *Remède de Fortune*, and the *Confort d'Ami*, Guillaume de Machaut was also a prolific producer of shorter love poems set to music: *ballades, lais, rondeaux*, and *virelais*. His *Messe de Notre Dame* is one of the best examples of early polyphony.

Guillaume de Machaut:
a select bibliography

Calin, Willam C. *A Poet at the Fountain: Essays on the Narrative Verse of Guillaume de Machaut*. Lexington, Ky., 1974.

Domling W. *Die mehrstimmigen Balladen, Rondeaux und Virelais von Guillaume de Machaut*. Tutzing, 1970.

———. "Isorhythmie und Variation. Uber Kompositionstechniken in der Messe Guillaume de Machauts," *Archiv fur Musikwissenschaft*, 28 (1971), 24-32.

Gombosi, O. "Machaut's *Messe Notre-Dame*," *The Musical Quarterly*. 36 (1905), 204-24.

Guillaume de Machaut. *La louange des dames*, ed. Nigel Wilkins. Edinburgh, 1972.

———. *Le livre de voir-dit*. Geneva, repr. Slatkine, 1969.

———. *Oeuvres de Guillaume de Machaut*, ed. E. Hoepffner. 3 vols. Paris, 1908-21; repr. Johnson, New York, 1965.

———. *The Works of Guillaume de Machaut*, ed. Leo Schrade. 2 vols. (*Polyphonic Music of the Fourteenth Century*, Vols. II-III). Monaco, 1956.

Levarie, Siegmund. *Guillaume de Machaut*. New York, 1954.

Machaby, Armand. *Guillaume de Machaut*. 2 vols. Paris, 1955.

Perle, G. "Integrative Devices in the Music of Machaut," *The Musical Quarterly*, 34 (1948), 169-76.

Lowinsky, Edward. "Music in the Culture of the Renaissance," *Renaissance Essays*, ed. Paul O. Kristeller & P. Weiner. New York, 1968.

Reaney, G. *Guillaume de Machaut*. Oxford Studies of Composers, 9. London, 1971.

———. "Guillaume de Machaut: Lyric Poet," *Music and Letters*, 39 (1958), 38-51.

Reese, Gustave. *Music in the Middle Ages*. New York, 1940.

———. *Music in the Renaissance*. New York, 1959.

Robertson, Alec, & Denis Stevens. *The Pelican History of Music*. Baltimore, 1970.

Seay, Albert. *Music in the Medieval World*. Englewood Cliffs, N.J., 1965.

Williams, Sarah J. "An Author's Role in Fourteenth Century Book Productions: G. de Machaut's Livre ou je met toutes mes choses," *Romania*, 90 (1969), 433-54.

Wimsatt, James I. *The Marguerite Poetry of Guillaume de Machaut*. Chapel Hill, 1970.

Winternitz, Emanuel. *Musical Instruments and Their Symbolism in Western Art*. New York, 1967.

MUSIC pervaded medieval medical practice and theory in astounding manner. Not only was music prescribed for good digestion and for bodily preparation before surgery, but also as a stimulus to wound healing, a mood changer, and as critical accompaniment to bloodletting. Specially composed medical music (the *shivaree*) graced the wedding chamber to assure erotic coupling at the astrologically auspicious moment. Music of the heavenly sphere was thought closely integrated with all human harmonies definable in a human being's horoscope (thus the significance of birth time and conception time) and was expressed rhythmically by bodily pulses. Pulse music was predictive in diagnosis, prognosis, and treatment by medicine and surgery. Such practical medical music expressed magnificently the harmonious medieval world order. Closely allied with astrology and ideas of time, medical music may have influenced the development of the mechanical clock.   MPC

Machaut's Medical Musical World

MADELEINE PELNER COSMAN

Institute for Medieval and Renaissance Studies
City College, City University of New York
New York, New York 10031

A FTER a fine dinner including asparagus and turnips beautifully cooked with chestnuts and cream, accompanied by music of lute, vielle, and psaltery, a fourteenth-century feaster would not be surprised if suddenly stimulated to amorous ideas. Listening attentively to the music, he would have eaten such foods thought to be aphrodisiacs fully aware of their expected effect. Fault for his sudden after-dinner appetite thus would not be his stimulating company but rather his menu and its music. If not interested in satisfying erotic impulses, that medieval diner would quickly quench amorous ardor by a cooling salad of lettuce tossed with rosemary, fennel, and rue,[1] munched to the accompaniment of cornets and shawms.

Similarly, medieval physicans would not perform elective surgery without the patient's prior preparation by diet and music therapy. If the patient's bodily type were choleric, the surgeon would attempt to regulate his excitable, irascible constitution and calm the rhythm of his pulse by insisting before operation that tranquil music accompany a diet including little wine, and much boiled rice, barley cream, boiled veal and chicken, and in their season, figs, grapes, damson plums, baked pears and apples.[2] However, if the medieval patient's constitution were basically melancholic, the good fourteenth-century practitioner would "prepare the body well for operation"[3] by regulating music of mood with a milk diet comparable to the modern ulcer sufferer's. The medieval melancholic would eat only occasional boiled, lean rather than roasted meats, along with plenty of light cheese, eggs, custard and dairy products[4]—best to the musical accompaniment of sensuous, joyful melody (FIGURE 1).

In Machaut's medical musical world,[5] medical theorists in their texts and practitioners with their patients remarkably utilized both musical ideas and music performance in diagnosis of disease, prognosis of cure or death, and treatment by medication or surgery. Hospitals, clinics, and health spas sounded with rhythm and melody. Music was considered a significant mood changer and antidote to poisons. Wound healing was

[1]

0077-8923/78/0314-0001 $01.75/2 © 1978, NYAS

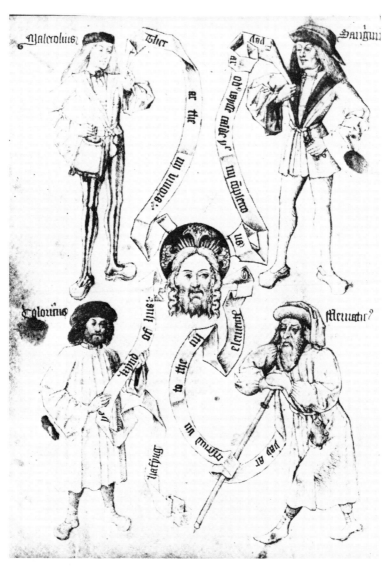

FIGURE 1. The four humors: blood, phlegm, choler, and black bile, cause, respectively (and clockwise), the *sanguine, phlegmatic, choleric,* and *melancholic* personalities, temperaments, and physiques, which are, in turn, associated with stages of life and seasons of the year. (From the *Guild Book of the Barber Surgeons of York,* the British Library, the British Museum, London, 15th century.)

stimulated by musical tranquilizers. Elaborate mathematical theories of human pulse music required of medieval physicians not only some knowledge of music theory but also a learned touch, a *tactus eruditus*, an ability to feel pulse music through sentient fingers, perceiving rhythm, proportion, meter, thus blood tempi. Music of the hunt and banquet were allied to medicine because food was thought to stimulate the four major bodily fluids—blood, phlegm, black bile, and choler—the four humors that determined health. Wedding feast melodies followed by musical processions to the nuptial bedroom, the erotic stimulating sounds of the *shivaree*, had their musical echoes in the medical-obstetric chamber. Such practical medical music magnificently expressed the harmonies of medieval world order.

Medical music is one of the more startling aspects of Machaut's world. Not only spectacular art form, mathematical science, and venerable expression of emotion, music affected the very workings of life itself, accompanying critical human events from conception of the child through burial of the dead. As Geraldus Cambrensis, the twelfth-century chronicler, insisted:

> The sweet harmony of music not only affords pleasures but renders important services. It greatly cheers the drooping spirit, smoothes the wrinkled brow, promotes hilarity. Nothing so enlivens the human heart, refreshes and delights the mind. Music draws forth the genius and by means of insensible things quickens the senses with sensible effect. Moreover, music soothes disease and pain. The sounds which strike the ear, operating within, either heal our maladies or enable us to bear them with greater patience. A comfort to all, an effectual remedy to many, there are no sufferings which music will not mitigate, and there are some which it cures.[6]

Medieval music as mood changer was no mere aesthetic frivolity. Practitioners well perceived the significance of mental health to physical well-being and to the body's longevity. Just as the moods and melodies proper for mourning the dead certainly would not be conducive to sexual activity; and just as flighty exuberance in thought or erratic rhythms could not appropriately accompany important strenuous labor, so each human action had its requisite attitude and rhythm, integrating spirit with physique. Disjunctures might lead to illness or death. As avidly as some medieval physicians prescribed elixirs, music, and dance against excessive melancholy,[7] such another practitioner as Gentile da Foligno recommended potions against infatuation by joy.[8] Music could manipulate mood with medical effect.

Mood music—melody or rhythm or mode used intentionally to influence emotions—of course has venerable tradition, appreciated by primitive societies as well as the most sophisticated. Greek and Roman notions of medical mood changers were mediated to medieval theorists and musicians by Isidore of Seville (d. 366),[9] Cassiodorus (born c. 485),[10] and, most importantly, Boethius (born c. 480).[11] The treatise *On Music* by Boethius was read in medieval universities, particularly in the medical school curriculum, which in certain Italian cities was called the faculty of arts and medicine.[12] Not only could music ennoble or corrupt character, an excellence and danger Plato feared,[13] but it could cure anti-social mental states such as fury.

Pythagoras, by means of a spondaic melody calmed and restored to self-mastery a young man who had become wrought-up by the sound of the Phrygian mode; so Boethius reminds his audience of music's potency.[14] One night, the youth's harlot was in his rival's house, with the door locked. The young man, in frenzy, unable either to reach his beloved woman or to destroy his hated rival, and having been listening to Phrygian music, was ready to set fire to the house. Deaf to his friends' pleas for restraint, he responded to reason immediately upon their changing the mode of the music, thus by rhythm and melody reducing fury to perfect calm.

However, musically stimulated anger could also have a positive medical effect. In such popular medieval health handbooks as the *Tacuinum Sanitatis*, anger, described as boiling of blood in the heart, had its optimum expression in engorging blood vessels, restoring color to a pallid face, and stimulating muscular activity.[15] Musically incited ire, therefore, was particularly useful for treating cases of paralysis.

Medieval medical theorists also recommended musical soporifics and energizers, on the authority of Boethius,[16] who maintained that the Pythagoreans freed themselves from cares of the day by certain melodies causing gentle, quiet slumber. Upon rising, they dispelled the stupor and confusion of sleep by other melodies, knowing that the whole structure of soul and body is united by musical harmony.

Beyond such medical effects of music upon mind, melody and rhythm had other therapeutic functions in hospitals, sickrooms, and health spas. Music was often recommended to palliate various orthopedic and rheumatic conditions; citing Boethius and classical precedent, medieval medical practitioners suggested the efficacy of melody against sciatica and lower back pain.[17] Moreover, music was sometimes thought an effective

FIGURE 2. The Great Ladder of Being or the Stair of Existence or Great Chain of Being portrays *homo*, man, in the middle between the inanimate stones and the spiritual greatness of angels. In God's House of Wisdom, man by spirit and intelligence may potentially ascend to the angels; but humankind falls by propensity of dross body to the level of brute beasts.

antidote to poisonous bites and stings of reptiles and insects.[18] Based on the medieval perception that human beings and animals were closely related in God's great scheme[19] (man in the middle, aspiring by spirit and soul to angel's heights, but drawn by dross body to animal's depths), the rhythms of each level of existence were believed inherently harmonized, each with those above and below (FIGURE 2). Therefore, for a patient bitten by a scorpion, a medical musician ought to emulate in music the offending biting beast's rhythms;[20] physician and patient might thus counteract harmful effects of the arachnid's poison. Scorning such ideas as ignorant superstition, William of Marra, in his fourteenth-century treatise on poisons,[21] nevertheless recommended music as treatment for the tarantula's bite because the effect of its poison is severe melancholy, whose best cure is joy—rejoicing stimulated, of course, by stirring melody (FIGURE 3).

Music therapy was significant for bodily preparation before elective surgical procedures such as ophthalmic or urinogenital operations. The sixteenth-century Italian plastic surgeon Tagliacozzi would not perform a rhinoplasty to reconstruct an absent nose (FIGURE 4) without first determining the patient's temperament, and establishing by diet accompanied by music that circumstance favorable to the operation's success.[22] Even post-surgical wound healing was thought stimulated by tranquil music (FIGURE 5).

Musical announcements allied pitch, sound, and rhythm to public health. Lepers and others with contagious diseases were required to blow horns or sound clappers to warn of their approach. While isolation wards and quarantine clinics for communicable diseases existed in the fourteenth century,[23] patients with incurable chronic diseases were required by law to sound their malady music, a leper often wearing his horn on his back (FIGURE 6). Epidemic outbreaks of syphilis, whose genital origin was well appreciated, were explained as resulting from disjunct heavenly music— malevolent, inharmonious conjunctions between such heavenly bodies as Jupiter and Saturn in Scorpio. Dürer, following current medical theory, explained through his famous graphic the 1484 venereal disease pandemic by such astrological disjuncture and disharmony (FIGURE 7).

Even more important in actual medical practice was music's association with a major modality of treatment: phlebotomy or bloodletting. For virtually every known medieval medical problem, as well as for general good hygiene, bloodletting was significant therapy by any of three major techniques: *fleeming*, simple incision in an artery or vein by scalpel, knife, or fleem; or *leeching*, application of live bloodsucking leeches to various

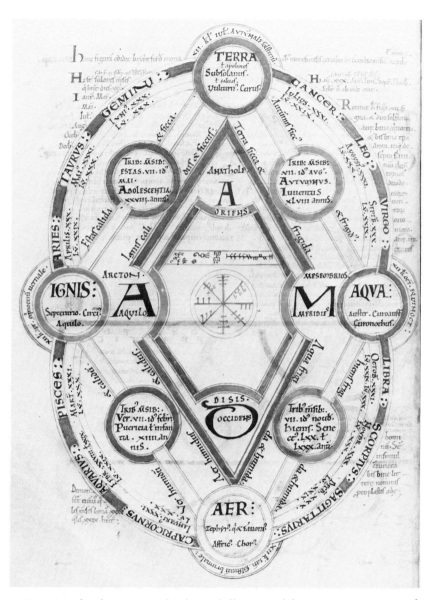

FIGURE 3. The glorious interrelatedness of all aspects of the universe are represented in this cosmological diagram in which (clockwise) earth, water, air, and fire are associated with the planets, the zodiacal signs, the four ages of man, the four seasons, and the "contraries" out of which the world is made: heat, cold, moisture, and dryness. (From *Byrthferth's Manual*, Saint John's College, Oxford, MS 17 folio 7v.)

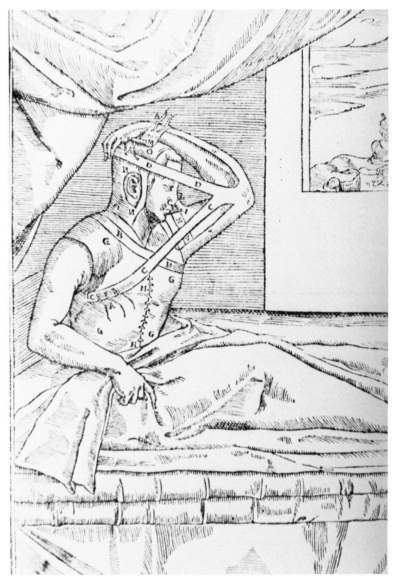

FIGURE 4. Plastic surgical reconstruction of a nose by raising a "pedicle" flap from the arm to replace the missing flesh follows preparation by diet therapy. The patient's left arm is held in position by an elaborate strapping and vest device. (From the *Life and Times of Gaspare Tagliacozzi, Surgeon of Bologna*, Martha Teach Gnudi and Jerome Pierce Webster, New York.)

FIGURE 5. A medical musician bows his viol, playing soothing melody to the sleeping patient.

bodily points; or *cupping,* application of glass or wood vacuum cups to the nicked skin's surface (FIGURE 8). Phlebotomy's purpose was to diminish blood volume, and to establish or re-establish an equilibrium amongst vital bodily humors. Music, both directly and indirectly, determined the efficacy of bloodletting.

Depending upon the patient's birthtime and natal zodiac constellation, as well as the time of onset of injury or illness, the medieval practitioner would let blood from veins at certain phlebotomy points (FIGURE 9). Delineated in various medical texts and bloodletting calendars, these points were to be opened at auspicious moments, the practitioner carefully avoiding malevolent phlebotomy times. Calculating such discrete times, the medieval physician utilized that exquisitely effective chronometer and biological slide rule, the astrolabe. The astrological sign governing the injured or ailing part of the body also determined the time and incision site for phlebotomy.[24] With the zodiacal sign Aries, governing the patient's head—its aches and injuries—Gemini, the arms, Sagittarius the thighs, and Pisces underpinning the aching feet, the practitioner would attempt by treatment to re-establish the patient's bodily harmony with the celestial rhythms whose disjuncture or disproportion initially caused the illness or injury (FIGURES 10 and 11).

More directly practical instrumental music accompanied phlebotomy and medical ministration in the health spas. To increase the efficacy of mineral waters for bathers, either soothing or stimulating melodies accompanied hydrotherapy (FIGURE 12). A women patient being bled into a bowl from an arm vein while soaking her sore leg, with her physician in-

FIGURE 6. A leper, or a patient with a serious skin disease represented by "spots," wears his warning horn strapped to his back, sounding it to avoid contaminating the unwary.

FIGURE 7. Albrecht Dürer depicts the 1484 outbreak of syphilis as associated with malevolent astrological conjunctions.

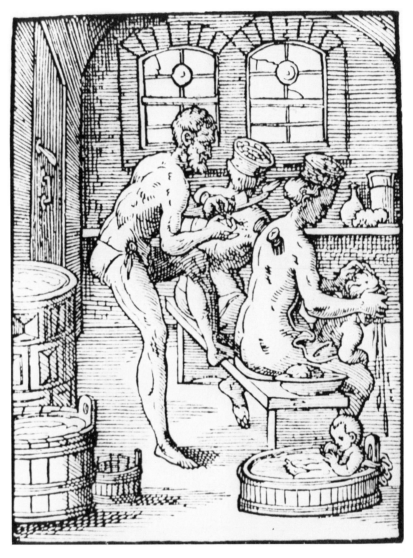

FIGURE 8. A woman is being phlebotomized by *cupping;* she sports two vacuum cups on her back, while she sits nude, except for a hat, washing the hair of one of her children, while yet another sits safely strapped in a small bathing vessel. A male bathing attendant aids these hygienic ablutions.

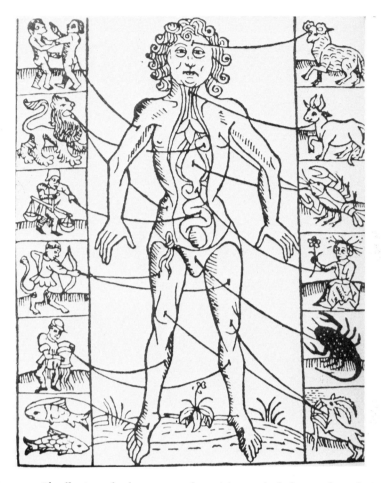

FIGURE 9. Bloodletting calendars suggested propitious and ideal times for reducing blood volume at particular phlebotomy points, each of which was governed by astrological sign.

specting her urine specimen in a urine flask, would enjoy truly sanguine music, stimulating the right mood for healing, the musical pulsing and throbbing rhythmically correlated either to her ailment or her temperament (FIGURE 13).

Other bathers uniting medicine and music added sexual dalliance.[25] Not only individual bathing couples would hear seductive music of the lute and horn (FIGURE 14), but larger communal bath parties, regaled by music,

FIGURE 10. An Astrological Man or *homo signorum* wears the names of the astrological signs governing his anatomical parts. Aries controls the head, Gemini the shoulders, Libra the stomach, Sagittarius the thighs, and Pisces the feet.

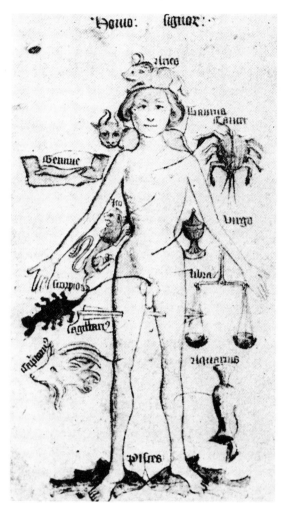

FIGURE 11. An Astrological Man, *homo signorum*, sports the insignia of the zodiacal signs governing the various parts of his anatomy; Aries the Ram surmounts his head; Cancer his chest; Scorpio, the genitalia; Sagittarius, the thighs. (From the *Guild Book of the Barber-Surgeons of York*, London, the British Museum, British Library).

FIGURE 12. In a sixteenth-century Germanic health spa, bathers are separated by a banquet board, while listening to a jester's playing on a viol.

would harmonize the delights of the health spa with those of the pleasure palace. Spectacular medieval bathing establishments offered bath vessels for nude couples desporting before bedding, sharing aphrodisiac food from the dining boards, hearing sensual stirrings of the lute (FIGURE 15).

Just as food, music, and health conjoined in the bath, so also in the banquet hall. Fourteenth-century fabulous feast music was astoundingly significant, essential both to ceremony and digestion.[26] Well-liveried servants in procession carrying numerous dishes and courses of food were governed by fanfares and trumpet sounds. The foods determined by the temperaments of the feasters, each dish might have its appropriate musical complement. Trumpets, pipes, and bells, lutes, horns, and rolling drums provided not only visual splendor with stirring sound but digestives. Discordant sound might cause incomplete food metabolism.

FIGURE 13. In an early sixteenth-century German health spa, the woman at left endures phlebotomy while the nude bathers await their food carried in covered tureens to the dressed guests regaled by jester and flute player. A physician, at right, scanning a urine flask, oversees the rites of hygiene.

Not only an aid to normal digestion, feast music was an important anti-dote to poison. The fourteenth-century theorist William of Marra said in his *Papal Garland Concerning Poisons* that joy derived from music might attract the spirits from within the body to the periphery, so preventing poison from penetrating to the vitals.[27] Therefore, when the extraordinary huntsman and patron of the arts, Gaston Phébus, commissioned great composers' works to accompany his midnight chicken-wing orgies, his concerts may have been planned as much for safety as aesthetics.

Medieval banquet scores musically balanced culinary courses and their service. Instruments alternating with solo song, chorus, and dance were interspersed amidst banquet foods and wines (FIGURE 16). For culinary music at the D'Este court of Ferrara, the chief cook Christaforo Da Messis-bugo[28] lists for each fish course, meat, and wine, its supplement by viols, voice, and choir. While the noble guests washed hands with perfumed water, a musical performance by six singers, six viols, a lyre, a lute, a zither, a trombone, recorder, flute, and numerous keyboard instruments accompanied the seventeenth course.

FIGURE 14. Under the zodiacal sign of Venus, nude lovers revel in a tented tub while listening to erotic feast music before they consume aphrodisiac foods and drinks set before them.

FIGURE 15. Erotic bath festivity entertains five couples in individual bathing vessels joined by the long banquet board athwart the gunnels. Wine lubricates the voluptuaries' passage from bath to bed.

Even the animals reaching the medieval banquet hall were hunted with specific horns and melodic calls for particular animals.[29] The more rhythmic the animal's catching and kill, the more healthful its effect as food. As most fourteenth-century hunting horns played only a single pitch, hunting calls were dependent upon rhythm and intensity of sound. Special calls, the *mote*, blown at the uncoupling of the hounds; the *rechete*, to recall hounds or urge them to the kill; the *mane* and the *pryse*, rhythmically uniting calls with action would portend the animal's functioning

Die sechszondachtzigist figur

FIGURE 16. Musicians signal the presentation of a course in an elaborate medieval noble banquet. (Woodcut from Wohlgemuths' *Der Schatzbehalter*, Nuremberg, printed by Koberger, 1491. New York, The Metropolitan Museum of Art, Rogers Fund.)

FIGURE 17. A *conception time mirror* surmounts the bed in this fifteenth-century marriage portrait. The oval pendant on the bed-backdrop behind the happy couple in the bedroom is so arranged as to reflect the nocturnal heavenly bodies, thereby the ideal astrological moment for sexual coupling. (From the *Histoire de Renaud de Montauban*, France, 1467-70. Paris, Bibliothèque de l'Arsenal, MS 5073).

without discord when eaten in the banquet hall. Harmonious hunting pre-figured healthful eating.

Hunt poems and hunt melodies inspired the fourteenth-century musical forms, the French *chasse*, Italian *caccia*, and English *catch*.[30] Usually canons, these significant early forms of polyphony utilized the melodies and rhythms of ceremonial pursuit of a keenly desired animal and transferred them to the hunt of the desired lover—hunt music indirectly affecting sophisticated love allegories and love songs.

Sounding the spectacle of a newly married couple's march to the bedroom was the *shivaree*, erotic, stimulating music to assure their consummate coupling. Once therein, heavenly harmonies were reflected in a mirror registering the moon and the stars of the night sky. This star rhythm would determine the exact propitious moment for sexual intercourse and thus for the conception of a remarkable child. Such a reflecting glass was called a *conception time mirror* (FIGURE 17). A startling fifteenth-century birth scene depicting a medieval Caesarean section, with the child being

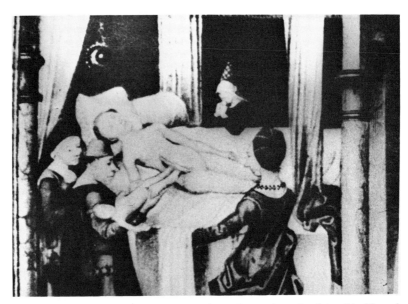

FIGURE 18. A *birth time mirror* hangs above the mother's head in this fifteenth-century painting of a Caesarean section, the newborn babe leaving, head first, through an abdominal incision. The exact birth moment would determine the child's astrological inheritance: physical, temperamental, intellectual, emotional, and professional.

removed head first from an incision made in the abdominal wall, shows the mother's head surmounted by a globe, a *birth time mirror* (FIGURE 18). Such birth and conception time mirrors would determine indirectly significant affairs of family and state. From the very moment a child was conceived or born, its qualities were affected by the constellation governing its birth: physical nature, temperament, personality, predilections, even the babe's likely adult profession were influenced.[31] Physicians and clockmakers were born under the sign of Mercury; poets, painters, aesthetes and cooks under jurisdiction of Venus.

Probably the most important medieval medical music was the music of pulse. Physicians took patients' pulse either at the brachial artery (FIGURE 19) or at the wrist. The physician calculated rhythm, tested pulse strength or irregularity, comparing this data against numerous pulse music treatises written by such venerable physicians as Galen, whose sixteen books on pulse distinguish among 27 separate varieties of human pulse.[32] Following Galen and Avicenna,[33] the medieval physician expected to identify musical proportion in pulse by touch (FIGURES 20 and 21). The close integration of music with medicine in the fourteenth century (Italian) universities suggests intriguing interpretations of ideas and practices governing health and the *Ars Nova*.[34]

Amongst the numerous medieval writers on pulse music, those most closely concerned with astrology most clearly invoked the significance of pulse to diagnosis and treatment,[35] and thus, the importance of music as one of the seven liberal arts essential to medical training. (Our modern insistence upon liberal learning for the medical profession may derive ultimately from that practical medieval concern for pulse music.) Among these pulse music writers associating with the best musical theorists of their day in faculties of arts and medicine (such as Bologna, Padua, and Perugia, centers for musical innovation)[36] was such a learned physician as Peter of Abano.[37]

His predilections for pulse are dramatic. Physician, astrologer, philosopher, professor both at Paris and Padua, Peter of Abano insisted that every self-respecting medical practitioner not only know music theory but feel it in pulse. In a widely publicized treatise, *The Conciliator*, he gave elegant musical descriptions (following Boethius and Isidore of Seville) of concordance, dissonance, and mathematical musical proportions, and semitones, the scale, the monochord, the Greek diatonic, chromatic, and enharmonic tetrachords! Peter of Abano listed as easily perceivable by touch those pulses from trochaic beats for infants, to iambotrochaic for the aged;

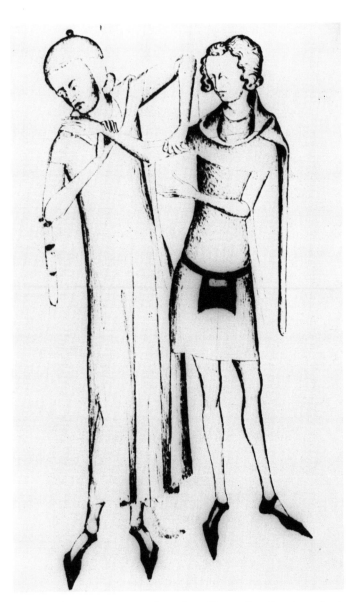

FIGURE 19. A physician wearing a short-sleeved long robe takes the patient's pulse at the axilla.

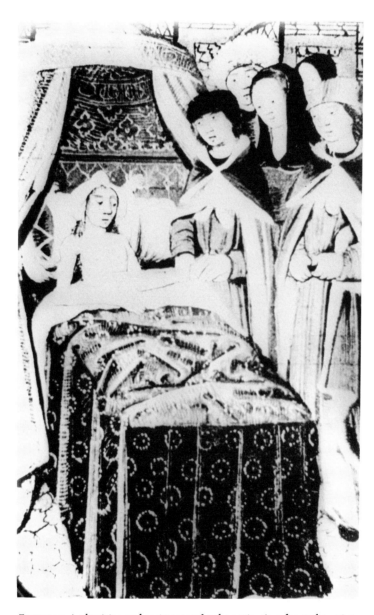

FIGURE 20. A physician and assistants take the patient's pulse at the wrist.

FIGURE 21. One physician takes the patient's pulse at the wrist, while three medical colleagues confer, studying the patient's urine, that valuable diagnostic and prognostic tool. Pulse rhythm would determine, in part, time for administration of medication and time for surgery; pulse would prognosticate cure.

from special pulses, caused by climate, pregnancy, and disease, as sluggish *pulsus formicans*, crawling slowly as an ant, to leaping, pounding *pulsus gazellans*, fleeting flying fast as the gazelle.

Pulse was thought to demonstrate the very nature of music.[38] As music consists of high and low notes arranged in proportion, so pulse consists of strokes of greater and lesser speed and intensity—both music and pulse characterized by rhythmic patterns of time intervals.

Medical music of fourteenth-century daily life was thought mere earthly echo of perfect heavenly harmonies. Medieval theorists adapting classical ideas of an ordered universe applied them to new Christian purpose. If God's total creation were perfect and divinely reasonable, then all aspects of it were interrelated, orderly, and harmonious. Just as the human body was the microcosm—the great created universe, the *macrocosm*,[39] in small —so medieval instrumental and vocal music was more than simple accidental melody made by human hand or voice for human ear. In heavenly harmonies of the spheres, *musica mundana*, originated those earthly echoes in the well-ordered world. That heavenly sphere music had its equivalent in the rhythms of the human body, *musica humana*, such as its periodic diseases, its cyclic fevers, its interrelated though malleable mind and physique. Last in descent of musical greatness is our paltry mortal musicianship, *musica instrumentis*.[40] Instrumental music that we play and sing sounds trivial echoes of celestial harmonies (FIGURE 22). Yet that least music for most of us is glory sufficient.

Remarkably, medieval medical music seems associated with far more than healing. Machaut's medical musical world may have developed the mechanical clock.[41]

Years ago a grateful patient gave my physician-father an extraordinarily medieval gift: a calendar watch that registered minutes, hours, day, and year, as well as configurations of stars and moon in the night sky. Engraved in minuscule letters was this legend: *My lucky stars plus your skill equals my life's time.* While the little mechanism was modern, the gesture was medieval: to unite for and with the physician the ideas of medicine, astrology, and time. In fact, the association amongst these three was so inextricable in medieval theory and practice that medieval medicine justly might be called Time Medicine, *Chronophysica*.

Medieval medical time associated zodiac configurations of time-of-injury or onset of illness with time of the patient's birth, time for treatment via medicine or surgery, time for prognosis of the patient's fate, and time for cure or time for death. For all these medical "times" the practitioner's

FIGURE 22. In this allegorical and practical portrait of "music" from a fourteenth-century manuscript of Boethius's *De musica* and *De mathematica*, the woman illustrating a fine pipe organ is surrounded by figures playing lute, psaltery, viol, tambourine, bagpipe, tabor and drums, horns, and castinets.

instruments were elaborate, effective timing devices such as the astrolabe and volvelle. Such timers determined favorable or inauspicious conjunctions between the heavenly bodies and the patient's body—uniting by mathematical calculations the patient's past, present, and future. Medieval medical time thus was associated with one critical human concern: life between birth and death ruled by good or bad health. So important was Time to medieval medicine and so compelling the practitioners' requirements for accurate timers that medieval medicine probably contributed to the development of the clock, one of the most significant machines in the history of the west, technology and social need converging at the medieval sick bed.

I dare arrogate this theory which unites music, medicine, and astrology, knowing that at this moment it is unprovable. But to me it satisfactorily explains what is otherwise as yet mysterious: why in the mid-fourteenth century (Machaut's world) timing so preoccupied so many great minds as to allow solution to the problems of clockmaking. Timing and rhythm, essentials for all medical and surgical techniques, affected bloodletting, diet therapy, cauterization, surgical operation. Patient and practitioner believed that success or failure—thus life or death—were calculable by chronometers.

Medieval medical practitioners' timers were astonishingly sophisticated. Astrolabes, quadrants, and volvelles were portable measures of hours and minutes, as well as computational devices for horoscopes. Among several types of astrolabes, the most important for physicians was the *planespheric astrolabe*.[42] Not only a reckoner of time, the astrolabe was a universal instrument and calculator. Its movable concentric discs, usually made of metal such as bronze, its predictable almucanters and azimuths, made it useful for observing and reckoning positions of heavenly bodies; for measuring heights, distances, and latitudes; and for horoscopy. (Modern wrist-watches certainly are less efficient!) The face of the astrolabe was a veritable analog computer for solving problems dealing with time, star positions, length of day or night, by mere rotation of the pierced metal *rete*, a stylized star map, essentially a stereographic projection of the skies. Beneath this *rete* was a series of *plates* or *tympans* upon which were incised images of the celestial spheres projected onto a plane parallel to the equator. Drawn for specific latitudes' the *tympans* would depict the horizon, zenith, altitude, meridian, and the Tropics of Cancer and Capricorn. The *tympans* in turn would rest in a flattened disc called the *mater* or the *mother*. Atop this pleasant pile of concentric discs would turn the *alidade,* a pivoting rule with a sighting bar. *Rete, tympans,* and *alidade* were flexibly anchored to the *mater* by the *horse* or *bird, a* pivot pin often whimsically shaped in animal form.

Though medieval patient, practitioner, and early scientist believed that hours determined fate, all astrolabes, even the most splendid medieval chronometers, were manual, not automatic. The physician-astrologer set by hand the discs, lines, parallels, angles, and correspondences. And only rarely did astrolabe chronometers allow plotting of time schemes more accurate than hours or quarter-hours. My reading of the fourteenth century medical evidence suggests that medical practitioners required a more exact timer—a calculator of minutes, for example—and a mechanism both more perpetual and self-propelling, which would thereby emulate on earth the independent predictability of heavenly spheres, and imitate in a machine the rhythmic regularity of human pulse.

Such medical interests would have been the most compelling social requirement for the new time technology of the Middle Ages, for physicians needed, daily, an instrument for plotting the primacy of time over influences of God, nature, and man's acts. Individual life was thought risked, the human fate weighed in time's balance at each chronometer computation. Such an instrument replicating automatic rhythms of human pulse, harnessing cyclic rhythms of the stars in a machine, Giovanni da Dondi created in 1348: an efficient, predictable, popular, mechanical clock.

Giovanni da Dondi,[43] physician, astrologer, and clockmaker, made a timer more accurate, more dependable, more magnificent than any before, not only with hours and minutes of the earthly day, but with all the planets in their orbits, heavenly bodies in their rotations, the feast days in the seasonal round, and even the embellishments of a dragon's head and tail. It was a machine so gorgeous and complex it was called in its time a wonder of the world.

Certainly Giovanni da Dondi was not the only physician-clockmaker working,[44] but his clock succeeded where others failed. (It still runs in a reconstructed version today in the Smithsonian Museum in Washington.) A specific constellation of events seems to have influenced the development of the mechanical clock in Machaut's lifetime in the fourteenth century— a series of dreadful cataclysmic circumstances requiring accurate timing for their amelioration and their future prevention: recurring epidemics of plague.

While 1348 marked the height of the Black Plague, epidemics terrorized Europe earlier.[45] In 1340, for example, the city of Florence buried one sixth of its citizens dead with plague. (Were we today suffering that scourge, of the 26,000 intelligent readers holding this essay in hand, 4333 amongst us would dramatically disappear). Such death quotas extraordinarily affected the living: their art, economics, philosophy, religion, and technology. If chronometers were instruments of prediction, of diagnosis, of prognosis, and of treatment for disease, then perfecting accuracy of the machine might extend life of the human being. I suggest then, that the medieval physicians and their scientist colleagues turned medical adversity into technological virtue.

Just as war can be credited as the impetus for technological developments, so might public health crisis. Plague did not cause the *invention* of the mechanical clock but rather its *development*, as accurate time measurer, stimulated by social medical requirement for such an instrument. I suspect it no accident that some of the most prodigious clockmakers were physician-astrologers, such as Giovanni and Jacopo da Dondi,[46] or that physicans such as Simon Bredon and Nicholas of Lynn made time-telling machines.

Healing, Hippocrates said, is a matter of time; but it is sometimes also a matter of opportunity. Plague may have been the opportunity to mechanically harness time for healing. Time was the great healer—so long as the healer timed. Medieval healing may have perfected a mechanical device for measuring time, making healing, yet again, a matter of time.

Music's timing and harmony resonated splendidly in Machaut's medical musical world. Musical instruments and voices sounded from the hunting ground to the health spa, the banquet hall to the bedroom *shivaree*. Mindful of celestial music, medieval people felt inexorable rhythms in their pulses, just as their physicians did. Such rhythmic insistences on man's relationship to the gigantic wondrous creation inspired both dignity and humility. If one is pulsing along on earth to the same beats as the heavenly planets and stars, then one must have some purpose beyond this insubstantial earthly existence. The medieval medical musical world celebrated rhythmic life as that long (or short) ceremony of melodic affirmations of universal order.

NOTES AND REFERENCES

1. On foods related to health, sex, disease, and treatment, see Madeleine Pelner Cosman, *Fabulous Feasts: Medieval Cookery and Ceremony* (New York, 1976), especially Chapter 5, "Sex, Smut, Sin, and Spirit"; C. Maillant, *Les aphrodisiaques* (Paris, 1967); and L. C. Arano, *The Medieval Health Handbook*, trans. A. Westbrook and O. Ratti (New York, 1976).
2. A convenient source for medieval diet therapy is the treatise on plastic surgery by Gaspare Tagliacozzi, as well as Alexander Read's partial translation thereof, in Martha

Teach Gnudi and Jerome Pierce Webster, *The Life and Times of Gaspare Tagliacozzi, Surgeon of Bologna, 1545-1599* (New York, n.d.), especially pp. 458, 649; and 67-79, 92, 93, 114, 123, 124, 136, 137, 205, 206, 208, 211-215, 249, 291, 309-14. Numerous significant works touching on the subject include P. O. Kristeller, "The School of Salerno," *Bulletin of the History of Medicine* 17 (1945); V. L. Bullough, "The Medieval Medical University of Paris," *Bulletin of the History of Medicine* 31 (1957); W. S. C. Copeman, *Doctors and Disease in Tudor Times* (London, 1960); H. E. Sigerist, *On the History of Medicine* (ed.) F. Marti-Ibanez (New York, 1960); H. M. Ferrari, *Une chaire de médecine au XVème siècle* (Paris, 1899); R. Dubos, *Man, Medicine, and Environment* (New York, 1969); A. Castiglioni, *Storia della medicina* (Milan, 1936); L. S. King, *The Road to Medical Enlightenment* (London, 1970); L. C. MacKinney, *Early Medieval Medicine with Special Reference to France and Chartres* (Baltimore, 1937); H. E. Sigerist, *Landmarks in the History of Hygiene* (Oxford, 1956).

3. On the relationship between music and medicine, see D. M. Schullian and M. Schoen, *Music and Medicine* (New York, 1948); Albano Sorbelli, *Storia della Università di Bologna* (Bologna, 1944); Guiseppe Ermini, *Storia della Università di Perugia* (Bologna, 1947); Oswei Temkin, *Galenism* (Ithaca, 1973); Nancy G. Siraisi, *Arts and Sciences at Padua* (Toronto, 1973).

4. Tagliacozzi,[2] pp. 459 and 650.

5. For the medieval musical context in which medical music fits, see Gustave Reese, *Music in the Middle Ages* (New York, 1940); and his *Music in the Renaissance* (New York, 1954); Albert Seay, *Music in the Medieval World* (Englewood Cliffs, 1965); F. W. Sternfeld, *A History of Music: Music from the Middle Ages to the Renaissance* (New York, 1973); François Gebaert, *Histoire et théorie de la musique de l'antiquité* (Paris, 1881); Curt Sachs, *The Rise of Music in the Ancient World, East and West* (New York, 1943); Manfred Bukofzèr, *Studies in Medieval and Renaissance Music* (New York, 1950); Richard L. Crocker, *The Early Medieval Sequence* (Berkeley and Los Angeles, 1977); Howard M. Brown, *Music in the Renaissance* (Englewood Cliffs, 1976); Sir John Hawkins, *A General History of the Science and Practice of Music* (New York, 1963); Paul Henry Lang, *Music in Western Civilization* (New York, 1963); Dom Anselm Hughes, *Early Medieval Music up to 1300, The New Oxford History of Music*, Vol. II (London, 1954-1967); Dom Anselm Hughes and Gerald Abraham (eds.), *Ars Nova and the Renaissance 1300-1540* (London, 1960-1969); Alec Robertson and Denis Stevens (eds.), *The Pelican History of Music, Ancient Forms to Polyphony*, Vol. I (Baltimore, 1960), and *Renaissance and Baroque*, Vol. 2 (Baltimore, 1963).

6. Giraldus Cambrensis, *The Historical Work of Giraldus Cambrensis. . .* , rev. and ed. Thomas Wright (London, 1863; repr. New York, 1968) pp. 127-130.

7. Anti-melancholy potions abound in the *Tacuinum Sanitatis*, the so-called "Tables of Health," the popular medieval health manuals. See L. C. Arano,[1] and Robert Burton, *The Anatomy of Melancholy* (London, 1932).

8. See Lynn Thorndike, *History of Magic and Experimental Science* (New York and London, 1934, 1966) III, pp. 233-502. And G. Girolami, *Sopra Gentile da Fuligno* [sic] *medico illustre del secolo XIV* (Naples, 1844); George Sarton, *Introduction to the History of Science* (Baltimore, 1948) 3, 1, pp. 848-852; F. Bonora and G. Kern, "Does anyone really know the life of Gentile da Foligno," *Medicina nei secoli* 9 (1972), pp. 29-53.

9. For convenience, see Oliver Strunk, ed., *Source Readings in Music History from Classical Antiquity through the Romantic Era* (New York, 1950); see Boethius *De institutione musica*, ed. Gottfried Friedlein (Leipzig, 1876).

10. Cassiodorus *Institutiones*, ed. R. A. B. Mynors (Oxford, 1937).

11. See Boethius, above, note 9.

12. Nancy G. Siraisi, *Arts and Sciences at Padua* (Toronto, 1973); Edward A. Lippman, "The place of music in the system of liberal arts" in *Aspects of Medieval and Renaissance Music*, ed. Jan la Rue (New York, 1976). Also compare, C. H. Haskins, *Studies in the History of Medieval Science* (Cambridge, 1924).

13. *The Republic*, trans. Paul Shorey, Loeb Classical Library, and ed. W. Heinemann (London, 1930), Book I, pp. 245-69, 287-95.

14. See Strunk,[9] pp. 79-86.

15. An example appears in Arano, *Medieval Health Handbook*, numbers 67, 176 (see note 1).

16. Boethius, *De musica*, pp. 178-89 (note 9); Strunk,[9] p. 83.

17. "Ismenias the Theban, when the torments of sciatica were troubling a number of Boeotians, is reported to have rid them of all their afflictions by his melodies." (Strunk,[9] p. 83).

18. See Lynn Thorndike, "Works on Poisons," in his *History of Magic and Experimental Science*, Vol. III, pp. 525-45.

19. The ordered Scheme of Being is best appreciated via A. O. Lovejoy, *The Great Chain of Being* (Cambridge, 1966); S. K. Heininger, Jr., *Touches of Sweet Harmony: Pythagorean Cosmology and Renaissance Poetics* (San Marino, 1974); John Hollander, *The Untuning of the Sky: Ideas of Music in English Poetry, 1500–1700* (Princeton, 1961). Three works by the great scholar Paul Oskar Kristeller are invaluable: "Ficino and Pomponazzi, On the Place of Man in the Universe," *Journal of the History of Ideas* 5 (1944) pp. 220-242; *Renaissance Thought* [1955] (New York, 1961); and *Catalogus translationum et commentariorum: Medieval and Renaissance Latin Translations and Commentaries* (Washington, 1960); See also Frances Yates, *The French Academies of the 16th Century* (London, 1947).

20. Lynn Thorndike,[18] Vol. III, p. 534.

21. For an accessible introduction see Thorndike,[18] pp. 525-40. See the Vatican Library's MS Barberini 306, Rome.

22. Tagliacozzi, ed. Gnudi and Webster,[2] and see Platina (Bartolomeo de Sacchi di Piadena), *De honesta voluptate: On Honest Indulgence and Good Health* (Venice, 1475; reproduced and translated, Malinckrodt Chemical Works, 1967); and see H. E. Sigerist, *Landmarks in the History of Hygiene* (Oxford, 1956).

23. Amongst numerous introduction to medieval and Renaissance epidemiology, some of the most useful are C. H. Talbot, *Medicine in Medieval England* (London, 1967); C. Morris, "The Plague in Britain," *The Historical Journal* 14 (1971); B. Benassar, *Recherches sur les grandes épidémies dans le nord de l'Espagne à la fin du XVIème siècle* (Paris, 1969); R. Ciasca (ed.), *Statuti dell'arte dei medici e speziali* (Florence, 1922); R. B. Litchfield, "Demographic Characteristics in Florentine Patrician Families," *The Journal of Economic History* 30 (1969); T. McKeown "A Sociological Approach to the History of Medicine," *Medical History* 14 (1970); M. Neuburger, *History of Medicine* (Oxford, 1925); E. Woehlkens, *Pest und ruhr im 16 und 17 jahrhundert* (Hanover, 1954); V. L. Bullough, "Population and the Study and Practice of Medieval Medicine," *Bulletin of the History of Medicine* 36 (1962); C. Salzmann, "Masques portés par les médecins en temps de peste," *Aesculape* 22 (1932); P. Ziegler, *The Black Death* (New York, 1969); Charles Mullet, *The Bubonic Plague and England* (Lexington, 1956); Frank Wilson, *The Plague in Shakespeare's London* (London, 1963); Anna Campbell,

The Black Death and Men of Learning (New York, 1931); and consult other references in note 2, above.

24. A comprehensive study with excellent bibliography is Harry Bober's "The Zodiacal Miniature of the *Très Riches Heures* of the Duke of Berry: Its Sources and Meaning," *Journal of the Warburg and Courtauld Institute* 1 (1948) p. 34f; and see R. T. Gunther's *Early Science in Oxford* (Oxford, 1925) and his *Early Science in Cambridge* (Oxford, 1937); see also Madeleine Pelner Cosman, "Medieval Medical Malpractice: The Dicta and the Dockets," *Bulletin of the New York Academy of Medicine* 49:1 (1973) pp. 22-47; "Medieval Medical Malpractice and Chaucer's Physician," *New York State Journal of Medicine* 72:19 (1972) pp. 2439-44; and "Medical Fees, Fines, and Forfeits in Medieval England," *Man and Medicine* (1975) pp. 133-58; as well as "Malpractice and Peer Review in Medieval England," *Transactions of the American Academy of Ophthalmology and Otolaryngology* (Vol. 80, 1975) pp. 293-97.

25. On the relationship between bathing and hygiene, see my "Fountain, River, Privy, Pot: Medieval London's Polluted Waters," *Fabulous Feasts: Medieval Cookery and Ceremony* pp. 94-101; and the bibliography on hydrodynamics, waters, and sanitation. On Italian baths in particular, see C. M. Kauffman, *The Baths of Pozzuoli* (Oxford, 1959). One of the more fascinating fourteenth-century manuscripts of bath habits is Petrus de Ebulo *de balneis Puteolanis*. (The Vatican. MS Rossiano 379, especially folio 21 ff). Intriguing materials are found in Lynn Thorndike, "Sanitation, Baths, and Street Cleaning in the Middle Ages and Renaissance," *Speculum* 3 (1929) pp. 192-203; G. Salusbury, *Street Life in Medieval England* (Oxford, 1948); Dorothy Hartley, *Water in England* (London, 1964) and Lynn White, Jr., *Medieval Technology and Social Change* (Oxford, 1962).

26. Banquet music is discussed by several of the authors listed in note 5 as well as by H. de Lafontaine, *The King's Musick* (New York, 1973); E. J. Dent, "Social Aspects of Music in the Middle Ages," *Oxford History of Music* (London, 1929); E. Winternitz, *Musical Instruments* (New York, 1967); W. Woodfill, *Musicians in English Society* (Princeton, 1953); H. M. Brown, *Instrumental Music Printed before 1600* (Cambridge, 1965); and his *Embellishing Sixteenth Century Music* (Oxford, 1975); and the bibliography to music of feasts in *Fabulous Feasts: Medieval Cookery and Ceremony*.

27. Lynn Thorndike, *History of Magic*: Vol. III, p. 534 (note 18).

28. *Banchetti Compositioni di Vivende* (Ferrara, 1549), ed. F. Bandini (Venice, 1960). See also L. Lockwood, "Music at Ferrara," *Studi musicali* 1 (1972); and H. M. Brown, "A Cook's Tour of Ferrara in 1529," a paper presented at University of Notre Dame Renaissance Conference (April, 1975).

29. An excellent literary example appears in *Sir Gawain and the Green Knight*; see the notes in the edition by J. R. R. Tolkien and E. V. Gordon, *Sir Gawain and the Green Knight*, (Oxford, 1955) esp. pp. 100-14. Most versions of the Tristan legend also deal with the music of hunt ceremony. For the relationship between the chase and love dalliance, see Marcelle Thiébaux, *The Stag of Love: the Chase in Medieval Literature* (Ithaca, 1974).

30. See Edward Lowinsky, "Music in the Culture of the Renaissance," in *Renaissance Essays*, ed. Paul O. Kristeller and P. Weiner (New York, 1968).

31. *Genethlialogy*, the medieval science of birth forecasting, is important to numerous astrological commentators, most of whom based their disquisitions upon Ptolemy's *Tetrabiblos*. A convenient edition is J. M. Ashmand (London, 1917); reprinted also

in *Health Research* (Mokelunne Hills, California, 1969). See J. R. Bram (trans.) *Ancient Astrology Theory and Practice, Matheseos Libri VIII by Firmicus Maternus* (Park Ridge, 1975).

32. Galen, *Opera omnia*, ed. C. G. Kuhn (20 volumes, Leipzig, 1821-33). See especially, as noted by Siraisi, below, *De pulsum usu*. 5:149-80; *De pulsibus libellus ad Tirones*, 8:453-492; *De pulsuum differentiis*, 8:493-765; *De dignoscendis pulsibus*, 8:766-961; *De causis pulsuum*, 9:1-305-430; *Synopsis librorum suorum de pulsibus*, 9:431-549; and *De pulsibus ad Antonium*, 19:629-642. Particularly important is Rudolph E. Siegel, *Galen's System of Physiology and Medicine* (Basel and New York, 1968); P. Capparoni, *Il "Tractatus de pulsibus" di Alfano I° Arcivescovo di Salerno (Sec. IX)* (Rome, 1936); and C. H. Haskins, *Studies in the History of Medieval Science* (Cambridge, 1924).

33. Avicenna, *Canon* (Pavia, 1510). See W. Puhlmann, "Die lateinische medizinische Literatur de frühen Mittelalters," *Kyklos* 3 (1930), 395-416; and A. C. Crombie, "Avicenna's Influence on the Mediaeval Scientific Tradition," in G. M. Wickens (ed.), *Avicenna: Scientist and Philosopher* (London, 1952).

34. The best, most comprehensive study of pulse music, which I gratefully acknowledge as inspiring my own, is Nancy Siraisi's superb "The Music of Pulse in the Writings of Italian Academic Physicians (Fourteenth and Fifteenth Centuries)" *Speculum* 4 (1975), pp. 689-710.

35. Gentile da Foligno, *Primus Avic. Canon. cum argutissima Gentilis expositione . . .* (Pavia, 1510; The New York Academy of Medicine). See the references in note 8; Siraisi,[34] and Frances Yates, *Giordano Bruno and the Hermetic Tradition* (London, 1964).

36. See Nancy G. Siraisi, *Arts and Science at Padua* (Toronto, 1973); Nan Cooke Carpenter, *Music in the Medieval and Renaissance Universities* (Normon, Oklahoma, 1958; reprint New York, 1972); and Edward A. Lippman, "The Place of Music in the System of Liberal Arts," *Aspects of Medieval and Renaissance Music*, ed. Jan La Rue, (New York, 1976).

37. Peter of Abano, *Conciliator* (Venice, 1496). See Lynn Thorndike, *History of Magic*, Vol. II: 874-947 (note 18); and his "Manuscripts of the Writings of Peter of Abano," *Bulletin of the History of Medicine* 15 (1944), pp. 201-19; G. Vecchi, "Medicina e musica, voci e strumenti nel "Conciliator" (1303) di Pietro d'Abano," *Quadrivium* 8 (1967), pp. 5-22.

38. So said Avicenna, *Canon* 1.2.3.1 [Pavia, 1510] cited by Siraisi, "Music of Pulse," p. 699.

39. See George P. Conger, *Theories of Macrocosms and Microcosms in the History of Philosophy* (New York, 1922); Rudolph Allers, "Microcosmus: From Anaximandros to Paracelsus," *Traditio* 2 (1944), pp. 319-407; and Leo Spitzer, *Classical and Christian Ideas of World Harmony* (Baltimore, 1963); as well as Edgar de Bruyne, *Étude d'ésthétique médievale* (Bruges, 1946).

40. On this Boethian tripartite nature of music, see his *De musica*, as well as David S. Chamberlain "Philosophy of Music in the *Consolatio* of Boethius," *Speculum* 45 (1970) pp. 80-97; and G. Pietzsch, *Der Klassifikation der Musik von Boetius bis Ugolino von Orvieto* (Halle, 1929). Compare P. Duhem *Le système du monde: Histoire des doctrines cosmologiques de Platon à Copernic* (Paris, 1913-1959) Volume 4.

41. I presented this theory in a lecture entitled "Heavenly Bodies: Medieval Medicine, Astrology, and Time," at the Metropolitan Museum of Art, Grace Rainey Rogers

Auditorium, 11 October, 1975, as the first lecture in a two-year series entitled "Time: Medieval, Renaissance, Perennial," co-sponsored by the Institute for Medieval and Renaissance Studies and the International Society for the Study of Time. The total series was made possible by a grant from the John Dewey Foundation and a matching grant from the National Endowment for the Humanities. Amongst the best background books are J. T. Fraser, *The Voices of Time* (New York, 1966), and his *Of Time, Passion, and Knowledge* (New York, 1975). The works of Lynn White, Lynn Thorndike, and Derek DeSola Price are critical: for example White's *Medieval Technology* (London, 1962); Thorndike's "Invention of the Mechanical Clock about 1271 A.D.," *Speculum* (1943) pp. 242-3; and Price's "On the Origin of Clockwork," *Smithsonian Bulletin* 218: *Contributions from the Museum of History and Technology;* and see S. Bedini and D. R. Maddison "Mechanical Universe," *Transactions of the American Philosophical Society* n.s. vol. 56:5.

42. A fine introduction to astrolabes is that which Geoffrey Chaucer wrote for his "little son," available in any decent edition of Chaucer's works. See also Andrew E. Brae (ed.) *The Treatise on the Astrolabe of Geoffrey Chaucer* (London, 1879); O. Neugebauer "The Early History of the Astrolabe," *Isis* 40 (1949); Lynn Thorndike, *The Sphere of Sacrobosco and Its Commentators* (Chicago, 1949); Lynn White, Jr., *Medieval Technology and Social Change* (Oxford, 1962); Ernest L. Edwardes, *Weight-Driven Chamber Clocks of the Middle Ages and Renaissance* (Altrinchan, England, 1965); J. Drummond Robertson, *The Evolution of Clockwork* (London, 1931); R. T. Gunther, *Early Science in Oxford* (Oxford, 1925) Volume 3; and Walter Clyde Curry. *Chaucer and the Medieval Sciences* (rev. ed., London, 1960); and see Madeleine Pelner Cosman, "Medieval Medical Malpractice and Chaucer's Physician," *New York State Journal of Medicine* 72 (October, 1972) pp. 241-43. Important studies of small units of time such as minutes include T. K. Derry and Trevor I. Williams, *A Short History of Technology from the Earliest Times to A.D. 1900* (London, 1956); and G. H. Baillie (ed.) *Britten's Old Clocks and Watches and their Makers* (London, 1956). A fine study of medieval clockwork is Sue Ellen Holbrook's "The Clock in Medieval Literature and Art: 'A Beautiful, Notable, Pleasant, Profitable Instrument" (unpublished dissertation; available from Dr. Holbrook at Temple University, Philadelphia, Pennsylvania 19122).

43. A pleasant introduction is Lynn Thorndike's *A History of Magic*, Vol. III. See the essay by Bert S. Hall in this volume, and the bibliographical references therein.

44. For the names and achievement of other clockmakers, see Derek DeSola Price[41] as well as the references listed in notes 41 and 42.

45. Lynn Thorndike gives an overview of the scientific and superstitious means of combatting plague terror in *History of Magic and Experimental Sciences*, vol. III, pp. 224-346. See the references in notes 2 and 23.

46. The father and son pair are well introduced by Thorndike (note 45), vol. III, pp. 386-97.

THE SPANISH JEWS who came to southern France in the second half of the twelfth century brought with them cultural values that they had acquired from their Muslim neighbors. Their translations of philosophic and scientific texts from Arabic into Hebrew were eagerly received by the Jews of southern France, who then proceeded to make their own contributions. Notable here are the works of Samuel ben Judah of Marseilles, Kalonymos ben Kalonymos, Levi ben Gerson, Immanuel ben Jacob, and Shelomo Davin de Rodez, all of whom wrote on scientific subjects in Hebrew in fourteenth-century France. Levi ben Gerson, the most original thinker among them, produced a new and successful lunar theory, and invented an observational instrument called the Jacob Staff that was used successfully in astronomy and navigation for several centuries.   MPC

The Role of Science in the Jewish Community in Fourteenth-Century France

Jewish Studies Program
University of Pittsburgh
Pittsburgh, Pennsylvania 15260

A FEW years ago L. V. Berman claimed that Samuel ben Judah of Marseilles, a fourteenth-century French Jewish translator of scientific and philosophic texts from Arabic to Hebrew, wished to transfer the Greek spiritual heritage into Hebrew.[1] As we shall see, this move succeeded to a significant degree; indeed, the Jewish community was able to build on this heritage, and not merely copy it. But should we be surprised at their interest in such secular subjects as logic and mathematics? Did the fact that Greek science came to them via Arabic sources have any deep significance, or is this simply a matter of linguistic concern? First, let us recall that Jews were not involved in scientific activity in the Biblical period or in Hellenistic antiquity. In effect, there was no relevant antecedent tradition. Second, it is often thought that Jewish interest in secular subjects invariably derives from contact with their non-Jewish neighbors. In the case of Jews in southern France this contact with Gentiles came after philosophy and science had already developed in the Jewish community. This anomaly is resolved by noting the dependence of these Jews on emigrés from Muslim Spain who in turn had participated in, and been influenced by, the high culture of the Arabs there. The first part of this paper will consider this process of transmission, and the second part will describe some of the scientific achievements of Jews in fourteenth-century France.

◄§ I §►

Soon after the Muslim conquests most Jews adopted Arabic as their vernacular, replacing Aramaic which had been the lingua franca of the Near East. Perhaps of greater importance, Arabic was used for a wide range of literary purposes. While the roots of the Judeo-Arabic synthesis lie in the East, the greatest achievements took place in Islamic Spain in the so-called "Golden Age" from the tenth to the twelfth centuries. The person who represented the ideal type for the Spanish Jewish élite was Samuel ibn Nagrela (993-1056), eulogized as "a prince, whose name has filled the earth; as a tower that has risen over the people of Israel, established aloft as a fort for his nation."[2] Samuel was a courtier who eventually became vizier over the kingdom of Granada, and leader of the Jewish community. He was also a wealthy merchant, a Rabbinic scholar, a noted poet, and an army commander. In his day, the Jews looked upon him as having the noble qualities usually associated with royalty. The Jewish courtiers— most of them might better be called civil servants—were the leaders of the community and developed their own set of values. Samuel was the model: he became a patron of Jewish scholars and courtiers just as the Muslim rulers were patrons of Arab scholars. His protégés composed liter- ary works in an elegant style that was patterned after that of their Muslim counterparts. In the scale of values of the time poetry ranked very high and this led to the development of Hebrew poetry that was based on Arabic meters but derived its idiom from the Bible and which was used as a vehicle for secular as well as religious expression. As a patron, Samuel encouraged his contemporaries to contribute to all branches of knowledge: rabbinics, linguistics, poetry, philosophy, science, and medicine. The courtiers considered themselves the nobles of the people with a sacred trust; they knew that their actions brought glory or shame to the entire community. As noble Jews they longed for the Messianic era when real power would be in their own hands ending the precarious state of affairs in which they lived.[3] They clearly felt prepared for this eventuality. The emphasis on the integration of secular and religious modes of expression was unique to the Spanish Jewish élite and is perhaps its most important legacy. It is also noteworthy that these Jews felt at home in Spain and looked upon it as the seat of culture.

Their world was not to last, however. Because of persecutions in the mid-twelfth century, the Jews of Muslim Spain were forced to flee to northern (Christian) Spain, and to a lesser extent to southern France and elsewhere. The Jews in Christian domains had a long tradition of Rab- binic scholarship in Hebrew, but no knowledge of Arabic. The refugees from Muslim Spain came to them with a cultural heritage of high status and roused their curiosity. The Jews in France had never participated in the governance of the country and had no tradition of courtiers. Moreover, their interest in secular culture did not result from contact with their Christian neighbors; indeed, they were excluded from the universities when they were founded. Whereas in Spain secular culture among the Jews had been part of a world view in harmony with that of the surrounding Arab society and was thought to prepare members of the community for their careers in the civil service, in France the devotion to secular culture had to be justified on completely different grounds. Basically, the Jews in France studied these disciplines for their own sake with the dedication previously reserved exclusively for Rabbinic texts, and because of "the nobility of the subject."[4] Other reasons were sometimes added, e.g. that science was useful for interpreting the Bible and the Talmud, and that Jews were respected by their Christian rulers for their secular learning (particularly medicine and astronomy).

Towards the end of the twelfth century, the Rabbis of southern France sent a letter to Maimonides, then resident in Egypt, inquiring about the meaning of some astrological terms. In his reply Maimonides tried to discourage them because he claimed that astrology was a form of paganism.[5] But Maimonides was happy to have his philosophic work, *The Guide for the Perplexed*, translated from Arabic into Hebrew in southern France for the benefit of the Jews there who could not read the original. Maimonides also makes it clear that science and philosophy are intimately related to religious values. His *Mishneh Torah*, a compendium of Rabbinic law written in Hebrew and widely influential, has a philosophical introduction and, at the beginning of the "Guide" he tells us that mathematics, logic, and astronomy are required before one can properly be introduced to philosophy.[6]

Translations from Arabic into Hebrew had begun to appear in the middle of the twelfth century. At first the translators concentrated on religious and philosophical works by Jews, but then they turned to other texts including scientific studies some of which were originally composed in Greek. By the thirteenth century, there were Jews in southern France who had become true enthusiasts for philosophy and the sciences and they were sometimes as fanatical as naïve believers. They asserted that unexamined piety was well beneath the intellectual experience that comes with philosophic study. Moreover, they looked to Maimonides as a model because he commanded the respect of Muslims and Christians, a goal they earnestly sought.[7] By the middle of the thirteenth century, a large enough set of works had been translated for the Jewish scholars to become intellectually independent. Though translations continued, the patterns were fixed, and a scholar who only knew Hebrew could make original contributions in a great number of fields.

This rationalist school met with opposition from the traditionalists who saw philosophy as potentially destructive of the religious beliefs of the community. A lengthy polemic led to a ban on philosophy proclaimed in Barcelona in 1305 which read in part:[8]

> From this day on and for the next 50 years no member of our community shall study Greek works on science and metaphysics, either in the original (*i.e.*, in Arabic) or in translation, before he will have reached the age of 25. . . .

Despite this ban on philosophy, it was still permitted to study medicine, astronomy, and all the works of Maimonides (including his philosophy!).

In another passage of the ban we are told that the philosophers "preach blasphemous homilies and scoff at the words of the sages. . . . Therefore, we excommunicate these transgressors, and place them under a curse and under the ban. . . ."

In the next year, 1306, Philip IV ordered the expulsion of the Jews from France (which did not include Provence at the time) and that seemed to end the controversy on philosophy for, surprisingly, there are no later traces of the ban either in France or Spain, and scholars went on studying philosophy.

The occasion for this ban was the scandal of a poor scholar named Levi ben Ḥayyim (d. after 1315) who wrote encyclopedic works on mathematics, science, ethics, theology, and philosophy. He was born in Perpignan and lived successively in a number of towns in southern France—at the time of the controversy, he was living in Narbonne under the patronage of Samuel de Scaleta, a wealth and pious Jew.[9] The leading Rabbi of Barcelona described Levi ben Ḥayyim as "worse than the gentiles who differ with us in their interpretation of a few verses. [However, he and] his colleagues do not spare even a letter of the Torah. . . ."[10] In reply to the suggestion that philosophy be banned, Jacob ben Makhir of Montpellier, a leading intellectual and scientist of the time, argued that secular study was not only permitted but highly desirable:[11]

> We would do well to learn the example of the most civilized nations who translate learned works from other languages into their own, and who revere learning. . . . Has any nation changed its religion because of this? . . . How much less likely is that to happen to us who possess a rational Torah. . . .

Modern scholars have been surprised at the treatment of Levi ben Ḥayyim because in examining his works they find nothing terribly radical; indeed, he seemed to be following in the footsteps of Maimonides.[12] Leo Baeck concluded that he was hounded not for his opinions, but because of his poverty and relative defenselessness. Halkin ends his study of Levi ben Ḥayyim's theory of the philosophical-allegorical interpretation of the Bible with the remark that Levi ben Ḥayyim was unjustly treated by his contemporaries and that his views were not at all subversive of Jewish beliefs.[13] Indeed, he argued strongly that the faith of Israel was superior to any other, and that the nobility of the Jewish character had been demonstrated by the steadfastness of the people in the face of persecution and suffering. Levi ben Ḥayyim also expressed the commonly held "myth" that philosophy originated in ancient Israel and was taken from them by the other nations, but because of exile this learning was lost by the Jewish community.[14] It is clear that this myth served an important apologetic role in the defense of science and philosophy against the traditionalists.

✍ II ☙

Though the French Jewish community in the fourteenth century was restricted to Provence, it was an intellectually vital group that made many original contributions to science and philosophy in Hebrew. The texts of Aristotle and Averroes were widely studied and commentaries were written on them. Significant advances were made in several branches of mathematics and astronomy that compare favorably with any in the Middle Ages, some of which still remain to be explored. From the fact that these Jews wrote in Hebrew, we learn that they wrote for the benefit of the Jewish community. They did not write in Latin or French, which would have been the case if their audience were primarily Christian. By way of contrast, in thirteenth-century Castile (Spain), King Alfonso X supported many scientists, including a number of Jews, and their works were written in Spanish.[15]

Samuel ben Judah of Marseilles, a member of a wealthy Jewish family, was born in 1284, and traveled widely in southern France. At the age of 18 he became a student of philosophy, as well as of astronomy and medicine, under the guidance of Sen Astruc de Noves—despite the ban on philosophy seven years earlier. He translated works on logic, political philosophy, astronomy, and natural science; his only independent work was a commentary on the Almagest. The last date for his activity is July

1340 when he finished his copy of a revised translation of Alexander of Aphrodisias' *De Anima*.[16] We learn a great deal about his method of working from the epilogues he appended to his translations. Two excerpts will illustrate this: the first from his translation of Averroes' *Epitome of Plato's Republic*, and the second from his revised version of Jacob ben Makhir's translation of Jābir ibn Aflaḥ's *Epitome of the Almagest*.[17]

(1) When I translated [Averroes' *Commentary on Plato's Republic*], I still did not have the commentary of Averroes on the *Ethics* of the philosopher Aristotle. . . . Therefore I put myself to much effort in order to procure a copy [of it] . . . and I roused myself to translate it. God, who is honored, in his kindness was partial to me so that I completed the translation of the commentaries on the whole of political science. . . . The work of revision and correction was completed [in 1320] in the citadel of Beaucaire, shut up and abandoned with the rest of my breathren, imprisoned in one of its fortresses. . . . Therefore, let not the serious student reproach me . . . [for my mistakes, since] there is no craftsman who does not, at times, err in his work or craft. . . .

(2) When I achieved a good understanding at that time of this honored science [astronomy] and all or nearly all of the other sciences, I realized from the words of Averroes in his book on this science that the good found in them was gleaned from the book of Ibn Aflaḥ. . . . We [I and my brother] exerted ourselves until we acquired a copy of the original Arabic. . . . We copied it very quickly, shut up two days in one of the houses of the seekers of wisdom who lived there [in Trinquetailles], nourished on a minimum of bread and water. Then we were forced to return the Arabic original to its owner. . . .

From these passages we can see how highly motivated Samuel was, for he was willing to endure many hardships. His diet of bread and water may mean that the owner of the Arabic text was a Christian, and that he was unwilling to eat the food of the house. It is also clear that Samuel was working on his own, and not as part of a coordinated effort. He did not translate books simply because they were there, or even because he had read them and liked them. He sought out books in certain subjects on the basis of allusions to them in texts he had read, and immediately set out to translate them. The astronomical work of Jābir Ibn Aflaḥ (twelfth century, Spain) influenced later writers in Arabic, Hebrew, and Latin; one copy of the Arabic text survives in Hebrew characters (British Library **Ms. Or.** 10,725, folios 92b-175b).

Some of the texts translated by Samuel do not survive in the original

Arabic. Among them is the *Treatise on Twilight* by Ibn Muꜥādh, an eleventh-century Spanish Muslim. The same text was also translated into Latin and was quite influential in the sixteenth and seventeenth centuries. The goal Ibn Muꜥādh set out to achieve was the computation of the height of the atmosphere based on the angle below the horizon that the sun must reach for twilight to end and night to begin. Ibn Muꜥādh takes that angle to be 19°, which is within the range of such parameters in Arabic sources, and finds the height of the atmosphere to be about 50 miles, a value generally accepted until the mid-seventeenth century.[18]

The text of Ibn Muꜥādh ends with a polemic that Samuel did not understand, and the final paragraphs are left in Arabic though written in Hebrew script. It is clear from this that Samuel's knowledge of Arabic was imperfect—in one case he failed to realize that a passage was from the Qur'an.

Another prolific translator of scientific texts during this period was Kalonymos ben Kalonymos (d. after 1328), who lived in Arles for the most part.[19] Of special interest is his translation of Ptolemy's *Planetary Hypotheses*, dated 1317; the extant Greek text is incomplete and the Arabic version survives in two copies. Ptolemy's system of cosmic distances was generally accepted in the Middle Ages, but before this text came to light some 10 years ago, there was no direct evidence to link it with Ptolemy. Though I published the Arabic version only, I had first investigated the Hebrew translation where I noticed the crucial passage for the recovery of the system for which Ptolemy is best known.[20]

The most original thinker to write in Hebrew in fourteenth-century France was Levi ben Gerson (1288-1344), who lived in Orange (at the time the Jewish population of that town was about 50–100 families).[21] He is best remembered for his philosophical works and commentaries on the Bible, but his contributions to mathematics and astronomy put him in the first rank of medieval scientists. Unlike Maimonides, Levi accepted astrology while keeping it quite apart from his astronomical research. His primary intention was to treat astronomy in a way that would be satisfactory from two points of view: that of mathematics and that of natural philosophy. In chapter one of his *Treatise On Astronomy* (136 chapters in all), he writes:[22]

> We found that [the ancient] mathematicians decided it was sufficient to produce a model [for planetary motion] from which there would follow that which is in close agreement with what is perceived by the senses, but they did not attempt to explain the model according to true principles [of natural

philosophy]. . . . In its perfection this investigation belongs to both sciences: to mathematics because of the geometric proofs, and to natural philosophy because of the physical and philosophical proofs. Since this is so, this investigation cannot be split in such a way that part of it would belong to a master of one science and the remainder to a master of the other, because the second would not know what is missing from the investigation of the first unless he already knew what the first had explained. . . .

On the basis of a very detailed analysis, Levi rejected Ptolemy's models and, in the case of the moon, he came up with a model that agreed more closely with observation than that of Ptolemy. In fact this lunar model is one of the most successful scientific theories of the Middle Ages, though it seems to have had little influence on subsequent developments.[23]

Levi indicates that he composed his astronomical tables at the request of "many great and noble Christians." Since a Latin translation of some chapters of his astronomy was dedicated to Pope Clement VI in 1342 (during Levi's lifetime), it is plausible to assume that Levi's astronomical research was of interest to, and perhaps even supported by, the papal court then in Avignon. Levi's brother was a physician,[24] and helped an Augustinian friar translate Levi's treatise on the astrological significance of a planetary conjunction to take place in 1345.[25] Levi's best known scientific achievement was his invention of the Jacob Staff, a simple but accurate instrument for astronomical obseravtions (FIGURE 1). It is widely used for astronomical navigation, with some refinements, from the late fifteenth century until the mid-eighteenth century. Levi's concern for accurate observations led him to introduce a diagonal scale for subdividing angles very finely that seems to have influenced Tycho Brahe's construction of astronomical instruments in the late sixteenth century.[26]

In the next generation, Immanuel ben Jacob of Tarascon, known as Bonfils, constructed an extremely popular set of astronomical tables, which was translated into Latin and Byzantine Greek even though it is based on the Jewish calendar.[27] Bonfils was aware of Levi's work and may even have been his student, but chose to depend for his lunar tables on a ninth-century Muslim astronomer, al-Battānī, rather than on Levi who had explicitly rejected the results that derive from his Muslim predecessor. During the summer of 1977 I came across a short text in a Hebrew manuscript (Casanatense, Rome, Ms. 204:115b) with the heading: "Table for the motion of the [solar] apogee arranged for years and months according to the calculation of *Ralbag* [Rabbi Levi ben Gerson], whose soul is in paradise, composed by Immanuel ben Jacob: the radix is according to the leap day [Feb. 29] of 1340 of the Christian era, noon of the last day of

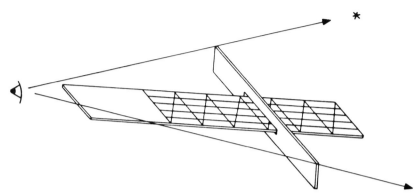

FIGURE 1. This drawing illustrates the Jacob Staff invented by Levi ben Gerson (1288-1344); the manuscripts contain a verbal description only. To use this instrument, the cross-piece is moved along the staff until two stars are seen as shown here. The angular distance between them may then be determined from the ratio of the length of the cross-piece to its distance from the eye.

the year." The motion of the solar apogee is given in this text as $1;22,33°$ in 60 years which agrees with the value in Levi's text of $1°$ in $43\frac{2}{3}$ years.[28] Levi included some 45 tables arranged for the Christian calendar in his astronomical text, but this is not found among them. Here we have an example of one of his successors, after his death, composing a table in his style and with his parameters.

In addition to scientific developments within the Jewish community, there were translations of works originally composed in Latin. One such text is a set of tables, radix 1368, that was translated into Hebrew from Latin by Shelomo Davin de Rodez, a pupil of Bonfils (Ms. Munich 343: 104b-167a). It includes a very detailed and unusual set of tables for planetary longitudes and latitudes, based on the Alfonsine Tables of the thirteenth century. As if by chance, John North just finished describing the Latin version of these very same tables based on manuscripts in Oxford, with radix 1348, thus permitting me to identify their origin.[29] As far as we know, no copy of this Latin text survives in France, and so the mode of transmission cannot be reconstructed at this time.

This scientific tradition continued in southern France through the fifteenth century until the expulsion of the Jewish community from Provence, but for the papal enclave around Avignon, at the end of that century.

&§ §&

ACKNOWLEDGMENTS

This study was generously supported by a research grant from the National Endowment for the Humanities. I am most grateful to the Institute of Microfilmed Hebrew Manuscripts at The National and University Library in Jerusalem which I visited in the summer of 1977 for providing access to their vast holdings. The film of the Casanatense (Rome) text that I examined there has the serial number: 42.

NOTES AND REFERENCES

1. See p. 289 of L. V. Berman, "Greek into Hebrew: Samuel ben Judah of Marseilles, Fourteenth-Century Philosopher and Translator," in *Jewish Medieval and Renaissance Studies*, ed. A. Altmann, (Cambridge, Mass.: Harvard University Press, 1967, pp. 289-320.

2. See p. 282 of G. D. Cohen, *The Book of Tradition by Ibn Daud*. Philadelphia: The Jewish Publication Society, 1967.

3. Cohen[2], p. 284.

4. Levi ben Gerson's Introduction to his Astronomical Tables, British Library Ms. Add. 26,921, folio 12b.

5. See pp. 463-473 of I. Twersky, *A Maimonides Reader*. New York: Behrman House, 1972.

6. See page 3 of Maimonides, *The Guide of the Perplexed*. trans. S. Pines. Chicago: University of Chicago Press, 1963.

7. See p. 204 of I. Twersky, "Aspects of the Social and Cultural History of Provençal Jewry." *Journal of World History* 11 (1968), pp. 185-207.

8. See vol 1, pp. 301-302 of Y. Baer, *A History of the Jews in Christian Spain*. 2 Vols. Philadelphia: The Jewish Publication Society, 1961.

9. Baer[8], Vol. 1, p. 292.

10. Baer[8], Vol. 1, p. 294.

11. Baer[8], Vol. 1, p. 296.

12. See p. 67 of A. S. Halkin, "Why was Levi ben Ḥayyim Hounded?" *Proceedings of the American Academy for Jewish Research* 34 (1966), 65-76.

13. Halkin[12], p. 76.

14. Halkin[12], p. 74.

15. See p. 365 of D. Romano, "La transmission des sciences arabes par les Juifs en Languedoc," in *Juifs et judaisme de Languedoc*. eds. M.-H. Vicaire and B. Blumenkranz. *Cahiers de Fanjeaux*, 12. Toulouse: E. Privat, 1967, pp. 363-386.

16. Berman[1], p. 298.

17. Berman[1], p. 299.

18. B. R. Goldstein, "Ibn Muᶜādh's Treatise On Twilight and the Height of the Atmosphere." *Archive for History of Exact Sciences* 17 (1977), pp. 97-118.

19. See pages 417ff. of E. Renan, "Les écrivains juifs français du XIVᵉ siècle," in *Histoire Litteraire de la France* 31 (1893), pp. 351-789, Reprinted 1969. Westmead, England: Gregg International Publishers.

20. Goldstein, B. R. "The Arabic Version of Ptolemy's Planetary Hypotheses," in *Proceedings of the American Philosophical Society*, N. S. 57 (1967). Philadelphia.

21. Shatzmiller, Y. "Gersonides and the Jewish Community of Orange in His Day," in *Studies in the History of the Jewish People and the Land of Israel*, eds. B. Oded et al. Vol. 2, pp. 111-126. Haifa, Israel: University of Haifa, 1972. In Hebrew with English summary. Guillemain, B. "Citoyens, Juifs et Courtisans dans Avignon au XIVe siècle," in *Comptes Rendus du 86e Congrès National des Sociétés Savantes*, Montpellier, 1961. Section phil.-hist. Paris: Gautier-Villars, 1962, pp. 147-160.

22. Paris, Bibliothèque Nationale, Ms. hebr. 724, folio 5b.

23. See pp. 53-74 of B. R. Goldstein, *The Astronomical Tables of Levi ben Gerson*. Hamden, Conn.: Archon Books, 1974.

24. Shatzmiller[21], p. 120.

25. See pp. 309-311 of L. Thorndike, *History of Magic and Experimental Science*, Vol. 3. New York: Columbia University Press, 1934.

26. Goldstein, B. R. "Levi ben Gerson: On Instrumental Errors and the Transversal Scale." *Journal for the History of Astronomy* 8 (1977), pp. 102-112.

27. Solon, P. "The Six Wings of Immanuel Bonfils and Michael Chrysokokkes." *Centaurus* 15 (1970), pp. 1-20.

28. Goldstien[23], p. 94.

29. North, J. D. "The Alfonsine Tables in England," in *Prismata: Festschrift für Willy Hartner*, eds. Y. Maeyama and W. G. Saltzer. Wiesbaden: Franz Steiner Verlag, 1977, pp. 269-301.

ENGLISH THINKERS and English thought, particularly early fourteenth-century Oxford philosophy, had astonishing impact upon the philosophical enterprises at the University of Paris. How did this English influence effect fourteenth-century Parisian learning? Fourteenth-century intellectual history requires our investigating how conceptions and techniques of analysis in logic and natural philosophy initially developed at Oxford (most notably at Morton College) were applied and extended in Paris. The effects of such "Anglicanae" can be found not only in the natural philosophy of the "Arts Faculty" at Paris, but also within the "Theology Faculty." The *Sentence Commentaries* of two seldom-studied theologians, John of Mirecourt and Pierre Ceffons, are especially informative instances of the character and extent of this Oxford influence in the history of Parisian ideas. MPC

Subtilitates Anglicanae in Fourteenth-Century Paris: John of Mirecourt and Peter Ceffons

JOHN E. MURDOCH

Department of the History of Science
Harvard University
Cambridge, Massachusetts 02138

Alas! by the same disease which we are deploring, we see that the Palladium of Paris has suffered in these sad times of ours, wherein the zeal of that noble university, whose rays once shed light into every corner of the world, has grown lukewarm, nay, is all but frozen. There the pen of every scribe is now at rest, the generation of books no longer occurs, and there is none who begins to assume the role of new author. They wrap up their doctrines in unskilled discourse, and are losing propriety of all logic, except that our English subtleties, which they denounce in public, are the subject of their furtive vigils.

T HE words are those of that curious fourteenth-century dilettante, Richard de Bury, written in the year 1344.[1] Although naturally not about Guillaume de Machaut himself, these words were intended as a description of a part—the academic part—of Machaut's world. To be sure, there has been much discussion of just how Richard de Bury's claims represent the historical facts. In particular, attempts have been made to weigh his assertion of the public denunciation, but private secret study, of those *subtilitates Anglicanae* against what we know of the Parisian condemnation in 1339 and 1340 of "dogmatizing the doctrine of William called Ockham."[2] Fortunately, it is presently not necessary to determine how accurately Richard's charges reflect any part of that condemnation, or even to enter into the very difficult matter of just what those Parisian edicts meant.[3] It is quite sufficient to note that, in spite of the undeniable exaggeration and English prejudice of these charges, Richard was correct in claiming that English subtleties were very much part of intellectual endeavor at Paris in the mid-fourteenth century. They were present in what

[51]

0077-8923/78/0314-0051 $01.75/2 © 1978, NYAS

Richard refers to in an earlier passage as the "quires of yesterday's sophisms,"[4] or, more exactly, in the conceptions and techniques that one finds in these English sophisms—for they were, basically, English—and in other works that emanated from Oxford circles.

It is to at least a fragment of this English influence, then, that I will address myself here. Without doubt, there was much appreciation and utilization in Paris of the content of English philosophy and science, and also much influence in the area of theological content. It is not this, however, that will be at the center of what I wish to say; specifically, I wish to examine the influence of methods that were characteristic of fourteenth-century Oxford natural philosophy and logic, particularly of that logic which was applied within natural philosophy. I wish also to reveal something of the extent of that influence upon Parisian learning taken more broadly, that is, not primarily within the area of natural philosophy and science, but within that of theology as well; such an account will, I believe, make the impact of *subtilitates Anglicanae* in fourteenth-century Paris more apparent.

Still, before broaching this broader topic, at least a passing word must be given to the rather obvious case of the Parisian practice of what Ockham and his followers had preached with respect to the extensive utilization of logic in the resolution of natural philosophical problems. It was not simply the case that Ockham quite frequently set the form in which a given problem was subsequently to be discussed,[5] but one is witness to more than a fair amount of his technique of resolving such problems by the invocation of what might most appropriately be termed "propositional analysis." That is to say, in place of investigating a given issue in natural philosophy by dealing directly with the things or events in nature involved in that issue, one analyzed the problem in terms of the propositions, and the terms within these propositions, that spoke of the things and events in question. Ockham himself fairly bristles with such analyses, and John Buridan, who can properly be regarded as something like the *chef d'école* for fourteenth-century Parisian natural philosophy, provides a clear example of a penchant for a similar kind of analytical approach and practice.[6]

There were, however, more specific phenomena within Oxford natural philosophy that were disseminated at Paris. They too formed part, even a greater part, of Richard de Bury's judgment. Involving as they did both mathematical and logical conceptions (often combined) at the methodological level, they were, moreover, phenomena that were quite characteristic of the English approach to things. I shall focus upon those aspects of these conceptions that will emphasize the "physical strand" of the impact of subtilitates Anglicanae, but even in this restricted area many of the more important features of this English influence will become evident. I have in mind in particular four conceptions or doctrines that were developed in the earlier fourteenth century at Oxford.

Two of these doctrines are to be found in Thomas Bradwardine, specifically in his *Tractatus de proportionibus velocitatum in motibus* (written in 1328). The first is that which is at the very basis of his renown in the history of medieval science: namely, his resolution of the problem of what relations obtain between speeds or velocities and the forces and resistances determining those velocities. His answer to this problem was that the proportion of velocities follows the proportion of forces to resistances which, in our terms, meant that an arithmetic change in velocities corresponds to a geometric change in force-resistance proportions. (Thus, for example, twice the velocity corresponds to the force-resistance proportion squared, three times to the relevant proportion cubed, and so on.)[7]

In the same work, Bradwardine expounded the second doctrine I have in mind: namely, that the speed of any body undergoing uniform local motion is to be measured by the distance traversed by the *fastest moving point* of that body.[8] This rule was especially relevant to cases of bodies in rotary motion, so that (for example) the motion of a rotating radius would be measured by the linear velocity of its outermost point, and it is to such instances that Bradwardine and his successor at Merton College, William Heytesbury, apply the rule.[9]

The third English doctrine that was to receive favor at Paris was concerned with the measurement of the distribution of a varying quality over a given extended subject: or, alternatively, with the measurement of motions of varying speeds over given intervals of time. Thus, if some quality (heat, for instance) was possessed by some subject body in the same intensity throughout, then that quality was said to be *uniformly* distributed over its subject. Similarly, a body undergoing local motion at a constant speed throughout a given time interval was said to be *uniformly* moved. Distributions or motions that were not uniform were termed difform. But there was a special class of difformly moved bodies or difformly distributed qualities: it comprised those cases of distribution or motion where the variation of the quality or motion in question was a uniform variation. Thus, if, beginning from rest, a body were to undergo constant acceleration, it was said to partake of a *uniformly difform* motion. Similarly, if the intensity of a given quality increased or decreased at a constant rate from one end of a subject to the other, the quality was claimed to be *uniformly difformly* distributed. What is more, these special uniformly difform qualities or motions were held to be measured by their mean degrees. That is, if, beginning from rest, a body moves with a uniformly difform motion over a given time interval at the end of which it has some terminal degree of velocity, it will traverse just as much space as if it moved *uniformly* throughout the same time interval with a velocity half the original terminal velocity (which is to say with a degree of velocity that is the [arithmetic] *mean* between rest, or zero degree, and its terminal degree). Similarly, a given subject which is uniformly difformly (say) hot, is just as hot as if it were uniformly hot throughout with the relevant mean degree of heat.[10]

The fourth and final doctrine was concerned with the setting of temporal limits to motions and changes or to the existence or non-existence of objects in time. The point of departure was Aristotle's argument that there is no first or last moment of a motion or process of change that belonged to the motion or process itself.[11] In scholastic terms, there is no first instant of being and no last instant of being for any motion. On the other hand, again following Aristotle, a proper temporal limit could be assigned for the onset of a motion by claiming that it had a last instant of non-being, that is, that there was a last instant to the time interval *before* the motion began; the beginning of a motion was therefore held to be *extrinsically* limited and the same was held to obtain for its termination. What was to become the standard treatise dealing with this matter of "first and last instants" was written by the English scholastic Walter Burley.[12] Furthermore, essentially the same material was expressed in a different form in treatises dealing with the logic of the terms "incipit" and "desinit," and once again the most influential works were those of English authors, most notably Richard Kilvington and William Heytesbury.[13]

It should be evident that all four of the doctrines I have mentioned had in one way or another to do with the *measurement* of things: of velocities, of forces and resistances, of qualities, and of the existence of things and processes in time. As such they formed the very heart of what has come to be known as the "calculatory tradition" at fourteenth-century Oxford. The point of present importance, however, is that all four of these doctrines, complete with much of the full development they received in all manner of works written by fourteenth-century English scholastics, were influential in Parisian circles where they often underwent new and different developments. Thus, not only do we find natural philosophers at Paris accepting Thomas Bradwardine's solution to the problem of relating forces, resistances, and velocities,[14] but we are witness to someone like Albert of Saxony taking pains to write a compendium of Bradwardine's work[15] or to someone like Nicole Oresme developing Bradwardine's "calculus of proportions" in a novel and important way.[16] Alternatively, the same Nicole Oresme establishes his own, fundamentally geometrical, system for the measurement of the variation of qualities and motions.[17] And although the doctrine of first and last instants and of "incipit" and "desinit" does not seem to have received the same attention by Parisian scholars, its traces too are evident.[18]

As I have already mentioned, however, a more revealing and, at the same time, unexplored picture of the extent to which these English techniques or doctrines of measure had penetrated Parisian quarters can be had if one turns from the natural philosophy of the Arts Faculty at Paris to the Faculty of Theology. As we shall see, at times only one of the four doctrines mentioned above will be applied in a given theological context, while in other cases more than one will be utilized. In any event, the substance of all four of these doctrines must be kept in mind in approaching the theological writings to be examined in what follows.

Although one could find considerable evidence of the penetration of these doctrines in any number of fourteenth-century Parisian theologians, especially informative are the cases of two, hitherto little studied, Cistercians: John of Mirecourt and Peter Ceffons.

Mirecourt lectured on the *Sentences* of Peter Lombard during the 1344-45 academic year at the Cistercian College of St. Bernard in Paris.[19] But his theological activities certainly did not end with that, for shortly thereafter 63 suspect propositions were extracted from his *Commentary on the Sentences*, an event to which Mirecourt replied with a first *apologia* explaining and justifying his contentions. The result was that a reduced roster of 41 condemned propositions was finally presented, an action that occasioned a second *apologia* from Mirecourt.[20]

Peter Ceffons, even less studied than Mirecourt, lectured on the *Sentences* in Paris in 1348-49, although the unique manuscript copy we possess of his commentary most likely represents it in its final redaction, which would have been prepared, it seems, sometime before 1353.[21] Thus both Mirecourt and Ceffons furnish us with instances of the Parisian theological enterprise shortly before and (perhaps) shortly after mid-century.

However, before we examine the guise assumed by the *subtilitates Anglicanae* in Mirecourt and Ceffons, we should realize that they were far from alone in applying such things to theological entities or in theological contexts. Any number of *Sentence Commentaries* of the period lead one to think that such an application was almost a mania. To be sure, the "measure tradition" is found lurking in all corners of the discussions of the infinite that were occasioned by the necessary facing of the issues of God's infinity, of His ability to create infinites, or of the possibility of the eternity of the world, all of these being standard problems in late medieval *Commentaries on the Sentences*.[22] Similarly, lecturing on Distinction 17 of the first book of Lombard's *Sentences* provided a natural occasion to try one's hand at the measurement of qualities and forms. For it was that very Distinction in Lombard which had earlier provided much of the context for the discussion of just how forms and qualities could undergo intension and remission in the first place, let alone have their different intensities measured.[23] But the Oxonian "measure tradition" also found application in other, less expected, theological areas. One of the most notable was that of the perfection of species or, to use more modern terms, that of the Great Chain of Being provided by the full scale of such perfections.[24] Perhaps even less expected is the application of techniques and doctrines of measure within a second area that was of utmost importance within fourteenth-century theology: that of the will and the problems of merit, reward, sin, and punishment tied to the actions of the will.[25] The application of calculatory material in these two areas is particularly thick in both Mirecourt and Ceffons, a factor that takes on a special significance when we come to realize that the notions of measure that they were there applying are to be found not only in English works on natural philosophy and logic, but also in English works on theology. And they were there often applied in identical theological contexts. The *Sentence Commentaries* of the likes of Adam Wodeham, Richard Kilvington, Thomas Bradwardine, and Thomas Buckingham are all excellent cases in point. Yet if we canvass those sources most frequently utilized by Mirecourt and Ceffons, the Commentaries of Roger Rosetus and Robert Halifax provide even better instances.[26] Ideas and whole passages are lifted from Halifax in particular. We must conclude, it would appear, that at least some of the English calculatory raw material used by Mirecourt and Ceffons in their deliberations was there in "ready-made" form in equally English theological writings.[27]

JOHN OF MIRECOURT

Let us now sample these deliberations: Mirecourt first. His most extensive application of the calculatory tradition occurs in seven questions of the third book on his *Commentary on the Sentences;* the context is that of the problem of measuring the acts of the will and the effects of these acts.[28] Mirecourt's approach is surprisingly systematic. He first establishes that not only can all acts of the will undergo intension and remission, but also all acts of the soul or intellect.[29] Secondly, this intension and remission occurs by means of the addition or subtraction of parts, a contention in which Mirecourt was tacitly appealing to something well established in the English tradition[30] and by which he was automatically to guarantee the application of notions of measure to the will's acts.[31]

Another foundational move was made by Mirecourt before he was able to apply these notions to the issues of interest to him; that is, he had to establish that the generation or corruption of any thing occurs or is produced *per motum vel mutationem.* (Although he does not confess it, he was here following Halifax word for word.[32]) But this meant in turn that there is no first instant of any intellective or sensitive act, nor any instant in which a luminous body first illuminates a medium or in which any natural agent first acts.[33] Thus far Mirecourt has spoken only of the generation and corruption of things; it remains to extend his contentions to the existence of things in time, something he accomplishes by appealing to the standard natural philosophical rules for first and last instants as applied to the existence of both permanent things (like Socrates) and successive things (like motion).[34] Specifically, he denies that even an indivisible thing can exist for but an instant, denies the existence of a last instant immediately before a permanent thing comes into existence, denies the existence of a terminal instant for its being, but affirms the existence of a last instant of non-being extrinsically limiting the onset of a successive thing like motion,[35] all of these being standard contentions in the literature dealing with temporal limits.[36]

Mirecourt is now prepared to apply these rules about limits to the issue of the will and its meritorious and non-meritorious acts. In a first question he establishes that, since any action of the will occurs successively over time and not instantaneously, its acting must obey the usual criteria for the existence of successive things and hence always lack a first instant of its being: there is no instant, that is, in which the will is acting before which it was not already acting.[37] In a second question, he fits the process of willing into equally standard notions relative to the limiting of continuous, successive entities: there can be no ultimate intrinsic instant of an action of the will; on the contrary, any such action is extrinsically limited by an instant of non-existence at both of its ends.[38] And in a final question he carries his application of natural philosophy to the will even further by introducing criteria stipulating which degrees of merit or demerit the will can, or cannot, obtain.[39]

There are, however, several difficulties that might be encountered in the setting of temporal limits to the successive action of the will, difficulties that would not arise in the case of most successive continua or motions. The first is that, because there is no last instant of the will's acting, we can always infer from its acting in any given instant that it will also be acting immediately after that instant.[40] But does this not imply a necessary, and not voluntary, continuation of the action of a will, something that would obviously be inconsistent with the freedom of the will? Not at all, Mirecourt replies: given any instant of the will's action, the inferred continuation of its acting is contingent, not necessary. One merely infers that the will does act after any given instant in which it is acting, not that it *must* act after any such given instant. The only "necessity" involved is that it would be impossible for the will to be acting in some given instant and never be acting after that instant; the will simply could not be acting under such circumstances.[41]

A second problem in specifying limits for the action of the will derives from the fact that it is capable of effecting contrary processes, that it can act meritoriously or can act non-meritoriously. But since meritorious action can successively follow non-meritorious action, and vice versa, we must be able to ascribe accurately the temporal limit that would separate two such successive actions on behalf of the will. It is possible, Mirecourt maintains, that one single instant can serve as the limit in which the will first ceases to act meritoriously and begins to act non-meritoriously. This possibility can be established, he argues, if we probe even further into natural philosophy. Aristotle had held that whenever a reversal of rectilinear motion occurs, a *quies media* must obtain. An object thrown upward, for example, must have an intervening moment of rest after its upward motion is completed and before its downward motion commences.[42] However, the scholastics revised Aristotle on this point and claimed that, although such a state of affairs might obtain if no other variables were introduced, if the object thrown upward were a bean and, while still moving upward, it met a descending millwheel, it surely would not enjoy a *quies media*.[43] Its downward motion would follow directly after its upward motion. It is to this that Mirecourt appeals, inter alia, in proving his conclusion concerning the successive contrary actions of the will; there too there is no *quies media*, a factor that allows the same instant to limit the cessation of one kind of acting and the beginning of its contrary.[44]

In sum, then, Mirecourt has applied substantial portions of the calculus of first and last instants in determining the appropriate temporal limits for natural processes that were of little concern to the natural philosopher, but of considerable concern to the theologian. For the will, taken in conjunction with the moral values pursuant to its acts, was a theological topic of appreciable importance in the fourteenth century.

Mirecourt's application of other aspects of English measure conceptions to the problem of the perfection of species is no less startling.[45] Indeed, the applying of measure was here rather problematic for a number of reasons. First of all, one had to account for relations of excess and defect between members of a single species. Secondly, one also had to have some kind of measure that would cover such relations between individuals belonging to radically distinct species (or between the radically distinct species themselves). And, finally, in order to have a system of measure that would cover the whole chain of being, one also had to take into account the relation had by all species and their constituent individuals to God. If nothing else, these final two requirements forced one to come to grips with the problem of measuring infinite distances. For not only are all created species and their individuals infinitely distant from God, but because different species are *radically* distinct they too must somehow be infinitely "apart" or incapable of comparison. For example, although both a man and an ass are infinitely distant from God, they are also somehow infinitely distant from each other.

With these kinds of "requirements" in mind, let us now turn to Mirecourt. He approaches these inherent difficulties by specifying two kinds of excess: one quidditative or essential, the other accidental. Quidditative excess is, of course, that kind of excess appropriate to individuals of diverse species (like our example of man vs. ass), whereas accidental excess cannot account for such specific differences but is applicable to such more ordinary cases of excess as Socrates' being taller than Plato or one object being hotter than another.[46] Furthermore, there are also two "latitudes" corresponding to the scale of perfections of all things: one essential, the other accidental.[47] One must go further, however, and establish that not only are these two kinds of excess and latitude distinct, but they are distinct in the sense that variations in one kind of excess in no way affect variations, or non-variations, in the other. In particular, quidditative excess or equality is in no way affected by the accidental excesses realized in comparing parts to wholes.[48]

This much accomplished, Mirecourt can then move to the delineation of just which kinds of excess are allowably infinite. Quidditative infinite excess cannot obtain between created species or their individuals; that is reserved for the relation of all such species and individuals to God.[49] On the other hand, infinite accidental excess is allowed only when it is measured with respect to something external to the exceeding subject.[50] Yet, whatever Mirecourt may have maintained with respect to these allowable infinite accidental and extrinsic excesses,[51] what was apparently of importance to him was the fact that, God and relations to God excluded, the quidditative excesses relevant to the scale of perfections could not be infinite.

In fact, the whole purpose of his investigation of this material was to construct a proper essential or quidditative latitude and to show how diverse species could be fitted into this latitude and hence be measured by it, the one to the other. Yet before allowing himself to concentrate on species alone, Mirecourt stipulates just what must be true of any latitude—essential or accidental, no matter—against which one measures perfections. It must begin from *non gradus* or zero degree of that property being measured and it must exclude the existence of any minimal degree.[52] Given this, any perfection being measured by means of such a latitude must be measured by its distance from *non gradus* and not by its distance from the *summus gradus* of the latitude, should there be such a degree.[53] His English predecessors had also debated just what *repère* should be employed in using a latitude for measure and had come to a similar, "distance from *non gradus*" conclusion.[54] Mirecourt does not, however, reflect their arguments for their choice, but instead, because he is basically concerned with species and essential perfections, gives a theological reason: the fact that God is the *summus gradus* of the relevant scale of perfections and all creatures are infinitely distant from Him; hence, to measure in terms of distance from such a *summus gradus* would be impossible.[55]

Mirecourt is now prepared to explain how diverse species fit into their thusly constructed essential latitudes. Since the perfection had by any species is unique, they possess indivisible degrees within the latitude (*non gradus* excluded), but, as such, constitute a chain of being that is continuous, since between any two species there can always be another.[56] However, this plenitude of the latitude of essential perfections is only potential; one *can* always have another species between any two given species; God *could* create such; but things are not that way in fact.[57] Furthermore, radically distinct species must correspond to different perfections in the scale, while things belonging to the same species must have the same perfection[58] and, accordingly, we should realize that any two distinct species can have degrees related to one another in any finite proportion.[59] However, we should also be aware that fixing the precise position of species on such a scale or latitude is only theoretically possible: we can never know the exact proportion of any one species to any other. All we can know is that any one species *does* exceed, or *is* exceeded by, any other.[60] That much follows from their radical distinction.

PETER CEFFONS

Peter Ceffons's *Commentary on the Sentences* is in many ways quite different from that of Mirecourt: in size (it would comprise some 2000 pages if printed) and in style and approach. However, since he does treat many of the same problems as Mirecourt and even uses much of the same natural philosophical material,[61] it might be useful to compare him with his fellow Cistercian in this regard before proceeding to the differences between their two commentaries. The best point of comparison is provided by the problem we have been examining of the perfection of species. Ceffons carries his investigation much further than Mirecourt and reveals his propensity for introducing as much mathematics and logic as possible into the analysis of an issue (something that was, of course, quite characteristic of fourteenth-century natural philosophy in general and of its exercise at Oxford in particular).

He approaches the problem in the same manner as Mirecourt,[62] but, unlike his predecessor, gets directly to the heart of what he takes to be the problem by excluding God from the picture and asking whether infinite excess can occur between finite things.[63] His resolution of this sub-question begins by citing the view of a certain Rogerus (who in fact is the English theologian Roger Rosetus) which denies the possibility of such an infinite excess.[64] But then, without naming him, Ceffons quotes Mirecourt's doctrine of essential versus accidental excess and, in terms of it, reports (again without name) other parts of Mirecourt's argument, in particular his conclusion that there can be no infinite quidditative excess or no infinite accidental and intrinsic excess between finite things.[65] At this point, however, he breaks with his unnamed predecessor and claims that the opposite view appears plausible to him.[66] He supports this by introducing what is perhaps the most extensive medieval investigation we have of curvilinear angles. In particular, he is interested in the angle of contingence or the horn-angle formed by the circumference of a circle and a straight line tangent to the circle (for example, the angle ABC in FIGURE 1). Such angles were much discussed in the Middle Ages since they provided

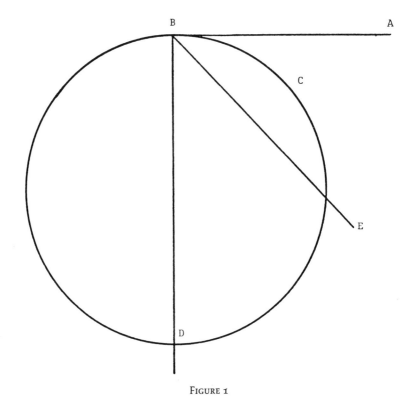

FIGURE 1

one of the few available intuitively clear examples of an infinitesimal:
it is a demonstrable fact that any horn-angle is less, indeed infinitely less,
than any rectilinear angle (for example, angle ABE) no matter how
small.[67] To this Ceffons adds the other curvilinear angle that fascinated
medieval scholars: the angle of a semicircle (which is the angle DBC in
FIGURE 1 and is infinitesimally less than the rectilinear right angle formed
by the diameter DB and the straight line BA tangent to the circle). The
result is that he has variables at his disposal that will allow him to formulate a whole battery of new rules about permissible cases of infinite excess:
the exceeding and exceeded objects are curvilinear and rectilinear angles.[68]
Ceffons is well aware of the novelty of the mathematics of the infinite he
was proposing and even troubles to cite (and name) Bradwardine on the
kinds of excess obtainable between proportions, and then shows that his
calculus of angles constitutes an exception to Bradwardine's rules.[69]

The point of all this is that Ceffons wanted a scale of measure that would prove effective when applied to the different perfections of radically distinct species and he believed that his calculus of angles provided him with such a scale. Indeed, the different kinds of angles in his calculus themselves belonged to radically distinct species. Thus, to cite only a few of the many examples given by Ceffons, we can have an infinite excess *ultra omnem proportionem* such as the excess represented by the relation of a rectilinear angle to a horn-angle (or of an angel to a fly) or we can have an infinite excess *citra omnem proportionem*, such as the excess represented by the relation of a right angle to an angle of a semicircle. And these angles are all *finite* things allowing of mutual infinite excess. So also, he suggests, are radically distinct finite species capable of mutual infinite excess.[70] What is more, the curvilinear angles can themselves be increased or decreased (by drawing smaller or larger circles to the relevant point of tangency or by adding curvilinear angles together) and they are still infinitely exceeded by rectilinear angles. This phenomenon corresponds, then, to an allowable increase or decrease of individuals *within* a given species while preserving the infinite distance between individuals of radically distinct species. And so on. Indeed, Ceffons carried his calculus of angles as applied to the perfection of species so far that he deemed it wise to apologize for his own excesses. Perhaps, he confesses, he had been too attracted by the mathematics of it all.[71]

Some of the characteristics of Ceffons's approach to problems are undoubtedly already evident. Hence, to speak only to our present interests, unlike Mirecourt (and others), he is not only willing to apply doctrines of measure to theological topics, but he is especially anxious to extend and develop the doctrines or techniques of measure themselves. Part of this would seem in some way to derive from his greater awareness of the English sources of many of these doctrines. In any event, he is more willing than most of his contemporaries to cite and to name these sources.[72] More than that, he has little hesitation in introducing matters tangential to the theological topics at hand, and will often even tell us why he has done so. He was, it seems fair to judge, an exceptionally talkative lecturer on the *Sentences.*

Thus, to turn to logic itself for a moment, Ceffons tells us that a very dear friend asked him to say something about *logicalia*, something that justifies, he feels, the ad hoc introduction into his *Lectura* of what amounts to a whole treatise dealing with the currently popular topic of the logic *de scire et dubitare* (that is, the modal contexts created by the presence of cognitive verbs).[73] His sources were, it seems, once more those *subtilitates Anglicanae*, since he cites, names, and argues, inter alia, with the *Sophismata* of Richard Kilvington.[74]

Moreover, in addition to introducing English-based topics by request (something he did, he tells us, *multa velocitate*[75]), there was another technique Ceffons utilized in order to import otherwise foreign matters into his theological works. It was one of the finding excuses. Hence, he justifies complying with another request he received for some logic by asking as the third question of Book II of his *Commentary* whether Blessed Augustine or even Master Peter Lombard or some other believing theologian could be led into an inconsistency by means of some insoluble proposition. The appropriate reply to this question is, as might be expected, the formulation of a whole disquisition on insoluble propositions or paradoxes. And, also as expected, the motivation is again English: Ceffons reproduces the substance of Roger Swineshead's *De insolubilibus* and appeals repeatedly to Thomas Bradwardine's treatment of the same logical topic.[76]

Still, while expounding, indeed while teaching, all of this logic to his fellow Cistercians, Ceffons did not refrain from indicating to his audience that the value of logic for doing theology was far from non-controversial. There are those, he reveals, who hold logic to be the cause of *curiositas* and even of damnation. They put logic together with—to cite one of Ceffons favorite couplets—the croaking of frogs and the cawing of crows.[77] But, Ceffons urges, this is all quite wrong-headed. Only those ignorant of logic defame it. It is most useful for theology and will often prevent one from falling into error. Indeed, although one may complain that he has already put too much logic into his *Sentences,* the truth is that such a complaint shows a lack of appreciation of its value in theological matters.[78]

The same techniques of introducing arts material into his *Commentary* appear when Ceffons turns to parts of natural philosophy and the quadrivium that were not so full of *Anglicana.* For example, he is able to deal with (perhaps, one imagines, is able to show off his knowledge of) astrology and astronomical considerations by working them into the context of the topic of creation at the beginning of Book II. He does this first by asking whether all things are created or produced, or come to pass, *de necessitate,* since this will open the door to a discussion *de eventibus necessariis astrorum.*[79] Secondly, because God has created the heavens, it is licit to debate whether one can prove by natural reasons that there are nine heavenly spheres.[80] Similarly, when it comes to natural philosophy, Ceffons may find it appropriate to discuss a matter traditionally treated in Book I of Aristotle's *Physics* simply because theologians are currently mentioning something vaguely relevant.[81]

On the other hand, he also employs his other more explicit, "on demand" technique of bringing in foreign material to introduce matters of natural philosophy. Thus, within the perfectly appropriate context of the problem of angelic motion he says that certain people have asked him to speak *de proportionibus, saltim modo grosso et rudi*. This gives rise to a whole separate *questio*, the beginning of which asks whether instantaneous action is inconsistent with Aristotle (and with the truth),[82] but whose substance is devoted to an exposition of Bradwardine's resolution of the problem of relating velocities, forces, and resistances.[83] Curiously, Ceffons does not here name Bradwardine, even though we know that he knew his *Tractatus de proportionibus*.[84] Yet in three suppositions he gives the substance of that at the basis of Bradwardine's resolution of his problem.[85] In the conclusions that Ceffons bases on these suppositions, it is clear that he is *teaching* Bradwardine's doctrines to an audience unfamiliar with the application of such a mathematics of proportions to natural philosophy. One can, he tells us, find doubling or halving variations in force versus resistance that will yield a double velocity, but one can also find such doubling or halving variations that will not give a double velocity. Values can be found, that is, that will establish apparently contradictory conclusions.[86] But that is just Ceffons's didactic point: the doubling of force (while resistance is held constant) or the halving of resistance (while force is held constant) does not *always* yield a double velocity, a state of affairs that makes it apparent that one should learn just what conditions do have to be satisfied in order to have a resultant velocity which is either determinately less than, equal to, or greater than twice the initial velocity. One should learn, that is, the relevant mathematics of proportions. And that is what Ceffons proposes to instill in his hearers.[87]

As if his treatment of *proportiones motuum* were not enough for his theological auditors, Ceffons goes on without pause to apply the mathematics of proportions he has thus far developed to the areas and volumes of circles, polygons, spheres, cylinders, pyramids, and even to a comparison of the sphere of the moon to the sphere of the earth.[88] And innumerable other things dealing with the proportion of proportions could be revealed, he claims. But what he has given should be sufficient, since the proportion of proportions is not often required *in scolis nostris*. Such things as irrational proportions do not matter much in theology. He should be excused, he avers, for having expounded as much as he did on proportions; but it was only the pleading of a "special friend" that caused him to do so.[89]

Nevertheless, only three questions later in his *Lectura*, Ceffons is back plowing English calculatory material into his commentary.[90] Here, however, there is no excuse provided by some friend's request nor even any theological "cover." The problem is put directly without any frills: whether any uniformly difform motion is just as great as any of its parts.[91] Ceffons has moved to another segment of Oxford measure preoccupation: namely, that concerned with the measurement of the uniform local motion of a body by its fastest moving point on the one hand, and that dealing with the mean degree measure of uniformly difform qualities and motions on the other. Indeed, Ceffons's first principal argument in reply to his *questio* quotes directly from Bradwardine's assertion of the fastest-moving-point rule for circular motion, although Ceffons adds the thought that God's ability to rotate the whole world provides an instance relevant to the kind of thing Bradwardine was treating.[92] This much would make it appear that the uniformly difform motion referred to in the original question should be interpreted as a motion uniformly difformly distributed over the *extension* of a rotating radius and that it was hence just as fast as one of its parts, namely, as fast as its fastest moving point.

However, Ceffon's argument *in oppositum* appeals to the quite different—but equally English—mean degree criterion for the measure of uniformly difform motions.[93] The problem seems to be, then, just how these two Oxonian rules of measure fit together. Perhaps it was a puzzle to his Parisian auditors and Ceffons was attempting to resolve it. In any event, we know that it was debated whether one might reject the fastest-moving-point rule for motions distributed uniformly difformly over space (as in the case of a rotating radius) and instead opt for a measure based on the linear velocity of the mid-point of the radius.[94]

Be this as it may, Ceffons's procedure is to begin resolution of the puzzle presented by his principal and *in oppositum* arguments by returning to didacticism with a vengeance. Uniform, difform, uniformly difform, difformly difform, and so on are first appropriately defined.[95] The most interesting thing, however, is not the definitions themselves, but his remark that the geometrical representation of motions, which we now usually associate with Nicole Oresme, was something commonly used by *scolares* and something that made the definitions of the different kinds of motion clearer.[96] The fact that Ceffons makes a point of this and goes on to expand his introductory definitional material to an extent beyond that needed for the problem at hand, suggests again that he was attempting to *teach* these calculatory matters.

It is only after all of this is appropriately in the minds of his hearers that Ceffons turns to the resolution of the problem with which this *questio* began. His reply is couched in a battery of conclusions that establish that uniformly difform qualities and motions are to be measured by their mean degrees, which in turn entails that they are not as intense as any of their parts.[97] In point of fact, the fallacy of applying a "most intense point" rule to uniformly difform qualities can be seen if one realizes that the highest degrees terminating such qualities do not belong to the qualities themselves; they are extrinsically, not intrinsically, terminated by these highest degrees.[98] But this does not mean that the fastest-moving-point criterion is never appropriate for the measure of uniformly difform things. For in reply to an objection, Ceffons agrees that in the case of a rotating wheel or radius we have an instance of a uniformly difform motion all parts of which exist simultaneously, a state of affairs that does at least allow measure by the fastest-moving-point rule.[99] Other puzzles, however, can be brought to bear against this suggested measure for a rotating wheel. What if, for example, we are dealing with an infinitely large wheel or a wheel that is undergoing continuous augmentation?[100] Although not necessarily exactly the same as what one finds in earlier English material, there is no doubt that such objections as these bear the stamp of the *sophismata de motu* that so flourished in the hands of Oxford scholars.[101]

As complicated as these brief examples from both Mirecourt and Ceffons of the invocation of the English calculatory tradition might be, it should be emphasized that they are only examples and quite limited ones at that. They are limited in the sense that they represent only a few selected areas in which this tradition was applied. And they are limited in the sense that only fragments of the application of calculatory material in even these selected areas have been presented. Yet I trust that this limitation has not been such as to prevent either appreciation of the fact that *subtilitates Anglicanae* did permeate Parisian learning at mid-century or the formulation of at least some idea of what forms this permeation then assumed.

Considering the display Mirecourt and Ceffons make of their utilization of English subtleties, it is extremely difficult to see them as instances of Richard de Bury's report that such matters were the "subject of furtive vigils." The condemnation of Mirecourt comes to mind as evidence of the fact that it might have been better had such things been more furtive and not so open; but even there, places in Mirecourt's *apologia* show that he was not at all chary to reveal the calculatory motives and conceptions behind some of the suspected propositions he had asserted.[102] It was certainly not for that kind of material for which he was called to account. It is also true that Ceffons frequently apologized for his introduction of the English calculatory tradition,[103] but it seems clear that his apologies were occasioned by the fact that he had gone on at such length and in such a loose way, and not by any subversiveness he thought what he had introduced might bear.

On the other hand, in succeeding years the kind of phenomenon represented by Mirecourt and Ceffons does seem to have been the subject of complaint, both personal and official. It appears, for example, that Ceffons may have been the culprit behind John of Ripa's impatience with those who use angles to explain the relations of species.[104] It is equally probable, moreover, that both Mirecourt and Ceffons and others of their sort were the object of the official admonition in the University of Paris statutes of 1366 to the effect that those reading on the *Sentences* should not treat logical or philosophical questions or matters unless they were strictly required by the text at hand.[105] This does not mean, of course, that the treating of *subtilitates Anglicanae* in particular was being criticized. But if one traces the appearance of such *subtilitates* in Parisian theology from the 1340s onward, it would not be a totally untoward hypothesis to presume that the excessive intrusion of logic and philosophy into theology that was officially discouraged was at least one of the fruits of the role played by "things English" at Paris.

NOTES AND REFERENCES

1. *The Philobiblon of Richard de Bury*, edited and translated by Ernest C. Thomas (London, 1888), pp. 211-212 (Latin text p. 89), with slight changes. The *Philobiblon* was presumably completed on 24 January 1345.

2. See the reference, for example, to the work of Franz Ehrle in Neal W. Gilbert, "Richard de Bury and the 'Quires of Yesterday's Sophisms'," in *Philosophy and Humanism. Renaissance Essays in Honor of Paul Oskar Kristeller*, ed. Edward P. Mahoney (New York, Columbia University Press, 1976), p. 233.

3. On this problem see Ernest A. Moody, "Ockham, Buridan, and Nicholas of Autrecourt," in his *Studies in Medieval Philosophy, Science, and Logic. Collected Papers, 1933-1969* (Berkeley, Los Angeles, London, University of California Press, 1975), pp. 127-160. Revisions of Moody's views on this matter have been suggested by T. K. Scott, "Nicholas of Autrecourt, Buridan, and Ockhamism," *Journal of the History of Philosophy*, 9 (1971), pp. 15-41 and Ruprecht Paqué, *Das Pariser Nominalistenstatut. Zur En-*

stehung des Realitätsbegriffs der neuzeitlichen Naturwissenschaft (Berlin, Walter de Gruyter, 1970), but without being thoroughly convincing, especially in the case of Paqué.

4. Thomas[1], p. 74. Cf. the article of Gilbert in note 2.

5. A good example of Ockham "setting the form" of a much discussed problem can be found in the debate that arose concerning a particular passage in Aristotle's *Physics* (Book III, ch. 1, 200b32-201a3). Initially, the problem was viewed as one of determining the appropriate category or genus to which motion or change belonged. On the other hand, Ockham's insistence upon the fact that only individual permanent things (*res absolute*) exist urged him to phrase the problem in terms of whether motion was some thing apart from or beyond such individual permanent things (his own view being that it was not). The point of present importance, however, is that in almost all instances after Ockham the problem was approached in his terms, even when its resolution differed from his. For the relevant references concerning this matter see Anneliese Maier, *Zwischen Philosophie und Mechanik* (=*Studien zur Naturphilosophie der Spätscholastik*, vol. 5), Rome, Edizioni di Storia e letteratura, 1958, pp. 59-144 and John E. Murdoch and Edith D. Sylla, "The Science of Motion" as in David C. Lindberg (ed.), *Science in the Middle Ages*, University of Chicago Press (in press).

6. Details concerning the practice of such "propositional" or "metalinguistic" analysis can be found in John Murdoch, "The Development of a Critical Temper: New Approaches and Modes of Analysis in Fourteenth-Century Philosophy, Science, and Theology," in Siegfried Wenzel (ed.), *Medieval and Renaissance Studies*, Number 7 (Chapel Hill, N. Carolina 1978), pp. 51-79 and "*Scientia mediantibus vocibus:* Metalinguistic Analysis in Late Medieval Natural Philosophy," to appear in the Acts of the VI International Congress of Medieval Philosophy, Bonn, 29 August-3 September 1977. The fact that Buridan "follows" Ockham in using propositional analysis extensively is not to deny important differences between them in other respects nor is it, especially, to claim that Buridan derived his "propositional approach." directly from a reading of this or that work of Ockham. Indeed, just what works of Ockham Buridan may have read is an open question.

7. See H. Lamar Crosby (ed.), *Thomas of Bradwardine. His Tractatus de Proportionibus. Its Significance for the Development of Mathematical Physics* (Madison, University of Wisconsin Press, 1955), p. 112 and J. Murdoch and E. Sylla, "The Science of Motion" (above, note 5). Note that what I am here calling, following medieval usage, a "proportion" would in modern terms be a ratio.

8. Ed. H. Lamar Crosby,[7] p. 130.

9. Bradwardine, (ed., H. Lamar Crosby[7]) *ibid.* and William Heytesbury, *Regulae solvendi sophismata* (Venice, 1494), fol. 37v; cf. fol. 198r. Heytesbury applied the rule to complex imaginative instances of the phenomenon in question, to such a case, for example, as the rotation of a body of which the outermost part is continually destroyed or undergoes corruption while the remaining, inner parts expand or are rarefied. In medieval terms, such a case was a "sophism" and provides a good example of *subtilitas Anglicana*.

10. This exposition of this doctrine is greatly oversimplified. For details see J. Murdoch and E. Sylla, "The Science of Motion" (above, note 5) and Marshall Clagett, *Nichole Oresme and the Medieval Geometry of Qualities and Motions* (Madison, University of Wisconsin Press, 1968).

11. *Physics*, Book VI, ch. 5 (cf. Book VIII, ch. 8).

12. Herman and Charlotte Chapiro, "*De primo et ultimo instanti* des Walter Burley," *Archiv für Geschichte der Philosophie*, 47 (1965), pp. 157-173. I have here ignored the existence of *permanent* things in time (such as the existence of a stone) that were distinguished from the existence of *successive* things like motions. Such permanent things *do* have, medieval scholars maintained, an initial, intrinsic instant limiting their existence, but, like successive things, only an extrinsic limit at the end of their existence.

13. See Curtis Wilson, *William Heytesbury, Medieval Logic and the Rise of Mathematical Physics* (Madison, University of Wisconsin Press, 1956), pp. 29-56 and J. Murdoch and E. Sylla, "The Science of Motion" (above, note 5).

14. Marshall Clagett, *The Science of Mechanics in the Middle Ages* (Madison, University of Wisconsin Press, 1959), pp. 440-44, 463-64.

15. H. L. L. Busard, *Der Tractatus proportionum von Albert von Sachsen*, Oesterreichische Akademie der Wissenschaften, Math.-Naturwiss. Klasse, Denkschriften, Bd. 116, Abhandlung 2 (Vienna, 1971).

16. Nichole Oresme, *De proportionibus proportionum and Ad pauca respicientes*, ed. Edward Grant (Madison, University of Wisconsin Press, 1966) and Edward Grant, *Nichole Oresme and the Kinematics of Circular Motion: Tractatus de commensurabilitate vel incommensurabilitate motuum celi* (Madison, University of Wisconsin Press, 1971).

17. Marshall Clagett, *Nichole Oresme . . .* (above, note 10).

18. Thus, Albert of Saxony includes 34 sophisms (Soph. 179-212) treating of 'incipit' and 'desinit' in his *Sophismata nuper emendata* (Paris, 1495). Our concern is here with the influence of English calculatory material on Parisian learning, but note should be made of its considerable impact upon later Italian learning in particular as well; see Marshall Clagett, *The Science of Mechanics . . .* (above, note 14), ch. 11 and J. Murdoch and E. Sylla, "Swineshead, Richard," in *Dictionary of Scientific Biography*, XIII, 209-213.

19. For the latest information concerning Mirecourt's career and the date of his *Lectura* see William J. Courtenay, "John of Mirecourt and Gregory of Rimini on Whether God Can Undo the Past," *Recherches de théologie ancienne et médiévale*, 40 (1973), pp. 226-228.

20. F. Stegmüller, "Die zwei Apologien des Jean de Mirecourt," *Recherches de théologie ancienne et médiévale*, 5 (1933), pp. 40-78, 192-204 and W. Courtenay, "John of Mirecourt . . ." (above, note 19), pp. 228-229.

21. The only substantial treatment, largely bio-bibliographical, of Ceffons is Damasus Trapp, "Peter Ceffons of Clairvaux." *Recherches de théologie ancienne et médiévale*, 24 (1957), pp. 101-154 (for the probable date of Ceffons's *Lectura* see pp. 109-110, 115-116, 127). Trapp suggests (pp. 103, 121) that the unique manuscript of Ceffons's *Lectura* (MS Troyes 62) may be an autograph copy, but the numerous errors to be found in the philosophical portions of the work I have examined render this most unlikely.

22. See Anneliese Maier, *Ausgehendes Mittelalter. Gesammelte Aufsätze zur Geistesgeschichte des 14. Jahrhunderts*, vol. 1 (Rome, Ed. di Storia et letteratura, 1964), pp. 41-85, and J. Murdoch, "*Mathesis in philosophiam scholasticam introducta*: The Rise and Development of the Application of Mathematics in Fourteenth-Century Philosophy and Theology," in *Arts libéraux et philosophie au moyen âge*, Actes du quatrième Congrès International de Philosophie Médiévale (Montreal/Paris, 1969), pp. 221-224.

23. This particular Distinction in commentaries on Lombard dealt with, inter alia, the problem of the augmentation of *caritas*, of how such augmentation could occur, and, as

such, afforded ample opportunity to investigate the intension of forms and qualities. If we realize that a good deal of the early discussion of the "ontology" of intension and remission took place in this theological context, and then note that this discussion was furthered and broadened in natural philosophical works (where the question of how to *measure* intension and remission was the most important addition to the debate), the subsequent continuing discussion of this matter in *Commentaries on the Sentences*, a discussion that incorporated many of the natural philosophical results, provides us with an excellent case of the "cooperation" of philosophy and theology in the development and resolution of a given problem. Much of the history of all of this is traced in Anneliese Maier, *Zwei Grundprobleme der scholastischen Naturphilosophie*, 3 ed. (Rome, Ed. di Storia e letteratura, 1968), pp. 3-109.

24. See J. Murdoch, "*Mathesis* . . ." (above, note 22), pp. 238-246.

25. See J. Murdoch, "From Social into Intellectual Factors: An Aspect of the Unitary Character of Late Medieval Learning," in J. Murdoch and E. Sylla (ed.), *The Cultural Context of Medieval Learning* (Dordrecht/Boston, D. Reidel, 1975), pp. 289-297.

26. Unfortunately, Rosetus and Halifax are thus far even less studied than Mirecourt and Ceffons. The two basic references are V. Doucet, "Le Studium franciscain de Norwich en 1337 d'après le MS Chigi B.V. 66 de la Bibliothèque Vaticane," *Archivum franciscanum historicum*, 46 (1953), pp. 88-93 [on Rosetus] and W. J. Courtenay, "Some Notes on Robert of Halifax, O.F.M.," *Franciscan Studies*, 33 (1973), pp. 135-142. A sampling of some of the modes of analysis in Rosetus and Halifax appear in J. Murdoch.[25]

27. See notes 32 and 61 below for examples of passages taken from Halifax. Note also might here be made of Ceffons utilizing a very extensive quote (MS Troyes 62, 73r-74v, where Ceffons himself notes that he is quoting "de verbo ad verbum") from Thomas Bradwardine's *De causa Dei* (ed. London, 1618; reprt. Frankfurt, Minerva Verlag, 1964) pp. 120-132. The context of this "borrowing" of Bradwardine is a discussion of the infinite. For the kind of thing Ceffons was citing from his English predecessor see J. Murdoch, "*Rationes mathematice*": *Un aspect du rapport des mathématique et de la philosophie au moyen âge* (Paris, 1962), pp. 17-20.

As far as I have been able to note, it is possible that Mirecourt's sources for his English calculatory material were all English theological works. But for Ceffons we know that he knew some of the English calculatory works themselves since, as we shall see (below, notes 69, 74, 76, 84, 88, 92), he cites Bradwardine's *Tractatus de proportionibus*, his *Insolubilia*, Richard Kilvington's *Sophismata*, and Roger Swineshead's *Insolubilia*.

28. Mirecourt, *Comm. Sent.*, III, QQ 4, 6-11 (MSS Prague Univ. III.B.10 (419), 85v-88r, 91v-105v; Paris BN 15883, 112v-133r). These *questiones* appear together in the Paris MS at the very end of Book III, with the changed order QQ 7, 8, 6, 4, 9-11. That Mirecourt was aware of the fact that he was indulging in a rather extensive examination of the application of "calculatory" material to theological issues is evident from the reference he makes in his first *Apologia* to these questions (F. Stegmüller, "Die zwei Apologien . . ." [above, note 20], p. 72).

29. MS Paris BN 15883, 118v-121r: *Quest.* 6: Utrum voluntas creata posset intendere vel remittere meritum suum vel demeritum. *Concl.* 1: Voluntas creata potest intendere meritum suum. *Concl.* 2: Potest intendere demeritum suum. *Concl.* 3: Potest meritum suum et demeritum remittere. *Concl.* 4: Quemcumque actum suum potest voluntas intendere et eundem remittere. *Concl.* 5: Non solum quilibet actus voluntatis sic

potest intendi et remitti, ymmo etiam quilibet actus intellectus et quilibet actus anime; *Concl.* 6: Quodibet accidens in anima receptum naturaliter vel supernaturaliter potest intendi et potest remitti. This *questio* is missing from the Prague MS, but appears, together with its *conclusiones* in a tabula at the end (112v) of the MS. Note should be made of the fact that the fundamental contention behind Mirecourt's conclusions was the all but universal belief that the will acts successively throughout some time interval and not instantaneously (see J. Murdoch, "From Social into Intellectual Factors . . ." [above, note 25], pp. 294-295).

30. See E. Sylla, "Medieval concepts of the latitude of forms The Oxford calculators," *Archives d'histoire doctrinale et littéraire du moyen âge*, 40 (1973), pp. 230-232.

31. MSS Prague III.B.10, 84v-88r; Paris BN 15883, 121r-123v: *Quest.* 4: Utrum voluntas per additionem partis ad partem vel per diminutionem partis a parte meritum vel demeritum suum possit intendere vel remittere: *Concl.* 1: Intensio actus meritorii ipsius voluntatis fit per additionem partis ad partem. *Concl.* 2: Intensio actus demeritorii voluntatis fit per additionem partis ad partem. *Concl.* 3: Cuiuslibet actus voluntatis intensio fit per additionem partis ad partem. Tres alie conclusiones possunt poni de remissione actus voluntatis conformiter tribus predictis.

32. MSS Prague III.B.10, 91v; Paris BN 15883. 112v: *Quest.* 7: Utrum alica creatura possit aliquod instantanee precise producere. . . . Ad evidentiam huius questionis pono duas conclusiones: Prima est quod omnis res quocumque modo de non esse ad esse producta—quod dico propter generationem Filii in divinis vel propter Spiritus Sancti processionem, quia ibi non est productio de non esse ad esse—illa res producta per motum vel mutationem producitur. Secunda quod omnis res quocumque modo corrupta, per motum vel mutationem corrumpitur. Et accipio motum vel mutationem proprie secundum quod Philosophus eas distinguit 5 et 6 *Phisicorum.* (Note that Mirecourt was taking *mutatio* to be a change involving time and not following its frequent scholastic use as implying a change that was *subito* or instantaneous.)

That Mirecourt was here quoting Halifax is evident from the parallel passage in the latter's *Comm. Sent.*; see J. Murdoch, "From Social into Intellectual Factors . . ." (above, note 25), p. 322-323, n. 83. For other borrowing by Mirecourt from Halifax see J. Murdoch,[25] op. cit., p. 331, n. 106; W. J. Courtenay, "Some Notes . . ." (above, note 26), p. 139; and K. Michalski, "Le problème de la volonté à Oxford et à Paris au XIVe siècle," *Studia Philosophica. Commentarii societatis philosophicae polonorum*, 2 (1937), pp. 346-347.

33. MSS Prague III.B.10, 91v-94v; Paris BN 15883, 112v-115v: *Quest.* 7 (vide supra), *Concl.* 1: Quod non potest dari instans in quo intellectus creatus primo intelligatur. *Concl.* 2: Quod non potest dari in casu (!) instans in quo sensus primo sentiat. *Concl.* 3: Quod non est dare primum instans in quo luminosum primo illuminet medium. *Concl.* 4: Quod non est dare primum instans in quo agens naturale agat aliquam actionem nec aliquid.

34. MS Prague III.B.10, 93v: *Quest.* 8: Utrum alica creatura posset per solum instans a Deo conservari [corrected *apud tabulam*, 112v]. Arguitur quod sic. . . . In oppositum: si alica creatura posset conservari per instans precise, illud instans foret primum instans sui esse et ultimum sui esse. Et hoc est falsum, tum quia in isto instanti inciperet esse et desineret esse, tum quia esset dare ultimum rei permanentis in esse, tum quia, si res habet primum et ultimum, igitur immediatum, et per consequens non durat precise per instans. De questione proposita supponendo modos communes loquendi secundum

quod (!) dicimus res esse in tempore vel instanti, pono aliquas conclusiones (see next note).

35. MS Prague III.B.10, 93v-96v, *corr. apud tab.* 112v: *Quest.* 8 (vide supra), *Concl.* 1: Quod nulla indivisibilis potest esse vel non esse precise per instans sic quod non per tempus. *Concl.* 2: Si alica res permanens divisibilis vel indivisibilis in hoc instanti non est, illa post hoc [MS modum!] adhuc non erit, et ideo hec non est possibilis: hec res non est in hoc instanti et immediate post hoc erit. *Concl.* 3: Quod hec propositio non est possibilis: hec res permanens est, sive sit divisibilis sive indivisibilis, et immediate post hoc nec ipsa nec aliquid ipsius erit, et ideo si res in hoc instanti est, ista vel aliquid ipsius immediate post hoc erit. *Concl.* 4: Quod tenens aliquam rem esse successivam que non est res permanens habet dicere quod alica res successiva non est in hoc instanti quod immediate post hoc erit, utpote dies, nox, albefactio, vel alica alteratio vel alica augmentatio.

36. Cf. Walter Burley, *De primo et ultimo instanti.*[12]

37. MS Prague III.B.10, 96v-98r: *Quest.* 9: Utrum voluntas creata mereri vel demereri possit in instanti precise per instans. *Concl.* 1: Si voluntas meretur in hoc instanti, voluntas ante merebatur; et ideo non est possibile quod alica voluntas precise mereatur per instans. *Concl.* 2: Si voluntas demeretur in hoc instanti, quod ipsa voluntas ante demerebatur.

38. MS Prague III.B.10, 99v-101r. *Quest.* 10: Utrum voluntas creata, que in hoc instanti non meretur vel in hoc instanti non demeretur, immediate post hoc instans possit mereri vel demereri. *Concl.* 1: Si voluntas in hoc instanti meretur, ipsa immediate post hoc merebitur. *Concl.* 2: Si voluntas in hoc instanti demeretur, immediate post hoc demerebitur. *Concl.* 3: Quod possibile est quod voluntas nunc non mereatur que immediate ante merebatur vel demerebatur, et similiter possibile est quod voluntas in hoc instanti non meretur vel non demeretur quod (!) immediate post hoc mereatur vel demereatur. *Concl.* 4: Quod pro nullo instanti quo quis meretur debetur aliquod premium cuius nulla pars prius debebatur, nec pro alico instanti quo quis demeretur debetur alica pena cuius nulla pars prius debebatur.

39. MS Prague III.B.10, 102v-104v: *Quest.* 11: Utrum voluntas creata ad quemcumque gradum meriti possit intendere vel remittere actum suum. *Concl.* 1: Voluntas creata per aliquod tempus potest eligere actus meritorios equalis intensionis. *Concl.* 2: Aliquis est gradus ymaginabilis meriti tam intensus quod hec voluntas cum gratia quam habet vel habitura est ex ordinatione Dei et in tempore quo potest esse in statu merendi ad illum non posset intendere meritum suum et aliquis est gradus ymaginabilis tam remissus quod hec voluntas cum gratia quam habet vel habitura est cum conatu suo quantumcumque modico non posset illum habere per se sic quod illo habito cessaret ulteriorem habere. *Concl.* 3: Nullus est gradus meriti ita intensus quod voluntas cum gratia habita vel habenda posset acquirere quin posset intensiorem acquirere nec aliquem gradum ita remissum quin possit remissiorem. *Concl.* 4: Non semper ad quemcumque gradum voluntas vult intendere vel remittere actum suum demeritorium intendit vel remittit suum demeritum.

40. See conclusions 1 and 2 cited in note 38.

41. MS Prague III.B.10, 99v: Si voluntas in alico instanti habeat aliquem actum, contingenter et non necessario continuabit illum post hoc instans. Si in hoc instanti habet (!) actum, potest illum non habere post hoc instans ex quo contingenter habet illum post hoc instans. Si voluntas in hoc instanti habeat actum, voluntarie continuabit illum

vel aliquid eius post hoc instans et non invice. Hec non est possibilis: in hoc instanti habet actum et nunquam post hoc habebit illum nec aliquid eius, et ideo si voluntas velit alicem actum habere in alico instanti et istum non habere nec aliquid eius post illud instans, vult sibi impossibile.

42. Aristotle, *Physics*, VIII, ch. 8, 261b27-263a3.

43. See, for example, John Buridan, *Questiones super octo libros phisicorum Aristotelis*, VIII, Q. 8 (ed. Paris 1509), fol. 116r-v

44. MSS Prague III.B.10, 98v; Paris, BN 15883, 126r; Napoli BN VII. C.28, 127v: *Concl.* 3: Quod possibile est quod in illo instanti in quo voluntas primo desinit mereri ipsa incipiat demereri vel econverso. Ista conclusio probatur, quia non est hoc minus possibile quam sit possibile mobile aliquod immediate postquam movebatur alico motu locali moveri motu contrario; secundum est possibile, igitur et cetera. Maior est de se nota et minor probatur primo, quia motus sursum et deorsum sunt oppositi, et tamen si faba proiiceretur sursum et occurreretur lapidi molari descendenti, non violentaret lapidem illum ad quiescendum in medio, cum ille lapis sit maioris gravitatis in magna proportione qua possit pellere fabam quam faba sit in agendo sursum, maxime cum actio istius fabe sit contra eius inclinationem, scilicet quod pellat aliquid sursum; et per consequens non quiescit faba postquam tetigerit lapidem molarem, sed statim descendet, et immediate antequam tetigerit illum lapidem ascendebat, igitur immediate postquam ascendit descendit.

Note should be taken of the fact that, in allowing the same instant to serve as the initial limit of a meritorious action and the terminal limit of a demeritorious action (or vice versa), Mirecourt apparently means to have that single instant belong both to the period of merit and the period of demerit. But this is inconsistent with his belief (see above, note 35, Concl. 4; note 37, Concl. 1-2; note 38, Concl. 1-4) that, as successive things, meritorious and demeritorious actions are *extrinsically* limited by, respectively, last and first instants of non-existence (since to have the same instant limit *both* successive contrary actions entails that it be *intrinsic* to at least one of them). But this was a problem that plagued almost all scholastics when they came to the problem of assigning temporal limits to two contrary motions or actions that immediately succeeded one another in time. Some stipulated that in such cases one was dealing with a *privatio* versus a *positivum* and, although standard extrinsic limits were to obtain for the *positivum*, the *privatio* was intrinsically limited at both ends (see Curtis Wilson, *William Heytesbury* [above, note 13], p. 35). But, as far as I have been able to discover, Mirecourt does not make such a distinction. Nor does he here tell us what he may mean by different kinds of 'incipit' and 'desinit', something that was often standard scholastic fare (see Wilson,[13] pp. 41-42). To judge from a subsequent reply Mirecourt makes to an objection, it seems clear that what he intends by having a single instant serve as the cessation of (say) a meritorious action and the beginning of a demeritorious action is that at that instant the will in question is undergoing contrary actions (MSS *cit., ibid.*): Dico quod in isto instanti faba movetur et non motu ascensus nec descensus, sed movetur quodam motu composito ex motu ascensus et descensus cuius quelibet pars terminata ad hoc instans fuit motus ascensus et quelibet pars incipiens ab hoc instanti erit motus descensus; quelibet autem pars cuius mutatum esse correspondens isti instanti est mutatum esse interminatum nec est totaliter ascensus nec totaliter descensus. Consimiliter dico quod in isto instanti talis non meretur nec demeretur, sed tantum movetur motu nec est totaliter meritivo nec totaliter demeritivo, sed partialiter.

45. *Comm. Sent.*, Lib. I, Q. 10 (MS Napoli, BN VII.C.28, 17r): Utrum cognitiones excedant se perfectione proportionaliter per excessum obiectorum. Mirecourt does not open his discussion by treating directly of the perfections of created species but instead approaches his topic by speaking of cognitions because this particular *questio* is part of the Prologue to Book I of his *Lectura* and, traditionally, such prologues dealt with the problem of knowledge, especially with the problem of what kind of knowledge claims could be made for theology as a discipline.

46. MS Napoli BN VII.C.28, 17r; Paris BN 15882, 36v-37r: Respondeo premittendo unam distinctionem quod unam excedere aliam intelligitur dupliciter: uno modo quidditative seu essentialiter, alio modo accidentaliter. Res una dicitur excedere aliam *essentialiter* quando sic excedit quod etiam quodlibet illius speciei excedit illud nec est possibile quod aliquid sue speciei sit et non sic excedit illud, verbi gratia, homo excedit asinum quidditative. . . . *Accidentaliter* vero dicitur una res excedere aliam quando sic excedit quod possibile est aliquam rem eiusdem rationis esse quam non sic excedet vel non sic excedet aliam (!), verbi gratia, si Sortes sit maior Platone, excedet Platonem in magnitudine, et possibile est quod aliquis alius homo non sic excederet, quia erit equalis Sorti in magnitudine. Excedere autem accidentaliter dicitur dupliciter, uno modo intrinsece, alio modo extrinsece. Res una dicitur excedere aliam *intrinsece accidentaliter* quando hoc sit propter aliquam rem que non est extra ipsam, sicut quando dicitur res maior vel forma intensior. Res dicitur excedere aliam *extrinsece accidentaliter* quando comparatur ad aliam vel comparari potest penes excessum propter rem aliquam que non est ipsa nec aliquid ipsius nec in ipsa, sicut una dicitur propinquior alteri propter maiorem distantiam alterius rei ab illa re.

47. MS Napoli BN VII.C.28, 18v: *Concl.* 6: Totalis rerum ymaginabilium perfectio duplicem habet latitudinem, prima est essentialis, secunda accidentalis et utraque est infinita.

48. MSS Napoli BN VII.C.28, 17v & Paris BN 15882, 37r: *Propositiones ad Concl.* 1: (1) Nulla res alicuius speciei potest excedere rem eiusdem speciei quidditative vel excedi ab eadem. (2) Ubi res alicuius speciei excedit aliam rem quidditative et quelibet pars eius est eiusdem speciei cum toto, in tanta proportione quelibet pars eius excedit eandem rem quidditative. (3) Nulla res alicuius speciei creata quantumcumque perfecta potest aliquid excedere quidditative magis quam excedat vel minus quam excedat, quia tunc posset esse quod rem eiusdem speciei inequaliter excederet. (4) Tantum excedit quidditative albedo quantumcumque remissa nigredinem quantumcumque intensam quantum albedo quantumcumque intensa nigredinem quantumcumque remissam. (5) Quod compositum ex infinitis partibus eiusdem rationis non minus excedit rem aliquam quidditative quam pars eius quantumcumque remissa. (6) Nullum totum excedit medietatem sui eiusdem rationis cum toto quidditative et essentialiter.

49. MS Napoli BN VII.C.28, 17r: *Concl.* 1: Quod nulla cognitio nec aliqua res creata excedit aliam rem creatam infinite quidditative vel essentialiter.

50. MSS Napoli BN VII.C.28, 17v; Paris BN 15882, 38r: *Concl.* 2: Nulla cognitio nec res creata finita permanens excedit vel excedere potest rem aliquam creatam infinite accidentaliter et intrinsece. An example of the kind of infinite accidental intrinsic excess among finite created things that Mirecourt denies is: Nullum continuum aliquam partem sui excedit infinite extensive, quia quecumque vel quantumcumque pars detur, habet proportionem ad illam finitam extensive. *Concl.* 3: Quod aliqua cognitio vel aliqua res creata excedit vel excedere potest aliam infinite accidentaliter extrinsece.

51. The allowable infinite accidental excess is extrinsic in this case—that is, derives from a comparison of the exceeding thing with something that is *extra ipsam* (see above, note 46)—and Mirecourt tells us what he has in mind by citing two examples: (1) Dilectio qua diligitur Deus super omnia infinite melior est quam dilectio qua diligitur creatura. . . . (2) Angulus contingentie (i.e., the curvilinear angle between the circumference of a circle and a tangent to it) in infinitum acutior est angulo rectilineo. On such angles, see note 67 below.

52. MS Napoli BN VII.C.28, 18v: *Prop. 2 ad concl. 6* (see above, note 47): Quelibet latitudo ymaginabilis istarum latitudinum incipit a non gradu perfectionis; hoc est, quocumque gradu perfectionis dato, contingit dare duplo minorem et sic sine statu, et tamen nullus est gradus minimus ymaginabilis.

53. MS Napoli BN VII.C.28, 18v: *Prop. 3 ad concl. 6*: Maior vel minor perfectio rei essentialis vel accidentalis attenditur penes maiorem vel minorem recessum a non gradu perfectionis; non attenditur maior vel minor rei perfectio essentialis penes maiorem vel minorem a summo recessum illius beatitudinis que est Deus, quia quelibet res ab eo distat in infinitum.

54. See Marshall Clagett, "Richard Swineshead and Late Medieval Physics," *Osiris*, 9 (1950), pp. 131-161; J. Murdoch and E. Sylla, "Swineshead, Richard" (above, note 18), pp. 188-190. Although Mirecourt reflects these earlier English concerns over the appropriate scale of measure, there is, it seems, no provable dependence of his contentions upon some specific English work.

55. See note 53.

56. MS Napoli BN VII.C.28, 18v: *Propositiones ad concl. 6*: (4) Nullius rei create perfectio essentialis est determinata ad non gradum perfectionis, quia ipsa est indivisibiliter tante vel tante perfectionis et nullius perfectionis essentialis (MS essentia vel) non potest esse creatura aliqua quin alia possit esse in duplo minoris perfectionis essentialis quam illa. (5) Quecumque res est hec species, ut homo vel asinus, indivisibiliter essentialiter est homo vel asinus, et quantumcumque fieret una res minus perfecta essentialiter, illa non esset homo, et quantum fieret una res magis perfecta essentialiter, illa etiam non esset homo; et ideo forte species rerum sunt sicud numeri, ut habetur 9 *Metaphisice*. (6) Quod punctus qualiter se habet ad continuum, gradus ad forme intensionem, mutatum esse ad motum, taliter se habet gradus specificus ad perfectionem essentialem. (7) Quod inter quascumque species possunt fieri non tot quin plures. (13) Cuiuslibet rei date contingit dare gradum perfectionis essentialis quo nec habet maiorem nec minorem.

57. MS Napoli BN VII.C.28, 18v: *Prop. 11 ad concl. 6*: Infinities infinite sunt ymaginabiles species quarum nulla est posita in esse, ymo inter quascumque species sunt ymaginabiles infinite species quarum nulla est posita in esse.

58. MS Napoli BN VII.C.28, 18v: *Propositiones ad concl. 6*: (10) Nulle due species possunt esse equaliter perfecte essentialiter. (14) Plures res eiusdem speciei possunt esse una maioris perfectionis accidentalis quam alia, sed non sic est de perfectione essentiali.

59. MS Napoli BN VII.C.28, 18v: *Propositiones ad concl. 6*: (8) Due species plus possunt distare quam alie species, sicut et gradus. (9) Due species possunt distare in proportione precise dupla, alique in proportione tripla, et huiusmodi.

60. MS Napoli BN VII.C.28, 18v: *Prop. 5 ad concl. 5* (which is: Quod una res potest excedere aliam in duplo quidditative vel in alia proportione finita vel infinita, et

tamen non oportet cognitionem eius excedere cognitionem alterius propriam et distinctam in duplo vel in eadem proportione): Nullus homo scit in qua proportione precise una res excedit aliam quidditative, potest tamen scire quod excedit.
61. Thus, to cite an instance other than that to be dealt with immediately below, Ceffons treats the same topic we have above seen in Mirecourt of the "temporal measure" of the acts of the will in his *Comm. Sent.* II, Q. 12 (MS Troyes 62, 111v-114v): Circa distinctionem quintam et sextam quero utrum alica creatura possit agere in instanti seu indivisibiliter. Et videtur primo quod sic. . . . In ista questione est quedam opinio tenens quod voluntas non potest aliquem actum producere in instanti, et premittit ad hoc probandum alicas suppositiones quarum prima est hec. . . . Deinde ponit alicas conclusiones quarum prima est . . . (all of which is to say that Ceffons is citing verbatim the relevant suppositions and conclusions of Robert Halifax, a factor he acknowledges at the end of his "borrowing"): Ex istis dictis patet quomodo respondendum est ad titulum questionis, scilicet quod nullus actus voluntatis potest esse subito productus a voluntate, ideo et cetera. Hec Eliphat de verbo ad verbum. Est autem opinio alicuius alterius moderni concordans cum ista (whereupon there follows an exposition of the view of another English scholar, Richard Fitzralph). For Mirecourt's treatment of the same issue and his utilization of the very same material from Halifax, see note 32 above and the reference to J. Murdoch there cited.
62. *Comm. Sent.* I, Q. 6 (circa prologum) [MS Troyes 62, 19v]: Consequenter queritur utrum necesse sit quod generaliter scientie se excedant in perfectione secundum quod subiecta. For Mirecourt's similar approach to the issue of the perfections of species through considering an aspect of the problem of knowledge traditionally treated in prologues to Book I of the *Sentences*, see note 45.
63. MS Troyes 62, 19v: In ista questione oportet primo videre de excessu unius rei super aliam, scilicet <si> sit possibile quod una res aliam in infinitum excedat. Et solet dubitari utrum aliquod finitum excedat aliud finitum in infinitum.
64. MS Troyes 62, 19v: Est hic quedam opinio tenens oppositum dicens quod una res finita non potest excedere aliam rem plusquam in triplo et sic in infinitum. Idem dicit Rogerus quod non potest unus angelus unam muscam excedere infinities in perfectione quin angelus esset infinitus. Et dicit Rogerus quod tot musce possunt esse per potentiam Dei quod ille quod equivalerent illi angelo in perfectione. The appropriate passage from the *Lectura* of Roger Rosetus (on which, see the reference in note 26) is from Q. 5, art. 2 (utrum alica creatura possit esse infinita), MS Bruges 192, 43r-v: Tertia conclusio est ista: quod nulla creatura potest aliam excedere in infinitum, scilicet plusquam in duplo, in triplo et sic in infinitum. . . . Dico quod angelus non excedit muscam in perfectione infinities, scilicet in duplo, in triplo et sic in infinitum, quia nescio quin foret infinite perfectionis, et ideo non concedo quod angelus infinities excedit perfectionem unius musce, ymo tot possunt esse musce per potentiam Dei quod equivalerent uni angelo in perfectione.
65. To be specific, Ceffons cites (MS *cit.*, 19v-20r) the substances or gives verbatim from Mirecourt that quoted in notes 46, 48, 49, and 50 above, where only Concl. 2 is cited from the last named.
66. MS Troyes 62, 20r: Hic tamen videtur mihi quod oppositum est probabile. Since what follows concerns the relations between rectilinear and curvilinear angles, in all fairness to Mirecourt it must be noted that he allowed infinite excess to obtain between such angles provided it was accidental and *extrinsic* (see notes 50 and 51). But Ceffons

cites only Mirecourt's denial of infinite *intrinsic* accidental excess (Concl. 2 in note 50) and does so directly before he claims that the opposite view seems *probabile* in his eyes. Of course, it is not certain that it is the opposite of no more than that conclusion of Mirecourt that Ceffons had in mind.

67. The *locus classicus* for the horn-angle in medieval mathematics was Euclid's *Elements*, Book III, Proposition 15 (=III, 16 of the Greek) together with the comment of Campanus of Novara on the same. Cf. J. Murdoch, "The Medieval Language of Proportions: Elements of the Interaction with Greek Foundations and the Development of New Mathematical Techniques," in *Scientific Change*, ed. A. C. Crombie (London, Heinemann, 1963), pp. 242-250.

68. Thus, Ceffons deduces no fewer than nineteen corollaries (MS *cit.*, 20r-20v) that spell out the relations that obtain among rectilinear and curvilinear angles. For the substance and text of some of these corollaries see J. Murdoch, "*Mathesis . . .*" (above note 22), p. 244, notes 100-101.

69. MS Troyes 62, 21r: The particular "rule" of Bradwardine that Ceffons has in mind is: Quod proportione equalitatis (MS equaliter) nulla est maior vel minor proportio (MS proposito) [Cf. Bradwardine, ed. Crosby,[7] p. 80]. Ceffons's argument against Bradwardine amounts to the claim that the kind of excess—or 'maior' and 'minor'— his rule speaks of is not applicable to rectilinear vs. curvilinear angles since one is there dealing with angles that differ in species. Thus, with respect to the kind of argument Bradwardine gives in support of his rule Ceffons claims: Dico igitur quod ubi fit operatio distinctorum specie non est necesse . . . licet etiam non negetur quin inter alica distincta species posset inveniri proportio. The kind of *proportio* he has in mind is one that: improportionaliter excedit, sicud angulus rectilineus (MS rectus) excedit angulum contingentie.

70. The relevant texts are too long to cite here, but a sample from them can be found in J. Murdoch, "*Mathesis . . .*" (above, note 22), p. 245, notes 103-105.

71. MS Troyes 62, 22v: Nota quod aliquando timui in hac materia quin errarem et deciperem per affectionem ad alica dicta mathematicalia, ne forte non esset bene assignata similitudo ubi quandoque assignavi; hoc autem discernendum experioribus relinquitur, nam alicubi (but where, we do not know) in hac materia satis curiose scripsi.

72. In addition to his English predecessors and, of course, more ancient authors, Ceffons cites some of his Parisian contemporaries, the most notable for our purposes being Nicole Oresme (MS Troyes 62, 72r): Ad secundum quicumque (!) sibi superposita nec excedunt nec etc. [scil. exceduntur] sunt equalia, dicitur primo quod falsa est de vi vocis, quia punctus superpositus puncto nec excedit nec exceditur et tamen non sunt equalia; dicitur etiam quod illa maxima non valet nisi quod unum non est maius alio; et si sint comparabilia sunt equalia; si non, <non> oportet, sicud linea angulo, nec est equalis nec maior nec minor . . . hec sunt dicta Oresme tenentis quod non <est> equale unum infinitum alteri infinito. I have cited but only one of the "dicta" that seem to be among those of Oresme. Yet it, and numerous others, allow us to identify the specific source Ceffons was using, viz., Oresme's *Questiones super libros physicorum*. Thus, the matching passage in Oresme for that quoted above is (MS Sevilla, Colomb. 7-6-30, 38r): Solutio: dico primo quod illa dignitas falsa universaliter esset intellecta, quia punctus superpositus puncto non excecit etc., et tamen non dicitur equale. . . . Secundo dico quod ex illa, si non excedunt nec exceduntur, non habetur nisi unum non est altero maius aut minus, quia excedere est esse maius et excedi esse minus; ideo si sic

comparantur statim sequitur quod sint equalia, et si non, non oportet, sicut linea superposita angulo non dicitur equalis, maior vel minor.

73. Book II, Q. 2 (MS Troyes 62, 87r): Quia postulas amice dilectissime, o Bernarde, ut alica de logicalibus in huius secundi libri principio diligenter annectam, idcirco alica logicalia, que dudum multa velocitate composui, que tibi in scolis non protuli, hic annecto, que tuo prospicati (!) reliquuntur examini, nec correctionis limam diligentis horrescunt. Et quoniam in hiis diebus nonnulli dubitari videretur de scire et opinari, quero utrum circa idem scire et opinari contingat.

74. MS cit., 94v-95r: Et sic qui contra hoc obiciet advertat quod obiciat contra intellectum quem habui, quia aliquando non ita proprie expressi sicut bene potuissem. Nam multum festinanter scripsi hoc. Potest autem esse dubium de quadam solutione quam ponit Kilenton ad illud sophisma quod a nonnullis reputatur sophisma difficile et mirabiliter sollempne; et est inter sophisma Kilenton 49 (MS 4) sophisma, et est istud: *A est scitum a te*, supposito quod A sit altera istarum: 'Deus est' vel 'nullum concessum a Sorte est scitum a te', et concedat Sortes istam et nullam aliam: 'A est scitum a te'; tunc subiungit Kilenton hec que sequuntur de verbo ad verbum. Probatur sophisma sic quod sit B et arguitur sic: B non est dubitandum a te nec distinguendum a te nec negandum a te, ergo B est concedendum . . . ideo dubito utrum C sit verum et ita C est mihi dubium. Hec Kilenton de verbo ad verbum. The relevant passage in Kilvington's *Sophismata* is (MSS Paris, BN 16134, 71v; Vat. lat. 3066, 24v-25r): *A est scitum a te*. Supposito quod A sit altera illarum: 'Deus est' vel 'nullum concessum a Sorte est scitum a te'. . . . Tunc probatur sophisma sic quod sit B; tunc arguo sic: B non est dubitandum a te . . . ideo dubito utrum C sit verum et ita C est mihi dubium.

75. See the text in note 73 and the similar remark made by Ceffons about his procedure in note 74.

76. Book II, Q. 3 (MS Troyes 62, 96r-101r): Quia petitur a me ut, si quidquam de insolubilibus novi, de ipsis aliquid hic pertractarem; idcirco in hac lectura secundi sententiarum quero utrum beatus Augustinus vel etiam magister Petrus Lumbardus vel aliquis alius theologus fidelis per aliquod insolubile potuerunt ad inconveniens deduci. . . . Hec autem ultima conclusio (scil. Nulla duo contradictoria sunt simul vera in sensu in quo sibi invicem contradicunt) est contra alios qui astruunt duo simul contradictoria fore falsa. Unde Rogerus Suinchet in quodam tractatu de logica ponit tres conclusiones. Prima conclusio <est: Alica propositio> est falsa que principaliter significat sicut est. . . . (*Concl.* 2): In alica bona consequentia et formali, ex vero sequitur falsum. . . . (*Concl.* 3): Contradictoria invicem contradicentia sunt falsa. . . . Ad hoc dicit Bradwardinus quod A est dubium; et cum arguitur, igitur dubium proponitur, dicit quod consequentia est bona et consequens (! *sed lege antecedens*) est verum; sed contra, ergo A est verum, dicit quod non sequitur, quia illud consequens solum significat dubium proponi. Ceffons cites Bradwardine numerous other times in this *questio*. For verification of the references to Roger Swineshead, see Paul Vincent Spade, *The Mediaeval Liar: A Catalogue of the Insolubilia-Literature* (Toronto, Pontifical Institute of Mediaeval Studies, 1975), pp. 102-105; Spade is currently editing Roger's *Insolubilia*. For the reference to Bradwardine, see Marie-Louise Roure, "La problématique des propositions insolubles au XIIIe siècle et au début du XIVe, suivie de l'édition des traités de W. Shyreswood, W. Burleigh et Th. Bradwardine," *Archives d'histoire doctrinale et littéraire du moyen âge*, 37 (1970), p. 316.

77. Book I, Q. 4 (of whole); MS Troyes 62, 14r: Tertio circa prologum quero utrum

logica sit utilis theologis in inquisitione theologicarum veritatum. Et arguitur quod non, quia illud quod est causa curiositatis et etiam dampnationis non prodest; sic autem (MS aut) est de logica, ut patet de illo qui apparuit cum capa plena sophismatibus et cetera, cuius dictione dictum est: linquo coax ranis, cras corvis, vanaque vanis, ad logicam pergo, que mortis non timet 'ergo'. The medieval couplet quoted by Ceffons is one of Serlo of Wilton. For the relevant references to Serlo as well as quotation of another passage in which Ceffons uses the couplet, see J. Murdoch, "From Social into Intellectual Factors," (above, note 25), p. 317, n. 32.

78. Book I, Q. 4 (MS cit., 14r): Concl. 1: Quod logica sit utilis ad investigationem theologice veritatis et etiam veritatum theologicarum. Concl. 2: Quod etiam theologis utilis est rethorica; de grammatica enim non est dubium, quia ipsa est necessaria etiam ad intelligendum dicti (!) theologie. Sit autem suppositio ad declarationem prime conclusionis quod logica docet bene et proprie disputare. . . . Secunda suppositio: quod theologi solent ad veritatis inquisitionem disputare, patet in exercitiis communibus theologie. Tertia suppositio: quod dyalectica seu logica docet artem respondendi et bene opponendi. . . . Book II, Q. 3 (MS cit., 96r, 101r): Hic tamen bene audeo astruere quod nunquam vidi peritum logicum qui logicam diffameret, ignaros logicos logicam contempnere vidi plurimum (MS pulchrum) . . . non est mirum si teneant fidem falsam, cum sint penitus ignorantes et propter ignorantiam logice multi in vanos prolabantur errores. . . . Multi etiam logicam deiciunt et infamant. Et forsan videbitur quibusdam quod nimis hic de logicalibus pertractavi.

79. Book II, Q. 1 (MS Troyes 62, 83v, 85v): Circa principium secundi sententiarum in quo tractat magister de productione rerum aliarum a Deo, quero utrum omne aliud a Deo de necessitate producatur seu de necessitate eveniat. . . . Item in eventibus necessariis astrorum et in eclipsibus et consimilibus communiter astrologi omnes, tam latini quam hebrei, arabes et greci, dicunt in suis prenosticationibus tali die erit eclipsis secundum naturam, si Deo placet; et cum istis est maior necessitas, igitur etc. Ceffons refers to numerous authors (e.g., Arazachel, Benissac, Messallach, Alphraganus) of astrological works in the sequel. Cf. D. Trapp, "Peter of Ceffons" (above, note 21), pp. 101, 105.

80. Book II, Q. 4 (MS Troyes 62, 101r): Circa primam distinctionem secundi sententiarum in qua loquitur magister de creatione celi vel spere, quero utrum possit ratione naturali probari novem speras esse. In his determination of this question Ceffons appeals to, inter alia, Ptolemy, Geber, Thabit, and Alfraganus.

81. Book II, Q. 18 (MS Troyes 62, 133r): Quia hodie theologi faciunt mentionem de toto et parte, ideo circa distinctionem ———— (MS habet lacunam) secundi quero utrum totum sit sue partes. The passage in Aristotle that served as the point of departure for all scholastic discussions of this question is Physics, I, ch. 2, 185b 11-16. For a fourteenth-century Parisian treatment of the questio Ceffons here investigated, see John Buridan, Questiones super octo phisicorum libros, I, Q. 9 (ed. cit., note 43), fol. 11v-13v.

82. Book II, Q. 14 (MS Troyes 62, 116v): Sed adhuc quia volunt aliqui ut aliquid dicatur de proportionibus, saltim modo grosso et rudi, dubitatur utrum illi contradicant sententie Aristotelis et veritati qui asserunt alicam creaturam posse agere in instanti sed sequitur (! vel quod?) oportet creaturam agere successive ex limitatione sue potentie, quia finita est. This questio appears as a "supplement" to the preceding II, Q. 13 (MS Troyes 62, 114v-116v): Utrum angelus possit moveri, which itself contains the dubium: Utrum grave moveatur a se quando descendat.

83. Book II, Q14 (MS Troyes 62, 116v) continuing text in preceding note: Et videtur quod sic, quia isti destruunt totam habitudinem proportionum quam velocitas consequitur secundum Philosophum in pluribus passibus librorum suorum; igitur contradicunt ei et veritati. Consequentia nota est. Et antecedens patet, quia habet consequenter dicere quod aliquod mobile alica velocitate movetur et, si aufferatur medietas resistentie. nequaquam velocitas fiet dupla, quod est contra Commentatorem exponentem Aristotelem quarto *Physicorum*. Et quod habeant hoc dicere et cetera patet, quia, si semper ammota mediete resistentie, velocitas duplicaretur, tunc limitatione et finitate potentie dempta resistentia non determinaretur ad certam velocitatem finitam. In oppositum: quia multi antiqui et moderni hoc tenent concordantes secum omnia dicta Aristotelis, hic eligo partem affirmativam huius dubii. Dico enim quod tales contradicunt Aristoteli et veritati quod patet primo, quia isti ponunt determinationem velocitatis necessariam et certam secundum limitationem potentie, seclusa etiam omni resistentia. Et illud non stat cum dictis Aristotelis volentis quod velocitas consequitur proportionem geometricam; quod enim contradicant processui Aristotelis quarto *Physicorum* facilissimum est videre, cum ipse improbans vacuum concludat quod, si nulla esset resistentia, motus fieret in instanti. (Ceffons then cites several points made by Averroes on the relevant passage in *Physics* IV and continues): Ex hiis possunt multa dici contra istos (scil. those maintaining action in an instant). Et hec apparent aliqualiter ex questione precedenti. (Ceffons then introduces his exposition of Bradwardine with the following sub-question): Utrum autem ex geminatione (MS generatione) potentie aut amotione medietatis resistentie sequitur universaliter aut saltim quandoque geminari velocitatem dubitant aliqui.

84. See note 69.

85. MS Troyes 62, 116v (directly following text in note 83): Et ad huius difficultatis solutionem pono tres suppositiones: (1) Prima quod velocitas attenditur penes proportionem geometricam ita quod ubi maior proportio potentie ad resistentiam ibi maior velocitas et ubi minor minor et ubi equalis proportio equalis. (2) Secunda suppositio quod, cum fuerint tres termini continue proportionales, proportio primi ad ultimum est dupla ad proportionem primi super secundum et secundi super tertium. Et cum fuerint quatuor tales termini continue proportionales, erit proportio primi ad ultimum tripla ad proportionem primi super secundum vel secundi super tertium vel tertii super quartum. . . . Et sic semper uno minus quam in terminis erit denominatio proportionis. Hec secunda suppositio trahitur evidenter de principio quinti *Geometrice* [i.e., Euclid's *Elements*], similiter ex *Arismetica* Jordani ex principio secundi ubi de proportionibus tractat ubi sic dicit: cum fuerint tres termini continue proportionales, dicetur primi ad tertium proportio primi ad secundum duplicata, ad quartum vero triplicata. (3) Et quia numero Deus impare gaudet, tertia et ultima suppositio erit ista—quam in forma ponit Jordanus ibidem: similes sive una alii eadem dicuntur proportiones que eandem recipiunt denominationem, maior vero que maiorem, minor vero que minorem. With respect to the first supposition stipulated by Ceffons, it should be noted that "attenditur penes proportionem geometricam" was a standard way of referring to Bradwardine's solution to the problem of relating forces, resistance, and velocities (Cf. J. Murdoch and E. Sylla, "Swineshead, Richard" [above, note 18], p. 201). Furthermore, it should also be noted that Bradwardine himself sets down suppositions equivalent to the second and third given by Ceffons, although he confirms his suppositions by reference to an anonymous *De proportionibus* rather than appealing to the *Arithmetica* of Jordanus of Nemore (Cf. ed. Crosby [above, note 7], pp. 28, 76). Ceffons's citations from Jordanus

are accurate, save that Jordanus is concerned with proportions of numbers (see MS Venice, San Marco, Fondo antico 332, 42v-43r).

86. MS Troyes 62, 116v (directly following text in preceding note): Ex istis tribus suppositionibus elicam alicas conclusiones. (1) Prima est hec: est assignare <poten- tiam> et resistentiam (MS resistentiarum): et duplata (!) potentia et eadem manente resistentia, duplabitur velocitas (Ceffons illustrates this doubling of the relevant velocity by an example in which the resistance is held constant at 2 while the corresponding force increases from 4 to 8). . . . (2) Secunda conclusio est assignare potentiam et resistentiam: et si medietas resistentie dematur eadem manente potentia, duplabitur velocitas (for example, when the force is constant at 4 and the resistance decreases from 2 to 1). . . . (3) Tertia conclusio est assignare potentiam et resistentiam: et si dupletur potentia eadem resistentia manente, redderetur velocitas minor dupla (for ex- ample, when the resistance is constant at 2 and the force increases from 6 to 12). . . . (4) Quarta conclusio est ista: est assignare potentiam et resistentiam: et si resistente medietas aufferatur eadem potentia remanente, fiet velocitas minor dupla (for example, when the force is constant at 16 and the resistance decreases from 4 to 2). The fact that one can elicit such force-resistance changes in which sometimes a doubling of velocity occurs and sometimes it does not fits with Ceffons beginning his investigation with a sub-question that asked whether the relevant doubling of velocity occurs "universaliter aut saltim quandoque" (see end of text in note 83).

87. MS Troyes 62, 117v: Ex hiis omnibus conclusionibus patet hoc: quod nec univer- saliter ex duplicatione potentie sequitur velocitas minor dupla nec precise dupla nec etiam maior dupla; nec ex ammotione medietatis resistentie universaliter sequitur velocitas minor dupla nec dupla precise nec etiam maior dupla. . . . Et ex hiis patet se- cundo in quibus dupla sequitur, in quibus minor et in quibus maior. Precise quidem duplatur velocitas duplicata potentia aut remota medietate resistentie quando terminor- um primorum fuit proportio ipsa dupla. . . . Cum autem termini sunt in proportione plusquam dupla, duplicata potentia aut ammota medietate resistentie, fit velocitas ad primam velocitatem minor dupla; nam non pervenit ad illam que est prioris propor- tionis proportio duplicata. . . . Ubi autem est terminorum proportio minor dupla, dupli- cata potentia aut resistentie medietate seclusa, fit universaliter ad priorem velocitatem velocitas maior dupla. . . . Et stat in hoc totum quia, si sit proportio dupla ad habendum terminum proportionalem, requiritur duplatio eiusdem vel minoris resistentie in duplo. Si autem sit maior dupla ad habendum proportionalem terminum, requiritur plusquam duplum vel quod [MS quam?] resistentia plusquam in duplo minuatur. Si autem sit minor dupla ad habendum terminum proportionalem, requiritur minor dupla vel quod resistentia minus quam in dupla minuatur. In establishing much of the foregoing, Ceffons brings his hearers back to the relation that his claims have with the supposi- tions he began with. Relative to Ceffons's attempt to teach Bradwardine, mention might be made of the fact that other fourteenth-century scholars also devoted themselves to such didactic presentations. This can be seen, for example, in Symon de Castello's *De proportionibus velocitatum in motibus* (ed. James McCue, unpublished dissertation, University of Wisconsin, 1961) or in Otto of Merseburg's *Questio de proportionibus velocitatum in motibus* (MSS Erfurt, Amplon. 4° 110, 54r-68r and 2° 380, 67r-83r; Cracow BJ 735, 42r-52v and BJ 739, 54v-72v).

88. MS Troyes 62, 117v-118v (directly following text in preceding note): Et ut ex uno multa concludamus, capiamus suppositionem secundam (see note 85) que fundamentalis

sit nobis; ex illa autem media suppositione paucis coassumptis alique questiones que interdum occurrunt faciliter solvi possunt. Si enim querat aliquis: habeo circulum cuius est dyameter 10 pedum, tu circulum cuius est dyameter unius pedis, quotiens iste continetur in isto seu in quota proportione minor exceditur a maiori. . . . Si igitur queratur de proportione totius spere activorum et passivorum seu quatuor elementorum ad speram terre, est sciendum prout reperi in 21ª differentia Alfragani: "longitudo propinquior lune a terra est trigesies ter tantum quantum dimidium dyametri terre et dimidium dimidii ac 20 pars eius." . . . igitur secundum hanc opinionem dimissis in (!) minutiis facile est scire proportionem totius spere corruptibilium ad speram terre. Throughout all of this additional material *de proportionibus* Ceffons cites the likes of Euclid, Archimedes, Ptolemy, Theodosius, Thabit, Jordanus, and Sacrobosco. In his reference to Alfragani, however (for which see al-Fragani's *Differentie scientie* astrorum, ed. Francis Carmody [Berkeley, 1943], p. 38), Ceffons's 'reperi' is slightly disingenuous, since the chances are that he was here poaching on Bradwardine's reference to the same (see ed. Crosby,[7] p. 134).

89. MS Troyes 62, 119v: Innumerabilia alia circa materiam de proportionibus ex hiis inferri possunt, sed pauca sufficiunt studioso. Materia etiam de proportionibus proportionum non frequenter requiritur in scolis nostris, specialiter in diebus. Taceo etiam de proportionibus irrationalibus et multis aliis habitudinibus que requirerent tractatum ordinatiorem, et de hiis in theologia modicum nunc curatur. Sed hoc qualitercumque composui addidique premissis ad petitionem cuiusdam specialis amici in principio nominati (Cf. supra, note 73); alias, si Deus dederit, poterunt hoc ordinatius profundiusque tractari; nam satis curiose in hiis scripsi. Ceffons's use of the technical expression *de proportionibus proportionum* need not derive from Nicole Oresme's work bearing that title (see above, note 16), since the expression can also be found in the works, for example, of John Buridan and John Dumbleton (see M. Clagett, *The Science of Mechanics* . . . [above, note 14], pp. 441-442). Similarly, in his reference to "irrational proportions," Ceffons need not have been aware of Oresme's *De proportionibus* (or his *De commensurabilitate*)—in which irrational proportions play an important role—since Oresme had already delineated the notion of an irrational proportion in his (undoubtedly earlier) *Questiones* on Aristotle's *Physics,* a work Ceffons apparently did know (see note 72). In this regard, see, for example, Oresme's *Quest. phys.* (MS Sevilla Colomb. 7-6-30, 67r-v): Secundum quamlibet proportionem et qualitercumque potest dividi continuum in duo media vel etiam in partes commensurabiles. Ex quo sequitur duo (!) quod qualibet proportione rationali data, per infinitum modicum fieret irrationalis aut econverso addendo vel minuendo. In any event, all of this strongly suggests that, however inexpert Ceffons may have been in "calculatory matters," he was certainly abreast of what was currently transpiring in this regard. On all of this, however, I have thus far not been able to discover a definite reference to Oresme. Cf. notes 72 and 96.

90. Indeed, he even turns to such material just two questions later: Book II, Q. 16 (MS Troyes 62, 122v): Utrum quelibet potentia terminetur per maximum in quod potest.

91. Book II, Q. 17 (MS Troyes 62, 129r): Utrum quilibet motus uniformiter difformis sit tantus quanta est eius alica pars.

92. Book II, Q. 17 (MS Troyes 62, 129r): (directly following text in preceding question): Et videtur quod sic, quia denominatio (MS denunciatio) fit a maiori communiter. Unde quelibet spera ita velociter movetur sicut alica eius pars et attenditur velocitas eius penes punctum velocissime motum. Unde secundum Bradwardinum in secunda

parte capituli quarti de proportionibus motuum (see ed. Crosby [ed. cit., note 7], p. 130):
"velocitas motus localis attenditur penes velocitatem puncti velocissime moti in corpore
moto localiter, quia velocitas est in eo quod mobile pertransit magnum spacium quies-
cens verum <vel> ymaginatum in parvo tempore; vel etiam velocitas motus est in eo
quod mobile pertransit aut pertransiet magnum spacium fixum, si ei esset applicatum,
in parvo tempore." Et hoc dicit propter ultimam speram de qua dicunt philosophi quod
ultra illam nihil est. Et similiter secundum theologos Deus potest circumvolvere totum
mundum, et tunc ultima convexitas mundi non describeret aliquod spacium.

93. Book II, Q. 17 (MS Troyes 62, 129r): In oppositum arguitur, quia quilibet motus
uniformiter difformis incipiens a non gradu correspondet suo medio gradui et est ei
equalis; igitur non quilibet motus uniformiter difformis est tantus seu ita intensus sicut
aliqua eius pars. For the "mean degree criterion" here appealed to, see note 10 above
and the exposition which it documents.

94. Such a mid-point alternative was held by Gerard of Brussels (see M. Clagett, The
Science of Mechanics . . . [above, note 14], pp. 185-189) whom Bradwardine cites and
criticizes (ed. Crosby [above, note 7], p. 128). Cf. Albert of Saxony's Tractatus proportionum (ed. cit., note 15), pp. 67-71.

95. MS Troyes 62, 129r-v: In ista materia primo videndum est quis sit motus uniformis
vel difformis aut uniformiter difformis vel etiam difformiter difformis. Deinde ponentur
alice conclusiones et postea moveri poterunt alique obiectiones aut etiam dubitationes.
Quantum ad primum: motus enim uniformis est satis notus. Dicitur enim uniformis
quasi tenens continue unam formam. Unde ille motus qui continue tenet gradum unum
sive alica intensione vel remissione dicitur uniformis; <igitur> est ille motus uniformis quo continue in equali tempore equale spacium pertransitur. . . . Difformis autem
motus est qui non continue sic retinet unam forman. sed ————— (Ms hab. lacunam)
ita quod nunc est intensior nunc remissior, . . . ille ergo est motus difformis quo in
temporibus (MS tribus) equalibus spacia pertranseuntur inequalia. Et iste est duplex,
quia quidam est uniformiter difformis, quidam difformiter difformis. Uniformiter
difformis est ille qui in difformitate sua retinet unam formam. Unde et ille (MS illud)
dicitur uniformiter difformiter (! MS, sed delendum est) intendere motum suum qui in
quacumque equali parte temporis equalem velocitatem acquirit, et uniformiter remittitur
motus talis quando in quacumque parte equali temporis equalem latitudinem velocitatis
deperdit. Difformiter autem difformis est qui in difformitate sua non retinet unam formam, sed in partibus equalibus temporis inequales latitudines velocitatis acquirit.

96. MS Troyes, 129r-v: Unde et reducendo ad figuras motus ipsos per ymaginationem,
motum uniformem solent communiter scolares ymaginari quasi unam superficiem
equaliter altam sicut unum quadrangulum equiangulum cuius linea superior est equidistans basi; et talis etiam superficies dici solet uniformiter alta, quando scilicet linea
altitudinis est equidistans basi. Motus autem difformis ponitur ad modum superficiei
difformiter alte, cuius scilicet linea altitudinis non est equidistans basi. Et hoc contingit
dupliciter, quia aut illa altitudo potest esse uniformiter difformiter alta aut difformiter
difformiter. Uniformiter difformiter est altitudo alta, sicud alii dixerunt, et bene quando
quelibet tres linee vel plures equaliter distantes inter se excedunt se secundum proportionem arithmeticam, id est, equaliter. . . . motus uniformiter difformis potest incipere
a non gradu vel a gradu alico; si incipiat a non gradu, tunc possumus ipsum ymaginari
sicut unum triangulum uniformiter difformiter altum (MS altam) cuius acuties incipit
a puncto in quo due linee concurrunt; si autem incipiat a gradu, tunc possumus ymagi-

nari modum illum quasi unam superficiem quadrangulam non rectangulam aliqualiter acutam cuius acuties est linea ultima minima in quadrangulo illo, ut patet in figura (no figures are given in the MS at this point). . . . Et hae declarationes motus uniformis et difformis aliqualiter dat intelligere clarius quam si non ad figuras ymaginatio flectetur. Ceffons does not mention Oresme at this point, but the geometrical representations of which he speaks could have been derived from the latter's works or teaching. However, Ceffons need not have had, and most likely did not have, Oresme's *Tractatus de configurationibus* at hand (it was written only after 1351). He could instead have taken at least some of his material from Oresme's earlier *Questiones super geometriam Euclidis*. Indeed, the above definition of a uniformly difformly high altitude resembles Oresme's definition of a uniformly difform surface in this earlier work (see M. Clagett, *Nicole Oresme* . . . [above, note 10], p. 550).

97. Book II, Q. 17, Conclusiones (MS Troyes 62, 129v): (1) Non contingit dare motum uniformiter difformem velocissimum nec etiam tardissimum. . . . (3) Quod nullum uniformiter difformiter calidum est ita calida sicut alica eius pars. . . . (4) Quod nullus motus uniformiter difformis est ita intensus sicut alica eius pars. . . . (5) Quod in omni motu uniformiter difformi incipiente a non gradu tota latitudo motus correspondet gradui medio. For the second conclusion see the following note.

98. Book II, Q. 17, Conclusiones (MS Troyes 62, 129v): Secunda conclusio quod etiam si sint alique qualitates que habeant aut habere nate sint summos gradus, nulla tamen qualitas uniformiter difformis terminatur inclusive ad summum gradum sic quod alica pars subiecti sit summe talis, puta summe alba aut summe calida aut huiusmodi.

99. MS Troyes 62, 130v, 131r: Secundo contra eandem conclusionem (scil., concl 2), quia esto quod aliquis motus uniformiter difformis sit, ipse tamen terminatur ad aliquem gradum inclusive, ut patet, quia in rota que circumvolvitur uniformiter super centrum immobile, motus extensus per rotam est uniformiter difformis, et tamen est dare punctum velocissimum seu velocissime motum. . . . Ad secundum . . . dico quod in motu uniformiter difformi cuius partes sunt simul ut in rota, non habetur pro inconvenienti quod terminetur ad gradum inclusive. Unde nos ibi ymaginamur ultimum punctum seu puncta in circumferentia; dicimus quod ita velox movetur ista rota sicut punctus qui est aut esset in circumferentia tota, quia rota movetur uniformiter super centrum suum et quoad uniformitatem suam non terminatur sicut motus uniformiter difformis. Unde ista rota ponitur moveri uno gradu velocitatis uniformi.

100. MS Troyes 62, 132v, 133r: Loco tamen obiectionum contra has conclusiones potest esse dubium utrum possit illud sustineri quod est statim concessum, quod scilicet rota ita velociter moveatur sicut aliquis eius punctus. Et forsan videretur alicui quod non possit sustineri, quia tunc, si esset dare infinitam rotam que movetur, illa moveretur infinita velocitate vel non moveretur. . . . Secundo quia aliquod est quod continue movetur et nullus est in ea (!) punctus velocissime motus; quia pono quod continue augeatur rota secundum omnem partem circumferentialem; et tunc in quolibet instanti erit verum dicere quod nullus est punctus velocissime motus, quia in quolibet instanti est assignare extrema puncta que (MS quia) nunquam prius movebantur. . . . Ad primum . . . dico quod rotam esse infinitam claudit contradictionem. . . . Ad secundum . . . dicitur quod ibi attenditur velocitas penes punctum qui quolibet puncto illius rote velocius movetur, per nullam tamen latitudinem velocius movetur quam aliquis punctus huius rote sic quod inter velocitatem suam et velocitates punctorum huius rote nulla esset velocitas media. Et ideo solent alii dicere quod velocitas illius rote que continue

non habet puncta extrema attenditur penes lineam quam describeret quidam punctus qui indivisibiliter velocius moveretur quam si (! *sed del. est*) aliquis in magnitudine illa data.

101. Compare, for example, the sophisms from William Heytesbury cited in J. Murdoch, "*Mathesis* . . ." (above, note 22), p. 234.

102. See, for example, F. Stegmüller, "Die zwei Apologien . . ." (above, note 20), pp. 47-48, 61-62, 71-73.

103. MS Troyes 62, 101r (directly preceding the last line quoted in note 78): Si quid in hac materia veritatis inveneris, Deo ascribas a quo omnis veritas est; si vero aliquid falsitatis repereris, ignorantie mee, quam supra confessus fui, audacter ascribas, et rogo quod in hoc non est defectus. Et hoc quantum ad materiam insolubilium pro nunc; quantum ad obligationes, puto quod eas hic non subiungam. See also the texts in notes 74 and 89.

104. See J. Murdoch, "*Mathesis* . . ." (above, note 22), p. 246.

105. *Chartularium Univ. Paris.*, vol. 3, p. 144: Quod legentes Sententias non tractent quaestiones vel materias logicas vel philosophicas nisi quantum textus Sententiarum requiret.

KINEMATICS is an aspect of dynamics, dealing particularly with the laws and actions of motion in a system of material particles without reference to the forces acting on that system. Fascinating resemblances and differences exist between techniques used in the Middle Ages for analyzing terrestrial and celestial kinematical problems. There is less interaction between these two domains of earth and sky than might be expected, because Aristotle had a particular, peculiar way of dividing up the universe. By reference to the astronomical tables of an anonymous fourteenth-century writer, and to those of John Killingworth of the next century, as well as to contemporary medieval solutions of intricate astro-mechanical problems, it becomes plain that the ingenuity of the philosophers and mathematicians who wrote in Oxford and Paris on the subject of terrestrial kinematics was matched by the achievements of the medieval astronomers.   MPC

Kinematics—More Ethereal Than Elementary

Rijksuniversiteit Groningen
Filosofisch Instituut
Groningen, The Netherlands

Moreover, rotatory motion is prior to rectilinear, because it is simpler and more complete ... Aristotle, *Physics*, VIII.9.

The number of [planetary] movements can be determined only by that mathematical science which is most akin to philosophy, that is, astronomy, which alone deals with substance that is perceptible but eternal, whereas the others—arithmetic and geometry—do not treat of substance at all. That the motions are more numerous than the bodies in motion is obvious to anyone who has given even moderate attention to the matter ... Aristotle, *Metaphysics*, Λ. 8.

THIS paradox—namely that celestial motions are simpler than rectilinear, and yet more complex (numerous)—lies at the very root of Aristotle's division of the universe. Circular motions were for Aristotle essentially simpler, and in that sense prior, because they were thought not to involve us in all the complexities of infinity, as the notion of unlimited rectilinear motion seemed to do.[1] Aristotle's finite universe was suitably divided into an inner spherical region (bounded by the sphere of the Moon) within which all was subject to continual alteration, growth and decay, and an outer spherical shell, the ethereal region, containing the spheres of the planets. Natural motions within the inner region were rectilinear but finite, trapped as they were within a finite boundary. They were directed either towards or away from the center of the universe. Celestial natural motions were circular, and on that account enjoyed a measure of perfection. These motions had an eternal character, according to Aristotle, and "a motion that admits of being eternal is prior to one that does not."[2] Needless to say, Aristotle will accept a temporal infinity where he would not accept the equivalent in space. This equivocation gave Averroes, Aquinas, and other loyal commentators certain problems,[3] but these are not my concern. I simply want to direct attention to that hard and fast

[89]

0077-8923/78/0314-0089 $01.75/2 © 1978, NYAS

boundary between two fundamentally different cosmic regions, each with its own type of motion. And I want to do so at the outset so that I may all the more plausibly distinguish between two different sorts of fourteenth-century scholar, each concerned with the analysis of one of those motions. I am of course referring on the one hand to those who—especially in Oxford and Paris—laid the foundations of terrestrial kinematics, and on the other hand to those who advanced the techniques of Ptolemaic astronomy.

Did neither group contribute to an understanding of the problems posed by the other? Was specialization such that the groups were made up of entirely different individuals? These questions have not been asked sufficiently often. If we are to judge by the sharp division between modern historical studies of the two groups, there was precious little interaction. I think this is broadly true, and by drawing attention to the Aristotelian distinction I have tried to explain why this fact should not surprise us. The Aristotelian distinction must surely bear at least part of the blame for any intellectual barrier we happen to find. Even so, we know of many medieval scholars who knew at least the rudiments of both Ptolemaic astronomy and terrestrial kinematics. If you put two sets of related ideas into one man's head, is it not reasonable to expect some sort of interaction? I cannot say that my study of this side of the problem is a very thorough one, but such evidence as I have found tends to suggest that first impressions are more or less correct, and that interaction was negligible across the relevant boundaries in the university curriculum. In other words, as far as original kinematic analysis is concerned, it seems that there were indeed two sorts of medieval scholar. It is therefore hardly surprising that each has produced its own sort of medieval historian. (And *there* is a barrier we have it in our power to abolish.)

In what follows, the most I can hope to do is draw together some of the loose threads, referring to arguments that have already been separately and expertly treated. For the fine detail I must refer to other commentaries. I shall collect together instances where interaction between the concepts of astronomy and terrestrial kinematics seems plausible, and I shall indicate a number of interesting, and sometimes intricate, examples of kinematic analysis that bear comparison with those most widely discussed by historians of kinematics. Above all, I want to show how astronomical kinematics was far from being a stagnant subject in the fourteenth and fifteenth centuries.

I shall begin with a few truisms, for the sake of those who are strangers

to medieval scientific thought. The fourteenth-century discussion of freely
falling bodies was not an empirical one. The difficulties of turning it into
an empirical study comparable with the study of planetary motions were
entirely practical. Even the Moon takes a leisurely month to move once
round the sky against the background of fixed stars, while Saturn takes
almost 30 years. On the other hand, a stone dropped from the highest
point of Merton College in Bradwardine's day would have reached the
ground in less than 2 seconds. (So far as I know, no one thought of investigating mathematically that other sub-lunar natural motion, namely the
rising of fire.) Secondly, the links between force (and resistance) and local
motion—in other words, between dynamics and kinematics—were without a counterpart for celestial motions. If the astronomer pondered the
cause for these, he fell back on intelligences and the Aristotelian prime
mover.

Those who advanced the study of kinematics at Oxford and Paris—and
the names of Thomas Bradwardine, William Heytesbury, Richard Swineshead, John Dumbleton and Nicole Oresme come most readily to mind—
may all be reasonably described as philosophers trained in mathematics.
The philosophical problem was one of explaining how forms or *qualities*
could vary in strength or *intensity*. This was the problem of intension and
remission of forms, which is to say, of increase and decrease in their intensity. The Mertonians treated this problem in general (in regard to intensity of color, heat, and so on) by analogy with the treatment of problems of motion in space.[4] The vocabulary used is to a casual observer
very reminiscent of that used in astronomy, with such words as *"motus"*
("movement," but often meaning "position," and even at times functioning as "velocity"), *"gradus"* (as in *"gradus motus"*—"degree of movement"), and *"latitudo"* (as applied to measurements on the velocity axis of
a velocity-time graph). It is difficult to decide whether these resemblances
are of any significance. Graphical methods were of course well established
in astronomy, not only for the simple representation of position through
the allocation of coordinates, and in transformations from one system to
another, but also for computation using geometrical loci with what would
now be called "analogue methods."[5] On the other hand, we should not
forget that what are to our eyes technical mathematical terms were in the
past simply common words—for movement, step, and breadth, in the
instances mentioned. Before the recognizably modern coordinate graph,
as in the work of Giovanni di Casali and Nicole Oresme, there is to be
found a two-dimensional *geometrical* analogy to contrast the intensity of a

quality and its quantity of extension.[6] There is no mention in any of these contexts of an astronomical association, and it seems to me to be quite mistaken to suppose that in kinematics the words "*latitudo*" and "*longitudo*" have any but their everyday meanings of length and breadth.

Although it should hardly be necessary to point out that astronomy has operated with kinematic concepts since Babylonian times, this is by no means to say that there was any conscious attempt to *define* the concepts so used. In his invaluable collection of materials on medieval mechanics, Marshall Clagett mentions a number of definitions from early astronomical writings.[7] Autolycos of Pitane (ca. 310 B.C.) defined uniform movement in terms of equal distances ("lines") in equal times (with the consequence drawn that distances were in the same ratio as corresponding times), and he applied the definition in some theorems on the uniform rotation of a sphere. Two centuries later, Geminos emphasizes the importance of geometry for astronomy, but by then this was something of a platitude in the Greek world. Even by the time of Autolycos, the extraordinary system of homocentric spheres as devised by Eudoxos had shown the great power of the geometric method. The treatise by Autolycos was translated into Latin by Gerard of Cremona in the twelfth century. Gerard of Brussels, however, who can be regarded as having launched medieval kinematics on its career in the thirteenth century, makes no obvious use of Autolycos. Gerard's geometric approach stems from Euclid and Archimedes. He had the same main object as Autolycos, namely to treat of the rotation of the parts of a sphere, and of other bodies—not a very pressing astronomical problem, as it happens. Throughout all this work there is lacking the concept of velocity as the ratio of essentially different quantities, viz., as a quotient of a length and a time. To speak of uniform motion in terms of equal distances in equal times is of course the counterpart of the astronomer's tables in which the *medium motus* is given for standard time intervals; for here "mean motion" is simply an angular distance measured from some base direction. The mental block was largely of Aristotle's making: lengths and times were of different species, and even a comparison of movement along a straight line and a curved was according to him out of order, since they too were supposedly of different species.[8] (This last idea Gerard of Brussels rejects, although he is squarely in the Aristotelian tradition.) Before they could readily progress towards the view of instantaneous velocity as a limit of a ratio of a length and a time, medieval writers on kinematics had to pass through the stage at which different "motions" (different velocities, as we might anachronistically say) were compared through the ratio of distances traversed in equal times. Gerard of Brussels, Bradwardine, and later writers have this idea explicitly; but of course it is implicit in the work of those astronomers who, like Ptolemy, tabulated motions and compared them (notably in eclipse calculations, which one may loosely describe as the problem of deciding whether two ships, of different sizes, on different (but intersecting) courses, and moving at different speeds, will collide). One should not underestimate the achievement of Ptolemy, or of those who later adapted his remarkable algorithm for this problem. Richard of Wallingford, for example, devised an instrument for the solution of this problem, as a component of his so-called "albion" (all by one). Just how ingenious this analogue device is can be seen by referring to his treatise on the albion and my commentary on his treatise.[9] For those who are not familiar with the problem, let me simply point out that its solution involves finding, not perhaps "instantaneous" velocities, but at least the hourly motion of the Moon as a function of a certain angle (the argument, so-called, when the Moon's epicycle is at apogee). Although, had they been challenged, Ptolemy and his followers would probably have taken the standard Aristotelian way of discussing velocity, in their working they came perilously near to a treatment of instantaneous velocities, and of instantaneous velocities in terms of quotients of small differences. It would be too pretentious to talk of differentials.

The sort of compound instrument designed by Richard of Wallingford —abbot of St. Albans and only a few years Machaut's senior—was part of a long tradition stretching back to antiquity. I cannot possibly begin to review this tradition here, but let me point out that it includes the graphical representation of planetary motions generally, and involves the generation of curves (usually circular arcs) on which position is correlated in some way with *time*. It is well known that three great problems of Greek geometry—the squaring of a circle, the duplication of a cube, and the trisection of an angle—led to the invention of curves involving movement, that is, using time as a parameter. There is no reason to suppose that the works of Hippias, Pappus, and Archimedes (on spiral lines) influenced in any way the computing tradition of which I spoke, but there too we find examples of curves generated by kinematic means. One notable example I might mention in particular is in fact to be found both in the computing tradition and in straightforward astronomical doctrine. I refer to the resultant deferent curve for the planet Mercury. Mercury was given a more complicated movement than any of the higher planets. Ptolemy supposed it to move on a circle (epicycle) whose center was to move on another circle (the deferent proper), with a center moving on yet a third circle. For all planets other than the Moon, two circles were enough, namely, a (moving) epicycle and a (fixed) deferent circle. Clearly it would be convenient if Mercury's second and third circular movements could be combined. In fact they were combined into a single (oval) curve on an eleventh-century instrument designed and described by the Hispano-Moorish astronomer known as Arzachel. Richard of Wallingford, possibly by a process of independent invention, makes use of comparable ovals twice over, for one of his scales transferring the time divisions from the oval Mercury deferent to a concentric scale, and for another of his curves producing what can be described as an approximation to the geometrical inverse curve. The latter was so designed as to have its radius vector proportional to a certain correction term ("equation") needed for the calculation of Mercury's position. It would, however, hardly be relevant if I were to pursue further the history of the Mercury deferent. I introduced it simply as an example of the way in which medieval astronomers used geometrical techniques for the "static" solution of problems involving *times*, but times which were no longer, as it were, thrown away when the parametric representation had done its work. The time-value corresponding to each point of an Archimedean spiral was of no great interest to the geometer of the ancient or medieval world. Not so the times corresponding to the points of curves analogous to the Mercury deferent, which therefore more truly belong to the subject of kinematics.

Another subject intimately connected at once with both astronomy and mundane kinematical principles is the analysis of the motions of trains of gear wheels. If I may be allowed to introduce him once more, Richard of Wallingford gives evidence not only of personal ingenuity but of traditions that we can now only dimly perceive. At first the geared planetarium, and later the mechanical and particularly astronomical clock, served as a meeting point of theory and practice. If we search only for anticipations of Galileo, the concept of velocity as a ratio, or the equation of free fall, we shall overlook entirely the calculus of gear ratios. We might take a high moral line over this matter. Certainly most trains were designed merely to preserve a constant ratio between the members of a set of constant mean angular velocities. Bearing in mind the very awkward ratios that obtain between the planetary motions, it is easy to see that the difficulties were rather greater than those of linking the hour-hand and the minute hand of a watch; but awkward as they were, and transported trains of wheels though they might have required, the motions in ratio were usually constant. Consider, however, Richard of Wallingford's solution to the problem of making the Sun's wheel move with a continuously varying velocity. This velocity was not to be equal to the velocity of the real Sun (i.e. in the ecliptic), but of the projection of the Sun on the celestial equator. He gave the solution, but without discussing the way of arriving at it, or even the formulation of the problem. The solution was to engage an *oval* contrate wheel (i.e. with teeth at right angles to its plane) with a pinion lying along a radius vector.[10] (The pinion must be of such a length that it will mesh with the contrate teeth at both greatest and least distance from the center.) How then did he decide on the very precise form of the oval contrate wheel? I have hazarded a guess in my edition of his works, and I suggest that as a part of his plan he effectively plotted the reciprocals of solar velocities (velocities in right ascension) in polar coordinates.

But there's the rub! Do I really mean *velocities*, or simply daily movements taken from an astronomical table? Of course in point of procedure I mean the latter, but there is surely a case to be made out for saying that by making something—namely a radius vector—a *function* of these small daily bits of "movement," Richard of Wallingford was at least groping for the idea of velocity as a richer conceptual entity than that of mean motion. But even granted that we have not caught the will-o'-the-wisp of instantaneous velocity, it must surely be admitted that—blacksmith's work or not—Richard of Wallingford's was a piece of kinematic wizardry comparable with that produced by the philosophers of his own university —philosophers who usually took only the crudest of astronomical examples to illustrate their classifications of motions. John of Holland, for instance, a mid-fourteenth century Oxford man, takes the so-called ninth sphere of the astronomers as an example of a sphere "moving uniformly with respect to time. . ." He makes no mention, however, of the movement with which so many of his astronomer contemporaries were obsessed, namely, the periodic (and therefore accelerative) motion of the eighth sphere. The steady movement of the stars with respect to the equinoxes was discovered by Hipparchos, in the second century B.C., but the kinematically complex explanation later offered for the movement (which was not looked upon as a precession of the equinoxes) seems to have originated with Thābit ibn Qurra of Harrān in the ninth century (Christian era). An admirable survey of this whole subject of precession has been given recently by Raymond Mercier, who incidentally touches upon the difficulties some astronomers have apparently had in trying to interrelate even steady periodic motions.[11] And while on the subject of precession, it should be noted that Hipparchos obviously had a clearer vision than most; one cannot help wondering whether the course of scientific history would have been in any way accelerated had his lost work on falling bodies been handed down to the Middle Ages.

I want to turn now to some of the more abstract parts of astronomy, namely planetary tables, in which kinematic ideas are often only to be inferred. I will try to give the barest outline of the sort of procedure followed by an astronomer who wished to calculate the position of a planet at a particular moment of time. Consider the planet Venus (P, FIGURE 1) which, according to Ptolemaic principles, moved steadily round an epicycle with center O, while O moved round the deferent circle, with center C.

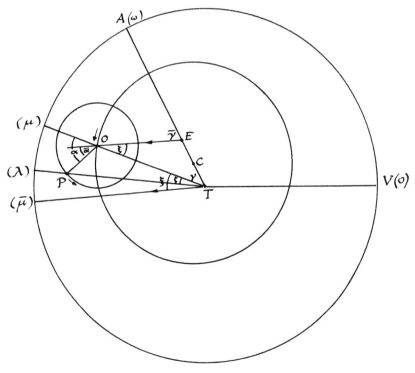

FIGURE 1

The Earth (T) was at some distance from C, and a point E, known as the equant, was the same distance to the other side of C. The equant was important because it was the EO, and not CO, which turned at constant angular speed. The resulting locus of P is a curve with a series of loops; but (if we overlook the complication that not all our circles were thought to be in quite the same plane) the planet would mostly be seen from the Earth to proceed in the same general direction as the Sun (in its annual path), but to turn back on itself from time to time, with the so-called "retrograde" motion. Its apparent place on the ecliptic, the path it seemed to follow through the background of stars, is at the point marked (λ). It is this point the astronomer wishes to find.

Now, once the relative dimensions of the circles and the direction of the line of symmetry TCE have been established, there are essentially only five further pieces of information needed to solve the problem. (We overlook the complication of precession.) We need to know where the radius vectors EO and OP were at some standard epoch, we need to know their constant angular velocities, and we need to know the time which has lapsed from the standard epoch. Given a scale diagram, or some instrumental equivalent—namely our "equatorium"—we can find the planet's apparent position with reasonable accuracy. Better still, given an electronic calculator and a knowledge of coordinate geometry, an even more accurate answer may be found. But the method that the Middle Ages learned from Ptolemy was a different one. From the position of EO with respect to the line of symmetry TCE, a small angular correction-term or "equation" EOT (marked ξ) was found. From this information, and the direction of OP, another equation PTO (marked ζ) was found. Combining by addition or subtraction the equations and the given angles, themselves linear functions of the time, the apparent position of the planet could be found. (The angle λ, the ecliptic longitude of the planet, was measured from the vernal equinox V.) In the course of this procedure, 10 or more different tables might have been used.[12]

What of these tables? They fall into two broad categories: (1) tables of mean motions, correlating angles with times; (2) essentially trigonometric tables, giving the "equations" as functions of other angles, and involving some highly ingenious processes of interpolation.

Although this rough classification does not begin to do justice to Ptolemy or his successors, I will try to restrict my attention to the problem of conceptualizing motion, although this can hardly be done by simply disregarding tables of type (2). It is the astronomer's grasp of the whole

system that counts. Consider, for example, the simplest "planetary" movement, that of the Sun, which was supposed to travel at constant speed
around the circumference of an eccentric circle. Not much cause for ex-
citement here! But when an almanac is compiled—i.e. a list of consecu-
tive solar positions, usually at one-day intervals—and provided, as was
often the case, with columns of first differences (i.e. daily motions), and
even of corresponding hourly motions, then the astronomer must have
begun to sense what it was to look at a table of changing velocities. We
know, for example, that Thābit ibn Qurra looked at the subject with a
keen analytical eye, and wrote a book on "the deceleration and acceleration
of movement along the ecliptic as dependent on distance from the eccentric
center".[13] Even allowing for the modern cast given to Thābit's words, on
which I am not qualified to comment, I have no doubt that when he discussed maxima and minima in the movement, and when he found the
point at which the "velocity" was equal to the mean value, he had a clear
understanding of the idea of acceleration. This has been shown true of a
yet greater astronomer, al-Bīrūnī, who early in the eleventh century
(Christian era) discussed the acceleration of the Sun in a perfectly explicit
way.[14] Bīrūni's discussion rests on his analysis of first and second differences of the true solar motion, this having been conceived of as comprising linear and non-linear components.

I have come across nothing so explicit in any writer in the medieval
West, although it is not unusual to find tables of differences; and these,
on closer inspection of associated texts, might prove to hide a deeper understanding than we yet appreciate. There are certainly matters of kinematic
interest in astronomical tables, especially in those that stemmed from
the tables of Alfonso X, King of Leon and Castile, which were assembled
at some time between 1263 and 1272.[15] The history of the tables after they
were first compiled is uncertain. They seem to have reached Paris a little
before 1320, and from there they spread to all quarters of Europe within
a decade or two. They were modified and rendered more convenient in
Paris and Oxford in particular, in Paris especially by John of Lignères and
his pupils John of Murs and John of Saxony, and in Oxford by an unknown astronomer, and later by John Killingworth. The aim of these
people was to lighten the task of using astronomical tables. Thus John
of Lignères had tables (tabulae magnae) which gave, by double-entry, a
single combined equation, rather than the two separate equations I mentioned earlier. Some Oxford tables, drawn up so as to be of use from 1348,
go much further than the tabulae magnae, and allow planetary longitudes

to be taken out more or less directly. (Correction for precession was necessary. Although the word "almanac" was sometimes applied to them, following John of Lignères' usage, they were not an almanac in the usual sense, of course, for it was first necessary to use tables for the mean motions with which to enter the final table.) The Oxford tables can be shown to have been grounded in the tables of John of Lignères, but where his ideas came from is uncertain. Earlier examples of double-argument tables are known from the Islamic world, but that there is a connection between them and the European tables at present seems unlikely. The Oxford tables of 1348 seem to have been well received. They are provided in one manuscript with an alternative set of canons, ascribing them to "Battecombe alias Bredon sive Bradwardynes." "Battecom" is alone mentioned at one later point and Bredon at another. I think we must simply be satisfied to accept that the tables were "in Oxonia constituta."

The fact that the 1348 tables were so close to an almanac or ephemeris meant that they could be easily used to carry information of a general kinematic sort. They showed whether motions were direct or retrograde, for instance, and they naturally therefore showed the planet's stationary points. A chapter of the canons for their use is given over to this general problem of planetary acceleration and deceleration. (One might even say that the writer was obsessed with this descriptive problem, for he refers not to the planets but to the "retrogradi"!) Another set of tables, meant to be used along lines that would take too long to explain, were drawn up by John of Murs, and effectively yield a picture of the relative motion of each of the planets with reference to the Sun, i.e. in terms of the age of the planet reckoned from the time of its conjunction with the Sun. These tables were rather difficult to use, and seem not have found much favor. They date perhaps from a little after 1320.

After the Oxford tables of 1348 I know of little important activity in this connection until the work of the Merton College astronomer John Killingworth, who died in his mid-thirties in 1445. A sumptuous copy of his tables, heavily interlined with gold leaf and done for Duke Humphrey of Gloucester, is to be found in the British Library (MS Arundel 66). The aim was to simplify the production of ephemerides, and not merely to yield isolated planetary longitudes. Without trying to explain, even in outline, John Killingworth's entire procedure, I will discuss one or two of his more interesting achievements. For a particular planet for much of the calculation he reckons mean motions as functions of the number of days from the noon when the planet was nearest the line of symmetry I alluded to earlier. (In my Venus example it was the line *TCE*. This can be called the "line of aux," or the "line of apogee.") The planets Jupiter and Saturn move very slowly, however, and the line of symmetry itself is moving slowly, and therefore he needs tables to allow us to make proper correction for this slow relative motion. This table obviously required clearer kinematic thinking than many medieval writers—Roger Bacon, for example—were capable of; but a third type of correction table was of an altogether higher order of originality.

As I have already hinted, John Killingworth reckoned the values of the constantly changing angles (μ and α in my figure) from the noon nearest the time the planet reached aux. Suppose, now, that we wish to find the planet's position at some later noon. Because we are working from noon to noon, an error is introduced into the calculation. The number of days between the two noons corresponds to a certain mean angular movement (μ_0), and this, added to the aux position, gives a position for the vector *EO* to which corresponds a certain equation (ξ_2, FIGURE 2). But the planet was

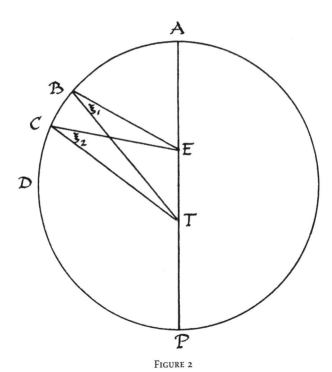

FIGURE 2

not at aux, at the first noon. It was at some other point, to which, when μ_0 is duly added, a different equation (ξ_1) corresponds. John Killingworth gives us therefore a table whereby we may find a correction factor proportional to the difference ($\xi_2 - \xi_1$), by which factor it is in due course necessary to correct both mean motions (μ and α). The trigonometric knowledge necessary for the tabulation of the correction factor was far from simple, but John Killingworth clearly understood what was needed. By reasoning on the basis of the differential calculus we ourselves can show that for errors as small as those ever encountered in this system, the error ($\xi_2 - \xi_1$) is directly proportional to the initial error, i.e. the displacement of the planet from aux at the first noon. John Killingworth somehow convinced himself of this proportionality, as can be seen from a closer examination of his method than I have been able to give here. He goes on, again in a way of his own, and making use of a new sort of double-entry table, to deduce the planet's longitude. However tedious even my bare outline might seem, the method is accurate and relatively easy to apply.

At the end of all this we have an ephemeris, looking much like any other ephemeris—that is to say, with no more continuity than in the readings on a digital clock. Where does John Killingworth stand, *vis-à-vis* the concepts of kinematics? Are we to say that those who know only digital clocks cannot know the concept of time? Astronomers like John Killingworth calculated with constant component velocities (if I may be allowed to use the word), but they compounded these to simulate the complex accelerated motions of the planets. (Indeed I do not need to remind you that the simplest circular motion entails an acceleration.) This was kinematics, not with the formal apparatus appropriate to freely falling bodies, but with an analytical structure of its own, a subject in its own right. There is no purpose in entering into a contest, as it were, between the two sorts of kinematics, but an extreme argument against the astronomical sort is not difficult to devise. "There is a sense in which Ptolemaic astronomy is not in any full sense kinematic" the argument might run, "for its technique is to begin with one geometric configuration and to pass to another by the addition of finite angles, which are simply multiples of the time interval; and from such a procedure you will at most extract the concept of velocity and even *that* only by a stretch of the imagination." As against this, it might be argued that *all* kinematics is concerned with separate events, and with devising a theoretical structure potent enough to connect them; and that compounding accelerated motions out of steady circular movements is one possible way to a solution of the problem of planetary motion—albeit a less potent way than the alternative later discovered by Kepler. With the advantage of hindsight we can see the Ptolemaic kinematics of the Middle Ages not as a subject giving conceptual clarity to the notion of linear acceleration, and not as having given rise to the Galilean theory of free-fall, but as the father of Kepler's description of planetary motion. And that was certainly no mean kinematic offspring.

NOTES AND REFERENCES

1. "The impossible does not happen; and it is impossible to traverse an infinite distance." Aristotle. *Physics*: 265a. Aristotle goes on to say that if something travelling along a straight line turns back on its path, the motion is composite, i.e. not simple.

2. See Aristotle,[1] and, showing the persistence of Aristotle's ideas in Copernicus' work, Copernicus, N., *De revolutionibus*. Book I, section 4.

3. See, for example, *Comm. in Phys.* 250 b.11 to 153 a.21.

4. For the general problem of intension and remission of forms, see Maier, Annaliese, *Zwei Grundprobleme der Scholastischen Naturphilosophie*, 2nd edit., Rome, 1951.

5. J. D. North, ed., *Richard of Wallingford*. An edition of his writings with translation and commentary. 3 vols. Oxford: Clarendon Press, Oxford University Press, 1976, passim.

6. M. Clagett, *The Science of Mechanics in the Middle Ages*. Madison, Wisc.: University of Wisconsin Press, 1959, p. 336.

7. See Clagett,[6] pp. 164-168; 178-182; 217-218.

8. Aristotle. *Physics*: 248a–249a.

9. As, for example, Richard of Wallingford. See ref. 5, vol. 2, especially pp. 206-209, and the corresponding sections of vols. 1 and 2.

10. See *Richard of Wallingford*, ref. 5, vol. 2, pp. 342-350.

11. R. Mercier, "Studies in the medieval conception of precession." *Archives Int. d'Hist. des Sciences*, 26 (1976). pp. 197-220; 27 (1977), pp. 33-71.

12. For a fuller account, see *Richard of Wallingford*, ref. 5, Appendix 29.

13. O. Schirmer, *Sitzungsber. der Med.-Phys. Soc., Erlangen*, 58 (1926), pp. 33-88.

14. W. Hartner, & M. Schramm, "Al-Bīrūnī and the theory of the solar apogee." In *Scientific Change*, ed., A. C. Crombie. London. pp. 206-218.

15. A detailed account of the matters raised in the last part of this paper will be found in: J. D. North, "The Alfonsine Tables in England." In ΠΡΙΣΜΑΤΑ. *Naturwissenschaftsgeschichtliche Studien. Festschrift für Willy Hartner*, eds. Y. Maeyama & W. G. Saltzer. Wiesbaden: Steiner Verlag, 1977, pp. 269-301.

JEAN BURIDAN AND NICOLE ORESME were two of the most brilliant, significant natural philosophers at the University of Paris during the fourteenth century. They shared a general conviction that *absolute truth* could only be found in faith and revelation; and that knowledge of the physical world was at best provisional, probable, and approximate. Moreover, both sought natural causes for observable events and effects. Yet their attitudes toward scientific knowledge were radically different. Relying on empirical generalizations, guided and controlled by natural reason, Buridan attempted to establish the firmest possible foundation for the acquisition of scientific truth. He believed that the best possible explanation which accounted for the phenomena qualified as *truth* and was sufficient for human purposes. Nicole Oresme followed a quite different path, emphasizing the uncertainties of natural knowledge. In resolving physical problems, his tactic was either to formulate equally plausible alternatives, none of which was demonstrably true, or to suggest tentative explanations that fit the data reasonably well. These opposed attitudes constituted a basic polarity in fourteenth-century scientific thought. MPC

Scientific Thought in Fourteenth-Century Paris: Jean Buridan and Nicole Oresme

EDWARD GRANT

Department of History and Philosophy of Science
Indiana University
Bloomington, Indiana 47401

O F the many natural philosophers at the University of Paris during the fourteenth century who explicated and disputed the innumerable problems of Aristotelian physics and cosmology, none were more illustrious than Jean Buridan and Nicole Oresme. These French scholastics were two of the most brilliant representatives of fourteenth-century science as a whole. Not only did they frequently exhibit originality of thought in the solutions they proposed for a host of specific problems, but they also represented two dramatically different attitudes that had developed during the fourteeenth century about the nature of science and the limits of human knowledge. Before describing and considering these attitudes, the main purpose of this paper, it will be well to identify Buridan and Oresme.

Taken in tandem, the life spans of our two authors were approximately parallel and equivalent to the life span of Guillaume de Machaut. Born around 1300, in the diocese of Arras,[1] Buridan studied at the University of Paris, where he earned a master of arts degree in the early 1320s. Twice rector of his university, in 1328 and 1340, Buridan was described in contemporary documents as a "very distinguished man" and "a celebrated philosopher." Apart from significant treatises on logic, all of Buridan's works were in the form of commentaries or *Questions* (*questiones*) on the treatises of the Aristotelian çorpus, which included all works relevant to medieval science.[2]

Buridan's almost exclusive preoccupation with the works of Aristotle was perhaps a consequence of the fact that he was, and remained, a secular cleric, who never obtained, and probably never sought, a theological degree. As a master of arts in the arts faculty of the University of Paris, he de-

0077-8923/78/0314-0105 $01.75/2 © 1978, NYAS

voted his entire intellectual life to the content and methodology of Aristotelian science. Following his death, sometime after July 12, 1358, when he was last mentioned, the great reputation he acquired as a perceptive interpreter and defender of that science outlived him for some two centuries. Not only were his works studied well into the sixteenth century at Paris, but they exerted a considerable influence in the new universities and centers of learning in Eastern Europe, especially in Vienna, Prague, Erfurt, and Cracow.

By contrast with Buridan, Nicole Oresme was a scholar of wide-ranging interests. Not only was he a commentator on the works of Aristotle, but he also wrote independent treatises on theology, mathematics, physics, cosmology, magic, money, and the evils of astrology, and was probably the first author to write anything of scientific significance in the French language.[3] Indeed, he helped fashion it into an instrument for the expression of scientific ideas.[4]

Of Oresme's life only a few dates and highlights are known with any certainty.[5] He was born around 1320 in Normandy, near the village of Caen. A student in the arts faculty at the University of Paris in the 1340s. where Jean Buridan was probably one of his teachers, Oresme, after a brief period of teaching in the arts faculty, became a master of theology in 1355 or 1356 and Grand Master of the College of Navarre in 1356. It was about this time that Oresme established connections with the French court, serving first King Jean II, and then, from 1364 on, his successor, Charles V, whom he had already served as teacher while Charles was dauphin. Mutual service to Charles V may even have brought Oresme into personal contact with Machaut, a suggestion that gains in plausibility from Oresme's evident interest in music[6] and his apparent acquaintance with Philippe de Vitry, the great champion of the *ars nova* in the first half of the fourteenth century.[7]

Service to Charles V undoubtedly enhanced Oresme's ecclesiastical and scholarly career. As dauphin and then king, Charles had secured for Oresme a series of ecclesiastical posts, culminating in 1377 with the bishopric of Lisieux, which was a reward for services rendered as a translator. In order that his councillors should better understand certain key works of Aristotle, so that they might govern the realm better and more wisely, Charles had commanded Oresme to translate from Latin to French Aristotle's *Ethics, Politics, Economics*, and the great cosmological treatise, *On the Heavens*.[8] Adding commentaries to the translations, Oresme carried out the royal command between 1369 and 1377. With the completion in

1377 of the translation of Aristotle's *On the Heavens*, which bore the
French title, *Le Livre du ciel et du monde*, Oresme's productive scholarly
career came to an end, although he lived on until 1382. The year 1377
thus marked not only the end of Machaut's life, but also the effective termination of Oresme's contributions to medieval science and scholarship.
No more fitting termination could have been hoped for, since the lengthy
French commentary that accompanied the *Le Livre du ciel et du monde*
was perhaps Oresme's greatest scientific treatise, incorporating and integrating the novel arguments and innovative ideas which had characterized
his thought for some thirty years.

   Although Oresme's scientific treatises were intellectually more daring
and wide-ranging than Buridan's, his subsequent direct influence was relatively minor compared to the enormous prestige and popularity enjoyed by
Buridan.[9] Apart from the publication of a few strikingly original mathematical treatises, in the early sixteenth century,[10] which, though difficult
to comprehend, had some influence at Paris, Oresme's works were virtually ignored. His most significant ideas were either too technical or contained attacks against magic and astrology, thus severely limiting his
appeal.

   With the completion of this brief summary of their lives and works,
we must now describe and evaluate the conflicting attitudes about the
nature of Aristotelian science that Jean Buridan and Nicole Oresme represented in the disputatious atmosphere of fourteenth century Paris. It is
hoped that the extent to which they were brilliant representatives of their
respective positions will become apparent as we proceed.

   The fact that Jean Buridan was only a master of arts—that is, strictly
a natural philosopher—whereas Nicole Oresme was that and a master of
theology as well is a significant datum on which to focus our comparison
of their conceptions of science. After 1272, and throughout the fourteenth
century, the faculty of arts at the University of Paris, required all its members, both bachelors and masters, to swear an oath that they would not
determine or dispute any purely theological questions.[11] Under penalty of
excommunication, they were, moreover, expected to resolve in favor of
the faith all questions that touched both faith and philosophy, which included, of course, natural philosophy. On at least one occasion Buridan expressed annoyance and frustration at this irksome restriction,[12] and at
other times he displayed a cautious and hesitant approach to physical
opinions and speculations that appeared to touch on theology.[13]

   But arts masters were frustrated from yet another source. In the third quarter of the thirteenth century, theologians challenged, and sought to erode, the confidence of natural philosophers in the principles that Aristotle had formulated to describe the structure and operation of the world. To weaken the intellectual grip of the natural principles and assumptions, some of which clashed with the demands of the Christian faith, the bishop of Paris was prevailed upon in 1277 to issue a condemnation of 219 articles that were no longer to be held on pain of excommunication.[14] As a consequence of this condemnation, which was in effect at Paris throughout the fourteenth century, all were compelled to concede that God could, if He wished, perform actions that were deemed naturally impossible in Aristotelian science.[15] By His absolute power, God could do anything He pleased short of a logical contradiction.[16] Attitudes toward the truth and validity of scientific knowledge were almost inevitably influenced by the condemnation, with masters of arts affected differently than theologians. As two of the best scientific minds at Paris during the fourteenth century, Jean Buridan and Nicole Oresme illustrate this profound difference with subtle clarity.

Although Buridan was no slavish follower of Aristotle, as a master of arts he had a professional commitment to uphold the general validity of Aristotelian natural science. Of course, he accepted the truths of revelation as absolute,[17] and frequently acknowledged that God could do anything that was considered naturally impossible. But God's power to do these things, ought not to imply that He had done so, or would do so. Thus, while Buridan was prepared to concede that "we hold on faith that just as God made this world, so could He also make another, or others,"[18] he preferred to believe that if God chose to create more creatures of the kind that appear in our world, He would simply enlarge our world to double, triple, or one hundred times its present size.[19] Similarly, God could create a finite or infinite space beyond the limits of our world, but we have no warrant to assume that He did, since the ordinary sources of evidence, namely sense experience, natural reason, and the authority of Sacred Scripture, fail to indicate the actual existence of such a space beyond our world.[20]

Rather than become preoccupied with supernatural possibilities, which, for a master of arts, could pose theological difficulties, Buridan devoted himself to the analysis and comprehension of the behavior of natural powers. He sought to defend Aristotelian science as the best means of understanding the physical world. Readily conceding that God could interfere at any time and alter the natural course of events, Buridan, nevertheless, assumed that "in natural philosophy, we ought to accept actions and dependencies as if they always proceed in a natural way."[21] Should a conflict arise between the Catholic faith and Aristotle's arguments, which, after all, are based only on sensation and experience, it is not necessary to believe Aristotle,[22] as, for example, in the doctrine of the eternity of the world. And yet, if we wish to confine ourselves to a consideration of natural powers only, it is appropriate to accept Aristotle's opinion on the eternity of the world, *as if it were true*.[23] Generally, Buridan was interested in arriving at truths about the regular operations of the physical world in the "common course of nature" (*communis cursus nature*).[24]

Was it possible, however, to establish a "common course of nature?" Since God could intervene at will in the causal order, how could scientific principles be absolute and certain to guarantee a "common course of nature?" On the assumption that scientific principles and causal laws require certainty, Nicholas of Autrecourt had argued that scientific knowledge was impossible. One could never be certain whether any particular empirical effect was the result of its natural cause or of God's direct intervention.[25] In response, Buridan insisted that fundamental and indemonstrable principles of natural science need not be absolute, but can be derived by inductive generalization—that is, "they are accepted because they have been observed to be true in many instances and to be false in none."[26] Since Buridan's methodology of science is predicated on the "common course of nature," God's intervention in the causal order becomes irrelevant,[27] and Buridan could declare that "for us the comprehension of truth with certitude is possible."[28] Using reason and experience, Buridan sought to "save the phenomena" in accordance with the principle of Ockham's razor—that is by the the simplest explanation which fit the evidence.[29]

Two illustrations will convey the manner in which Buridan combined empiricism, reason, and Ockham's principle of economy to save the appearances. In denying that the heavens, or celestial region, could be said to have matter, Buridan observed that the only method Aristotle had for determining that bodies in the terrestrial region possessed matter was substantial transmutation. In such transformations, which are observed by all, Aristotle assumed that one and the same subject remained in a body that had been generated and then corrupted, since he took it as axiomatic that something could not be made from nothing. On the common assumption that the heavens are not naturally generable or corruptible, and assuming further that generability and corruptibility are the characteristic features of anything we would want to call "matter," Buridan concluded that it would be "absolutely in vain, and without cogent reason, that we should posit such matter in the heavens."[30] In agreeing with Aristotle that something could not be made from nothing, Buridan was only too well aware that this was contrary to the Catholic faith. God, of course, could create something from nothing supernaturally, as when He created the world. But in natural philosophy one ought to adopt laws and principles that truly reflect the "common course of nature." That something could not be made from nothing was one of the most basic of such natural principles.

One of Buridan's most famous discussions concerned the possible axial rotation of the earth.[31] Using arguments based on relative motion, Buridan argued brilliantly that the earth's diurnal rotation from west to east (with the sphere of the fixed stars remaining stationary) was astronomically as plausible as the universally received Aristotelian opinion of the diurnal rotation of the fixed stars (with the earth motionless at the center of the universe). When supplemented by appeals to the greater harmony that would result if the earth rotated, it seemed that Buridan might opt for a rotating earth. Instead, he formulated an empirical argument which, as he put it, was "more demonstrative" (*magis demonstrativa*) than any other in convincing him of the correctness of the traditional opinion. If the earth really rotated with a daily motion from west to east, an arrow shot vertically upward should fall to the ground noticeably to the west of its launching site. Since such a perceptible discrepancy, which would result from the earth's eastward rotation as the arrow was in the air, is never detected, Buridan rejected the earth's rotation. But he was well aware of another explanation that could seemingly save the arrow phenomenon and yet be consistent with the earth's axial rotation. For if it were assumed that the arrow shares a common rotatory motion with the earth, the surrounding air, and the observer, the arrow would indeed fall back to the approximate place from which it was shot. In rejecting this argument, Buridan invoked his famous "impetus theory." When the arrow is projected, a quantity of impetus, or incorporeal force, is impressed into it, thus enabling the arrow to resist the lateral motion of the air as the latter accompanies the earth in its rotatory motion. As a consequence of its resistance to the motion of the air, the arrow should lag behind the motion of the earth and fall noticeably to the west of its launching site. Since this is contrary to experience, Buridan concluded that the earth does not possess a daily axial rotation. In this typical mode of argumentation, Buridan combined reason and experience to arrive at valid scientific knowledge, the kind of knowledge that could form the basis of our understanding of the physical world.

How very different was Nicole Oresme's approach. In apparent imitation of Socrates, Oresme declared at least twice that with respect to natural knowledge, all he knew was that he knew nothing.[32] Although he was deeply interested in scientific questions and would occasionally opt for one interpretation from among two or more, Oresme is best understood as belonging to that class of theologians who stressed the truths of faith to the detriment of natural knowledge. In stark contrast with Buridan, there can be little doubt that one of Oresme's objectives, and perhaps even pleasures, was to underscore the inability of human reason to arrive at certain knowledge about the physical world.[33]

To achieve this, it was essential to proceed as the masters of arts, or natural philosophers, did and adopt the mode of "speaking naturally" (loquendo naturaliter): that is, to operate within the prevailing framework of assumptions and principles customarily assumed in Aristotelian science. Oresme wisely refrained from invoking God and theology to discredit the arguments of natural philosophy, a move that undoubtedly increased the credibility of his position. The pretentions of science to exact knowledge were best combatted by formulating alternative scientific arguments that were equal or better than those traditionally accepted in natural philosophy. By suggesting sound alternatives to a variety of well-entrenched opinions, Oresme hoped to demonstrate that experience and natural reason were incapable of determining physical truth convincingly and unambiguously. Only faith could furnish us with absolute truth. In developing these alternatives, Oresme drew upon a profound knowledge of Aristotelian science and his considerable skill as a mathematician. Well armed, he sought to defeat the claims of natural philosophers on their own battle ground. As with so many sceptics in the history of western thought, his profession of ignorance was not an act of humility, but rather of arrogance, a transparent attempt to conceal the self-confidence of a brilliant and learned mind.

One of the most powerful tools that Oresme employed for the subversion of claims of exact knowledge about the physical world was the doctrine of the probable incommensurability of celestial and terrestrial motions.[34] Based on a mathematical proof that any two unknown ratios that might be proposed are probably incommensurable, and extending this to embrace any two continuous magnitudes of the same kind, such as velocity and time, Oresme was prepared to argue that any two celestial, or terrestrial, motions are probably incommensurable and form an irrational ratio.[35]

Oresme drew devastating consequences from the probable incommensurability of continuous magnitudes in nature. For if any two celestial motions are probably incommensurable, prediction of the precise disposition or positional relationships of two or more celestial bodies would be impossible. Indeed, if two or three celestial bodies entered into conjunction, or opposition, or into any other astronomical aspect, they could never again enter into that precise relationship. Knowledge of exact past relationships and future dispositions would thus be impossible. Celestial events are inherently unique and non-repetitive.[36] The implications of this doctrine for astrology should have been alarming to its practitioners. On the common medieval assumption that the celestial motions influence, and indeed govern, terrestrial events, and that the same cause always produces the same effect, the impossibility of the repetition of celestial aspects and dispositions signifies that since the same celestial cause does not occur twice, neither will the same terrestrial effect. Since Jupiter and Mars, for example, could only conjunct in a given point through all eternity, any other alleged conjunctions at that same point could only be approximate and would produce slightly different effects. Despite claims to the contrary, it now appeared that astrology and astronomy were inexact sciences, from which exact predictions were inherently impossible. At best only approximate results were attainable.[37]

Along with the refuted claims of the predictive powers of astrology and astronomy, the popular Stoic doctrine of the Great Year, which had been condemned in 1277,[38] also fell victim to celestial incommensurability. As defined by Cicero, a Great Year is said to occur everytime the sun, moon, and five planets have returned to the same positions relative to each other,[39] at which point all events of the previous Great Year are repeated. On the assumption of celestial incommensurability, however, Oresme argues that repetitions of this kind are impossible.[40]

Perhaps the most significant refutation of a natural principle stemming from the concept of probable incommensurability was Oresme's rejection of Aristotle's fundamental concept that whatever had a beginning, must have an end; and what has no end, cannot have had a beginning.[41] To refute this erroneous principle, which was associated with Aristotle's doctrine of the eternity of the world, Oresme assumed the eternity of the world, an assumption in keeping with his desire to operate within the framework of the commonly accepted principles of natural philosophy.[42] Oresme's general procedure was simple. He had only to imagine one or more unique celestial events, each of which terminated one cosmic condition that had existed from all past eternity and which immediately thereafter began a new cosmic condition that would last through all future eternity. The sun's actual motion provided more than one such illustration. Since the sun's resultant motion was assumed by astronomers to be the consequence of at least three different motions any one of which is probably incommensurable to the others, it followed that the sun's center, and therefore the apex of the earth's shadow, which depends on the sun's position, could never occupy the same celestial point twice. Thus the apex of the earth's shadow must continually, and necessarily, occupy one point after another in which it had never before been and to which it would never again return. In every point it occupied, then, some portion of the sunlight in the sky, which had existed there from all eternity, and was, therefore, without any beginning, would be darkened and come to an end. And at the moment when the apex of the shadow shifted to the next point, sunlight would once again shine on the previously darkened place and exist thereafter through all eternity, so that something which had a beginning need not necessarily terminate.[43] Indeed, on the universally held assumption that the planets and stars govern and cause physical events in the terrestrial region, Oresme argued the possibility that a unique celestial aspect, say a conjunction or opposition, might cause the existence of a new species never known before, which might then continue to exist through all future eternity.[44]

But Aristotle had posed an objection to events of this kind. Why, he asked, should something that had lasted through an infinite past suddenly cease to exist at one particular time rather than another? And why should something that will last through an eternal future come into existence at one particular time rather than another?[45] Indeed, it was precisely his inability to assign an apparent cause to explain the time of such events that led Aristotle to deny the possibility of their occurrence.[46] But Oresme believed he could provide a causal explanation. Any unique celestial event, or its unique terrestrial effect, is a direct consequence of the particular arrangement of the heavens and the constant velocities of the celestial spheres, which are guided and controlled by celestial intelligences. That a particular unique event should occur at one time rather than another is thus predetermined by natural necessity[47]: that is, by the inexorable, orderly, and controlled movements of the vast celestial machinery that was capable of operating with such clock-like precision that every event had, of necessity, to occur when it did.[48]

There is delicious irony here. Oresme, the theologian, defends the plausibility of his extraordinary doctrine of unique events by appeal to natural necessity, one of the most fundamental concepts in Aristotelian natural philosophy, but one which, in this particular instance, could not have been invoked by Aristotle for fear of jeopardizing another basic principle, that of the eternity of the world.

With the doctrine of celestial and terrestrial incommensurability, Oresme sought to refute significant and widely held beliefs in natural philosophy about exactitude and eternity. More often, however, he was content to provide an equally plausible alternative to basic Aristotelian cosmological conceptions. Under such circumstances, neither reason nor experience could demonstrate the truth leaving only extrascientific reasons to decide between them. Two significant problems in this category are the possibility of a plurality of worlds and the possible diurnal rotation of the earth.

In favor of the necessity of a single and unique world, Aristotle had formulated a number of arguments. Only one cosmic center and circumference were possible, thus necessitating that all heavy bodies move naturally toward the center and light bodies toward the circumference. Moreover, since all the matter that could possibly form a world already exists within ours, no surplus exists to form another. Indeed, nothing could exist beyond our world, neither body, nor place, nor void, nor time.[49] Oresme responded to all of these arguments.[50] By redefining the concepts of up and down, he showed that many worlds could exist separately and simultaneously in an infinite void space. Each world would have its own center and circumference, and heavy and light bodies would behave just as they do in our world, with no inclination to move towards our center or circumference. As for the paucity of matter to create these other worlds, Oresme insists that God could create matter *ex nihilo* for as many worlds as He pleased. But, since he wished to avoid the invocation of God's power as a resolution of problems in natural philosophy, Oresme quickly adds that even if we assume with Aristotle that things could only be made from already existing matter, Aristotle has no more demonstrated the impossibility of other worlds than he has demonstrated that the world has neither beginning nor end. Although convinced that natural philosophy cannot decide this important physical question, Oresme, nonetheless, concludes his discussion with the assertion that "there has never been nor will there be more than one corporeal world. . . ." But his point had been made. His decision in favor of a single world had no demonstrative, and, therefore, no scientific, basis.

Oresme treated the possible diurnal rotation of the earth in much the same manner.[51] He showed by a variety of arguments that the earth's rotation could just as well save the astronomical phenomena as did the traditional opinion. Indeed, on physical and nonphysical grounds, the earth's rotation was deemed simpler, nobler, and more conducive to cosmic harmony than the alternative. Ignoring the impetus theory, Oresme had no difficulty in reconciling the earth's rotation with the observed return of an arrow to the place from which it was shot. Whether at rest on the earth, or shot vertically into the air, the arrow shares the earth's rotatory motion. To an observer on earth, who also shares this circular motion, the arrow will seem to possess only a vertical component of motion as it falls to the place from which it was shot. From his lengthy argument, Oresme concludes that one could not choose between the competing alternatives on experiential or rational grounds. No demonstrative argument could decide the issue. And yet, at the conclusion of his discussion, Oresme, once again, opts for the traditional opinion, declaring that "everyone maintains, and I think myself, that the heavens do move and not the earth." Indeed, the earth's rotation seemed to him as much against "natural reason" as did articles of the faith. But if one must accept the articles of faith as true despite their conflict with natural reason, the same kind of reasoning need not be applied to physical problems. In the absence of demonstrative arguments, it is plausible, apparently, to follow natural reason, which, for Oresme, indicated the earth's immobility.

Perhaps we are now in a position to appreciate and understand the conflicting objectives that motivated these two great French scholastics, who, curiously, shared a general conviction that absolute truth was only to be found in faith and revelation and that knowledge of the physical world was at best provisional, probable, and approximate.[52] Both sought natural causes for observable events and effects. Indeed, Oresme even believed, and Buridan would probably have agreed, that supernatural, magical, and demonic causal explanations were to be avoided in accounting for allegedly marvelous phenomena. All such events ought properly to be explicable by natural causes.[53] And yet their attitudes toward scientific knowledge were radically different.

Buridan, the natural philosopher, sought to confer on natural knowledge all the power and dignity that was possible. Although experience and

reason produced truths and generalizations that were, by their very nature, provisional and tentative, because capable of disproof by a single
counter-instance, those empirical generalizations, guided and controlled by
natural reason, were a sufficient foundation on which to erect a solid scientific edifice. Scientific truth in such a context was sufficient for human
purposes.

Nicole Oresme followed a dramatically different path. As a theologian,
he sought to humble reason and experience and to undermine confidence
in their ability to yield truths about nature. Alternative solutions could
be formulated for even relatively straightforward physical problems. But
no matter how few or many natural causes could be suggested for saving
any particular natural phenomenon, reason and experience seemed incapable of choosing which, if any, was the true explanation.[54] Thus did
Oresme use reason to confound reason. But if the basic tools of natural
philosophy, reason and experience, could not furnish physical truth, how
much less could they comprehend the unintelligible truths of faith? In the
end, it was his faith that Oresme sought to protect from the probing
analysis of natural philosophy. To accomplish this, he found it necessary
to cast doubt on the principles and capabilities of natural philosophy.

Buridan and Oresme thus represent the polarities of fourteenth-century
scholastic thought. Buridan sought to establish the firmest possible natural foundations for the acquisition of truth about the physical world and
always tried to save the phenomena with the best possible explanation,
which then qualified as truth. Oresme emphasized the uncertainties of natural knowledge and, whenever possible, formulated equally plausible
alternatives, none of which was demonstrably true; or he simply suggested tentative explanations which fit the data reasonably well. For as
Oresme would have it, only God may know the true cause of a particular
event.

NOTES AND REFERENCES

1. For the details of Buridan's life and works, I have relied on Ernest A. Moody's
article, "Buridan, Jean," in the *Dictionary of Scientific Biography*, ed. Charles Coul-
ston Gillispie, Vol. 2 (New York: Charles Scribner's Sons, 1970), pp. 603-608.
2. The scientific treatises on which Buridan commented, or on which he wrote *questiones*, include the *Physics, On the Heaven (De caelo), Metaphysics, On the Soul (De
anima), Posterior Analytics, On Generation and Corruption, Meteorology*, and the
Short Natural Treatises (Parva Naturalia). Moody (*Dict. of Scientific Biography*, Vol. 2,
p. 608) cites early and modern editions of Buridan's works as well as a useful list of
secondary works. For a comprehensive list of editions and manuscripts of Buridan's
commentaries on the works of Aristotle, see Charles H. Lohr, S.J., "Medieval Latin

Aristotle Commentaries, Authors Jacobus-Johannes Juff," in *Traditio*, Vol. 26 (1970), 161-183. A lengthy list of articles on Buridan appears on pp. 161-163. Despite his significance, only a few of Buridan's works have received modern editions.

3. For the most recent list of Oresme's works and an excellent summary of his scientific thought, see Marshall Clagett, "Oresme, Nicole," in *Dictionary of Scientific Biography*, Vol. 10, pp. 229-230. The manuscripts and editions of Oresme's Aristotelian commentaries, as well as a lengthy bibliography on his life and works, appear in Lohr, "Medieval Latin Aristotle Commentaries, Authors: Narcissus-Richardus," *Traditio*, Vol. 28 (1972), 290-298.

4. On the neologisms which Oresme introduced into *Le Livre du ciel et du monde*, his French translation of the Latin text of Aristotle's *On the Heavens* (*De caelo*), see *Nicole Oresme: Le Livre du ciel et du monde*, edited by Albert D. Menut and Alexander J. Denomy, translated with an introduction by Albert D. Menut (Madison, Wis.: The University of Wisconsin Press, 1968), p. 13.

5. For a sketch of Oresme's life, see *Nicole Oresme: "De proportionibus proportionum" and "Ad pauca respicientes,"* edited with introductions, English translations, and critical notes by Edward Grant (Madison, Wis.: The University of Wisconsin Press, 1966), pp. 3-10; Clagett, "Oresme," *Dictionary of Scientific Biography*, Vol. 10, p. 223.

6. Oresme found occasion to discuss music in various works. Perhaps the most extensive discussion appears in his *De configurationibus qualitatum*, which has been edited, translated, and commented upon by Marshall Clagett, *Nicole Oresme and the Medieval Geometry of Qualities and Motions: A Treatise on the Uniformity and Difformity of Intensities Known as "Tractatus de configurationibus qualitatum et motuum"* (Madison, Wis.: The University of Wisconsin Press, 1968), pp. 37-39, 304-336 (also see the General Index, under "music" and "musical intervals and scales").

7. For Oresme's dedication of his *Algorismus proportionum* to Philippe, see Edward Grant, "Part I of Nicole Oresme's *Algorismus proportionum*," *Isis*, Vol. 56 (1965), p. 328 and n.5. Whether Oresme was personally acquainted with de Vitry cannot be determined from the dedicatory prologue. On the *ars nova*, see Willi Apel, *Harvard Dictionary of Music*, second edition, revised and enlarged (Cambridge, Mass.: Harvard University Press, 1969), pp. 58-60.

8. See Menut's discussion in *Oresme: Le Livre du ciel et du monde*, p. 3.

9. Oresme's mathematical techniques for graphing variations in intensities, or configurations, of qualities, which he developed in his *De configurationibus qualitatum et motuum*, were indirectly of enormous influence. Although by the fifteenth century, Oresme's name was rarely mentioned in connection with the graphing techniques he had devised and the variety of proofs and theorems which he had developed to illustrate the doctrine of the configuration of qualities, all this had been adopted by others, almost always without acknowledgment, and in this manner was eventually disseminated throughout Europe, ultimately influencing even Galileo and other seventeenth-century scientific luminaries. For the details, see Clagett, *Nicole Oresme and the Medieval Geometry of Qualities and Motions*, pp. 73-111.

10. I refer here to the *De proportionibus proportionum* and *Ad pauca respicientes*. In my edition of these two treatises, cited above, see pp. 130-132.

11. For a translation of the statute, see Lynn Thorndike (ed.), *University Records and Life in the Middle Ages* (New York: Columbia University Press, 1944), pp. 85-86; reprinted in Edward Grant (ed.), *A Source Book in Medieval Science* (Cambridge, Mass.: Harvard University Press, 1974), pp. 44-45.

12. This occurred in connection with Buridan's consideration of whether a vacuum, the natural existence of which Aristotle had denied, could be brought into being by the actions of any agent, including the deity. The concept of vacuum touched theology because it raised the question whether God required an empty space in which to create the world and because all Catholics had to concede the possibility that God could, if He wished, create a vacuum. In his discussion, Buridan acknowledged that he had been reproached by theological masters for intermingling theological matters in some of his physical disputations. He insisted, however, that despite the oath he had taken, it was necessary to invoke some theology in order to meet the demands of the question, for otherwise he would have to perjure himself by omitting relevant material from the argument. In the end, he chose to introduce a modest degree of theological discussion. For the relevant part of this question, see the translation in E. Grant, *A Source Book in Medieval Science*, pp. 50-51.

13. Here we may cite two instances. In the first, Buridan suggested that the celestial spheres might not be moved by intelligences, but rather by internal force, or *impetus*, as he called it, which God had impressed within them at the creation. Aware of the theological implications of his suggestion, Buridan quickly added that he was speaking only tentatively and that in this matter he sought guidance from the theological masters In the second instance, Buridan first upholds Aristotle's claim that no body whatever exists beyond the world, and then advises the reader "to have recourse to the theologians [in order to learn] what must be said about this according to the truth or constancy of faith." For the references and passages, see Grant, *A Source Book in Medieval Science*, p. 51, n. 4.

14. The Latin text of the articles in their original order appears in H. Denifle and E. Chatelain, *Chartularium Universitatis Parisiensis*, Vol. 1 (Paris, 1889), pp. 543-555. A regrouping by subject matter was made by Pierre Mandonnet, *Siger de Brabant et l'Averroisme Latin aux XIIIᵐᵉ siècle*, deuxième édition revue et augmentée: IIᵐᵉ Partie: Textes inédits (Louvain, 1908), pp. 175-191. Mandonnet's rearrangement of the articles was translated into English by Ernest L. Fortrin and Peter D. O'Neill in *Medieval Political Philosophy: A Sourcebook*, eds. Ralph Lerner and Muhsin Mahdi (New York: The Free Press of Glencoe, 1963), pp. 337-354; it was reprinted in Arthur Hyman and James J. Walsh (eds.), *Philosophy in the Middle Ages: The Christian, Islamic, and Jewish Traditions* (Indianapolis: Hackett Publishing Co., 1973), pp. 540-549. Selected articles relevant to medieval science have been translated in Grant, *A Source Book in Medieval Science*, pp. 45-50.

15. Article 147 condemned the opinion "That the absolutely impossible cannot be done by God or another agent.—An error, if impossible is understood according to nature." See Grant, *Source Book*, p. 49. As special cases, it had to be conceded that God could create a single world (articles 87 and 98 condemned the eternity of the world), or many worlds (article 34); that He could move our allegedly immobile, spherical world with a rectilinear motion (article 49); and that He could create an accident without a subject in which to inhere (articles 140 and 141). Grant, *Source Book*, pp. 48-49.

16. That not even God could produce a logical contradiction was widely accepted in the Middle Ages. For Aquinas's support of this position, see the translation of his *De aeternitate mundi* in *St. Thomas Aquinas, Siger of Brabant, St. Bonaventure, On the Eternity of the World* (*De Aeternitate Mundi*), translated from the Latin with an

Introduction by Cyril Vollert, Lottie H. Kendzierski, and Paul M. Byrne (Milwaukee: Marquette University Press, 1964), p. 22.

17. See Anneliese Maier, "Das Prinzip der doppelten Wahrheit," *Metaphysische Hintergründe der spätscholastischen Naturphilosophie* (Rome: Edizioni di Storia e Letteratura, 1955), p. 27.

18. "Tamen sciendum est quod licet per naturam non sit possibile esse alium mundum ab isto, tamen simpliciter hoc est possibile; quia tenemus ex fide quod sicut deus fecit istum mundum, ita posset adhuc facere alium vel alios plures." *Questions on De caelo*, Bk.1, Question 18 in Ernest Moody (ed.), *Iohannis Buridani Quaestiones super libris quattuor De caelo et mundo* (Cambridge, Mass.: The Medieval Academy of America, 1942), p. 84; see also p. 89, for the same sentiment.

19. After conceding that one could not demonstrate that beyond the world there is no magnitude and space, Buridan declares: "Tamen ego opinor quod illic non sit spacium vel magnitudo vel alius mundus et ad hoc adducit Aristoteles rationes naturales in primo Celi, que illic tractande sunt. Ideo solum ad hoc pono persuasionem talem quia non est verisimile quod Deus ibi fecit alium mundum vel alios mundos quia si plures creaturas mundanas voluisset fecisse quam fecerit, non oportebat facere alios mundos quia potuisset istum mundum fecisse in duplo vel centuplo maiorem; et si non fecit ibi Deus alium mundum vel alios mundos, non apparet ratio quare fecisse illic spacium quia illud de nichilo deserviret ultra istum mundum et apparet esse frustra." *Questions on the Physics*, Bk. 3, Question 15 in *Acutissimi philosophi magistri Johannis Buridani subtilissime questiones super octo Phisicorum libros Aristotelis diligenter recognite et revise magistro Johanne Dullaert de Gandavo antea nusquam impresse* (Paris, 1509; reprinted in facsimile under the title *Johannes Buridanus, Kommentar zur Aristotelischen Physik* by Minerva G.M.B.H., Frankfurt a.M., 1964), fol. 57v, col. 2.

20. "Secundo etiam dico quod non est ponendum modo supernaturali spatium infinitum extra caelum sive extra istum mundum, quia non debemus ponere quae non apparent nobis per sensum vel experientiam aut per rationem naturalem aut per auctoritatem sacrae scripturae, sed per nullum istorum apparet nobis quod sit spatium infinitum extra istum mundum. Bene tamen esset concedendum quod extra istum mundum posset deus creare spatium corporeum et substantias corporeas quantascumque sibi placeret, sed non est propter hoc ponendum quod ita sit." *Questions on De caelo*, Bk. 1, Question 17, Moody, *Iohannis Buridani Quaestiones super libris quattuor De caelo et mundo*, p. 79.

21. "Modo in naturali philosophia nos debemus actiones et dependentias accipere ac si semper procederent modo naturali; . . ." *Questions on De caelo*, Bk. 2, Question 9, edition of Moody, p. 164. Also cited by Maier, *Metaphysische Hintergründe*, pp. 18 and 328, n. 22.

22. "Et potest responderi ad rationes Aristotelis, quod ipse multa posuit contra veritatem catholicam, quia nihil voluit ponere nisi posset deduci ex rationibus ortum habentibus ex sensatis et expertis; ideo non oportet in multis credere Aristoteli, scilicet ubi dissonat sacrae scripturae." *Questions on De caelo*, Bk. 2, Question 6, edition of Moody, p. 152.

23. In considering "whether every corruptible thing is corrupted from necessity" ("Utrum omne corruptibile de necessitate corrumpetur"), Buridan offers three distinctions, of which the second is: "Alia distinctio est, quod possimus loqui de potentia divina, et secundum ea quae tenemus ex fide; aliter possumus loqui secundum potentias

naturales, *vel ac si esset vera opinio Aristotelis de aeternitate mundi, . . ." Questions on De caelo*, Bk. 1, Question 24, p. 118. Also cited by Maier, *Metaphysische Hintergründe*, p. 17. The italics are mine. Buridan was, however, not averse, on occasion, to use the divine power to his own advantage. In discussing the validity of the Aristotelian rules of motion, Buridan acknowledged that, although constant forces are required for their general validity, uniform motions and forces are not observed in nature. "And from these things it seems to me it must be inferred that these rules are rarely, or never, found to produce their effect. Nonetheless, these rules are conditional and true, for if the conditions stated in the rules were observed, everything would occur just as the rules assert. For this reason it ought not to be said that the rules are useless and fictitious because although these conditions are not fulfilled by natural powers, it is nevertheless possible, in an absolute sense, for them to be fulfilled by the divine power." My translation as it appears in Edward Grant, "Hypotheses in Late Medieval and Early Modern Science," *The Voice of America Forum Lectures, History of Science Series,* 3 (1964), p. 5. For the Latin text, see Buridan, *Questions on the Physics,* Bk. 7, Question 7, *ed. cit.,* fol. 108v, col. 1. The passage is also quoted by Anneliese Maier, *Die Vorläufer Galileis im 14. Jahrhundert. Studien zur Naturphilosophie der Spätscholastik* (Rome: Edizioni di Storia e Letteratura. 1949), p. 101, n. 41, and discussed by William A. Wallace, *Causality and Scientific Explanation* (2 vols.; Ann Arbor: The University of Michigan Press, 1972, 1974), Vol. 1, pp. 107-108. For Buridan, the rules of motion are logically consistent and would produce the predicted effects provided that the conditions are realized in nature, which might occur if God chose to intervene. It is a significant argument for the acceptance of idealized scientific laws, which though not realized in nature, ought yet to be accepted as valid.

24. Buridan uses this expression in his *Questions on the Metaphysics,* Bk. 2, Question 1 ("Utrum de rebus sit nobis possibilis comprehensio veritatis") in *In Metaphysicen Aristotelis, Questiones argutissimae Magistri Ioannis Buridani in ultima praelectione ab ipso recognitae et emissae . . .* (Paris, 1518; reprinted in facsimile under the title *Johannes Buridanus, Kommentar zur Aristotelischen Metaphysik* by Minerva G.M.B.H., Frankfurt a.M., 1964), fol. v, col. 2. In the title page added by Minerva, the edition is mistakenly dated 1588.

25. See Moody, "Buridan," *Dictionary of Scientific Biography,* p. 605. It was probably Autrecourt's opinion which Buridan had in mind when he offered the following as an argument against the possibility that we can comprehend the truth: "Et difficultas augmentatur multum per ea que credimus ex fide quia Deus potest in sensibus nostris formare species sensibilium sine ipsis sensibilibus et longo tempore potest eas conservare. Et tunc iudicamus ac si essent sensibilia presentia. . . . Immo cum nihil scias de voluntate Dei, tu non potes esse certus de aliquo." *Questions on the Metaphysics,* Bk. 2, Question 1, *ed cit.,* fols. 8r, col. 2-8v, col. 1.

26. Translated by Moody, "Buridan," *Dictionary of Scientific Biography,* p. 605. For Buridan's text, see his *Questions on the Metaphysics,* Bk. 2, Question 1, *ed. cit.,* fol. 9r, col. 1.

27. Moody, "Buridan," *Dictionary of Scientific Biography,* p. 605.

28. "Immo concludendum est quod querebatur, scilicet quod nobis est possibilis comprehensio veritatis cum certitudine." *Questions on the Metaphysics,* Bk. 2, Question 1, *ed. cit.,* fol. 9r, col. 1. On Buridan's nominalism and his insistence that "knowledge is to be objectively grounded in particular existents," see T. K. Scott, Jr., "John Buridan

on the Objects of Demonstrative Science," *Speculum*, Vol. 40, Nr. 4 (1965), 654-673; for the quotation, see p. 659.

29. In discussing whether the heavens can be said to have matter, Buridan declares that "in nature nothing ought to be assumed in vain; and yet it is vain to assume more when all appearances could be saved by fewer [assumptions]." (". . . quod in natura nihil debet poni frustra, et tamen frustra ponuntur plura quia omnia apparentia possent salvari pauciora." *Questions on De caelo*, Bk. 1, Question 11 ("Utrum caelum habeat materiam"), edition of Moody, p. 52. The same sentiment is expressed even more clearly in Bk. 2, Question 22, where Buridan explains that "just as it is better to save the appearances by fewer than by more [assumptions], if this is possible, so it is better to save [the appearances] by the easier path than the more difficult path." (". . . sicut melius est salvare apparentia per pauciora quam per plura, si hoc sit possibile, ita melius est salvare per viam faciliorem quam per viam difficiliorem." *Questions on De caelo*, Bk. 2, Question 22 ["Utrum terra semper quiescat in medio mundi"], *ibid.*, pp. 228-229.) In the very same question (p. 232), while discussing whether the earth rests in the milddle of the world, Buridan declares: "si terra quiescat, ideo frustra moveretur si moveretur; et nihil est ponendum frustra in natura." *Ibid.*, p. 232. See also the concluding sentence of the Latin text cited in note 24, above.

30. *Questions on De caelo*, Bk. 1, Question 11, edition of Moody, pp. 52-53. Instead of matter, Buridan assumed that the celestial region was composed of a simple, uncompounded substance, which is, nevertheless, subject to magnitude. It is the magnitude which provides extension to the substance, and is also subject to motion and other accidents.

31. For the Latin text, see *Questions on De caelo*, Bk. 2, Question 22, edition of Moody, pp. 226-233. The major part of the question has been translated by Marshall Clagett, *The Science of Mechanics in the Middle Ages*, pp. 594-598 and has been reprinted in Grant, *A Source Book in Medieval Science*, pp. 500-503. A summary account of Buridan's arguments appears in Edward Grant, *Physical Science in the Middle Ages* (New York: John Wiley & Sons, Inc., 1971), pp. 64-66.

32. The two passages appear in the unpublished, and as yet unedited, part of Oresme's *Quodlibeta*, and are quoted twice by Marshall Clagett. See his "Some Novel Trends in the Science of the Fourteenth Century," Charles S. Singleton (ed.), *Art, Science and History in the Renaissance* (Baltimore: The Johns Hopkins Press, 1968), p. 280, n. 12 and *Nicole Oresme and the Medieval Geometry of Qualities and Motions*, p. 12 and n. 6. A possible source for Oresme's statement is Walter Burley's *Liber de vita et moribus philosophorum*, which was probably written in the early 1340s. In describing the life of Socrates, Burley says: "Et licet esset sapientissimus nihil se scire putabat. Unde et illud sepe dicebat, ut ait Hieronymus ad Paulinum: hoc unum scio, quod nescio." Hermann Knust (ed.), *Gualteri Burlaei Liber de vita et moribus philosophorum*, mit einer Altspanischen Übersetzung der Eskurialbibliothek. *Bibliothek des Litterarischen Vereins in Stuttgart*, Vol. 177 (Tübingen, 1886), p. 110. In note *f*, in addition to St. Jerome's 53d letter to Paul, Kunst cites a number of other authors who quoted or paraphrased this famous Socratic pronouncement, the ultimate source of which is Plato's dialogues, especially the *Apology* and *Phaedrus*.

33. George Molland observes that, although Oresme sought to advance science, he was sceptical about attaining empirical scientific knowledge ("Nicole Oresme and Scientific Progress," *Miscellanea Mediaevalia Veröffentlichungen des Thomas-Instituts*

der Universität zu Köln, ed. Albert Zimmermann, Vol. 9: *Antiqui und Moderni, Traditionsbewusstein und Fortschrittsbewusstein im späten Mittelalter* [Berlin/New York: Walter de Gruyter, 1974], pp. 206-220).

34. The three lengthiest and most fundamental discussions occur in Oresme's *De proportionibus proportionum, Ad pauca respicientes* (for the title of my edition of these two works, see above, n. 5). and *Tractatus de commensurabilitate vel incommensurabilitate motuum celi* (for the edition, see *Nicole Oresme and the Kinematics of Circular Motion, Tractatus de commensurabilitate vel incommensurabilitate motuum celi.* Edited with an Introduction, English Translation, and Commentary by Edward Grant [Madison, Wis.: The University of Wisconsin Press, 1971]). Oresme also discussed incommensurability of celestial motions in his *Questiones super de celo, Le Livre du ciel et du monde, Quodlibeta, Questiones de sphera,* and *Quaestiones super geometriam Euclidis.* See *Nicole Oresme and the Kinematics of Circular Motion,* pp. 56-76, nn. 89-114.

35. The mathematical proof appears in *De proportionibus proportionum,* Ch. 3, Proposition X and the application to physical magnitudes in Ch. 4, Proposition XII (see my edition, pp. 247-55 and for discussion pp. 40-42; also pp. 303-309). For a summary, see *Nicole Oresme and the Kinematics of Circular Motion,* pp. 73-76, n. 113. Oresme was well aware that the incommensurability of celestial or terrestrial motions was not empirically demonstrable, "for by the part of a movement which would be imperceptible to the senses, even if it were a hundred thousand times larger, two such movements or similar motions could be incommensurable and yet appear to be comsurable." *Le livre du ciel et du monde,* p. 197. Oresme was, however, convinced that he had demonstrated the probability that any two given motions were incommensurable.

36. See *Nicole Oresme and the Kinematics of Circular Motion,* pp. 54-55, 142.

37. See *Nicole Oresme and the Kinematics of Circular Motion,* pp. 319-321.

38. Article 6, which reads: "That when all celestial bodies have returned to the same point—which will happen in 36,000 years—the same effects now in operation will be repeated." ("Quod redeuntibus corporibus celestibus omnibus in idem punctum, quod fit in xxx sex milibus annorum, redibunt idem effectus, qui sunt modo." *Chartularium,* Vol. 1, p. 544.

39. *De natura deorum* 2.20.51-52 in the translation by H. Rackham, Loeb Classical Library, London and New York, 1933), p. 173. In the Middle Ages, the period of the Great Year was often taken as 36,000 years, a figure determined by Ptolemy's value of 1° per hundred years for precession of the equinoxes. For a discussion of the Great Year in Greek antiquity and the Middle Ages, see Grant, *Nicole Oresme and the Kinematics of Circular Motion,* pp. 103-124.

40. In the fourth chapter of his *De proportionibus proportionum,* Oresme rejects the Great Year as an error in "philosophy and faith." The rejection is based on the probability of celestial incommensurability. See Grant, *Nicole Oresme "De proportionibus proportionum" and "Ad pauca respicientes,"* p. 307.

41. For Aristotle's discussion, see *De caelo,* Bk. 1, chs. 10, 12.

42. All this is incorporated into the following statement by Oresme: "Afterwards Aristotle tries to prove that everything, whether substance or accident or any tendency whatsoever which had a beginning, will have an end and will cease of necessity and cannot possibly last forever; and that it is likewise impossible that anything which will ultimately perish can always have been there without a beginning. Since this is not true and is, in its first part, against the faith, I want to demonstrate

the opposite according to natural philosophy and mathematics. In this way it will become clear that Aristotle's arguments are not conclusive. In the first place, I posit with Aristotle, although it is false, that the world and the motions of the heavens are eternal by necessity, without beginning or end." Menut, *Le Livre du ciel et du monde*, pp. 195-197. Beginning on p. 217, Oresme considers Aristotle's claim that what has no end cannot have had a beginning. In the *De proportionibus proportionum*, Oresme explained that when he assumes an eternity of future motion, he is "speaking naturally" (*naturaliter loquendo*). Indeed, for the sake of the discussion, he assumes all the principles enunciated by Aristotle in the second book of *De caelo* and elsewhere. See Grant, *Nicole Oresme and the "De proportionibus proportionum,"* pp. 305-307. Oresme's discussion on things beginning and ending in the context of eternity is taken up briefly by A. Maier, *Metaphysische Hintergründe*, pp. 27-31.

43. *Le Livre du ciel et du monde*, p. 199. In his *Questions on De caelo*, Bk. 1, Question 24, Oresme applied similar reasoning to lunar eclipses, showing that the earth's shadow will never return twice to the same point on the lunar surface. See *The "Questiones super de celo" of Nicole Oresme*, edited and translated by Claudia Kren (Ph.D. dissertation, University of Wisconsin, 1965), pp. 421-424, and Grant, *Nicole Oresme and the Kinematics of Circular Motion*, p. 63, n. 97. On the assumption of incommensurability, Oresme also conceived a perpetual circular motion which had a beginning but no end (*Le Livre du ciel et du monde*, p. 203). The fall of a heavy body through a successively more resistant medium was so devised that the motion, which had a beginning, never reaches its terminus (*ibid.*, p. 205).

44. See *Livre du ciel et du monde*, p. 243 and Grant, *Nicole Oresme and the Kinematics of Circular Motion*, p. 57 and all of note 90.

45. Aristotle, *De caelo*, 1.12.283a.11-24.

46. Aristotle further believed (*De caelo*) that if something could come into existence after an infinite past time or that something could endure through an eternal future after coming into being, then such a thing would simultaneously have the power of being and not-being. For Buridan's agreement with Aristotle, see his *Questions on De caelo*, Bk. 1, Question 23, edition of Moody, pp. 112-116.

47. *Le Livre du ciel et du monde*, p. 241.

48. In *Le Livre du ciel et du monde* (p. 289) Oresme actually likened celestial regularity to the workings of a clock. Just as a man could make a clock and let it run by itself, so also could God assign proportions of force and resistance in the heavens so as to produce regular and harmonious movements. See also Marshall Clagett, 'Nicole Oresme and Medieval Scientific Thought," *Proceedings of the American Philosophical Society*, Vol. 108, No. 4 (August, 1964), p. 300.

49. All of these arguments appear in *De caelo*, Bk. 1, chs. 8, 9.

50. The arguments described below appear in *Le Livre du ciel et du monde*, pp. 171-179, and are reprinted in Grant, *Source Book in Medieval Science*, pp. 550-554. Oresme distinguished three types of plurality of worlds, of which only the third and more commonly discussed variety is considered here. For a summary of Oresme's opinions on plurality, see Grant, *Physical Science in the Middle Ages*, pp. 74-75. Medieval discussions on a plurality of worlds, which includes Oresme, are detailed in Steven J. Dick, *Plurality of Worlds and Natural Philosophy: An Historical Study of the Origins of Belief in Other Worlds and Extraterrestrial Life* (Ph.D. dissertation, Indiana University, 1977), pp. 71-108.

51. For Oresme's consideration of the possible diurnal rotation of the earth, see *Le*

Livre du ciel et du monde, pp. 519-539. Most of Menut's translation has been reprinted in Grant, *Source Book in Medieval Science*, pp. 503-510. A brief description of Oresme's arguments is given in Grant, *Physical Science in the Middle Ages*, pp. 66-70.

52. Oresme favored the idea of the probable incommensurability of the celestial motions because this made knowledge of future events impossible and also enabled us to obtain some knowledge, while yet always leaving some things unknown for further investigation. See Grant, *Nicole Oresme and the Kinematics of Circular Motion*, pp. 319-321.

53. This is the major theme of Oresme's *Quodlibeta*. See *Nicole Oresme and the Marvels of Nature, A Critical Edition of his "Quodlibeta" with English Translation and Commentary* by Bert Hansen (Ph.D. dissertation, Princeton University, 1974), p. 32; for Oresme's statement, see p. 85.

54. This attitude is nicely illustrated in the *Quodlibeta*, where, according to Hansen (Ph.D. dissertation, Princeton University, 1974, pp. 34-35), Oresme "sometimes even reminds us that no one but God alone can render causes for particular cases. Quite often this agnostic approach results in his being content merely to indicate the kinds of causes or factors operating, without even suggesting how these parameters might be related." Indeed, even in his more rigorous *De configurationibus qualitatum et motuum*, Oresme was usually tentative in his causal explanations (see Clagett, *Nicole Oresme and the Medieval Geometry of Qualities and Motions*, p. 35). Oresme was "agnostic" in his approach to natural causes even before he became a theologian. As a theologian, however, he came to stress the uncertainty of natural knowledge, an attitude which is quite apparent in his last work, *Le Livre du ciel et du monde*.

Two BASIC TYPES of technological documentation are still extant from the middle ages: (1) sources that accidentally provide us with information about medieval technology, and (2) those which self-consciously intended to inform their readers about various arts and crafts. Modern sociological arguments concerning divisions between the learned men and the craftsmen offer intriguing explanations for the quantity and the structure of technological writings. Particularly noteworthy is the appearance of a new technical literature after 1400; fourteenth-century technical treatises preparing the way for this new development include Giovanni de'Dondi's *Tractatus astrarii* (ca. 1364) and Guido da Vigevano's *Texaurus* (ca. 1335).

Giovanni de'Dondi (dall' Orologio), Italian physician, astronomer, and clockmaker (1318 to 1389) was the son of a clockmaker, Giacomo Dondi (1298 to 1359) whose remarkable clock decorated the tower of the palace at Padua, not only striking hours, but detailing the sun, planets, days, months, and annual holidays. The son Giovanni's clock was even more complicated, dependable, and splendid; its description and diagrams appear in his *Tractatus.*

Guido da Vigevano (1280 to 1350), a physician at the French court and writer of medical treatises, composed the *Texaurus* or *Treasury of the King of France for the Recovery of the Holy Land Beyond the Sea,* intended for King Philip VI, who was planning a crusade. The treatise is bipartite: (1) a guide for health; and (2) a technological treatise on warfare including ingenious battle-machinery, which was light, transportable, and flexible. Ideas therein were copied and repeated by later inventors such as Leonardo da Vinci.   MPC

Giovanni de'Dondi and Guido da Vigevano: Notes Toward a Typology of Medieval Technological Writings

BERT S. HALL

Institute for the History and Philosophy of Science and Technology
Center for Medieval Studies
University of Toronto
Toronto, Canada M5S1A1

IT is now some 40 years since the study of technology in medieval Europe emerged as a serious scholarly occupation, and during that time we have learned a great deal about the importance of technology in the changing patterns of medieval life.[1] Along the way we have uncovered a substantial amount of information about both general patterns of technological development in the period and also about the documentary remains that reflect —with greater or lesser degrees of accuracy—the knowledge of medieval people about matters technical. Many desiderata remain to be fulfilled in this enterprise at present, among them a satisfactory scheme of categories within which various documents can be placed for further study, i.e. a typology of medieval technological writings. Unfortunately, the field is rather untidy and a great deal of work will have to be done before a full framework can be created. I would like to make a preliminary contribution to this task, however, by trying to distinguish two major types of technological writing in the middle ages. In keeping with the theme of this conference, two works written during Machaut's lifetime have been selected to serve as model examples of their types: Giovanni de'Dondi's *Tractatus astrarii* and Guido da Vigevano's *Texaurus Regis Francie acquisicionis Terre Sancte de ultra mare*. This paper describes both works, makes some remarks about their authors, and seeks to place each within a tradition of similar writings. Naturally, we cannot expect such a procedure to give us a comprehensive view of technological writings during the mid-

0077-8923/78/0314-0127 $01.75/2 © 1978, NYAS

dle ages, but for purposes of a preliminary categorization, two examples
will suffice.

SCHOLARS AND CRAFTSMEN

Early in his career, the student of medieval technology must confront the
slightly embarrassing fact that medieval people left remarkably few didactic writings on technology, despite their considerable range of technological
accomplishments. We have become aware of these accomplishments largely by relying on sources that accidentally give us information about technology: casual references in chronicles; works of literature or philosophy;
artistic representations showing tools or machines in common use; or the
surviving artifacts themselves, which can yield a rich harvest of information when properly analyzed. None of these was consciously intended to
give the user or reader much insight into technology as such, and the
historian must engage in a substantial effort of reconstruction when he
uses them as sources. Technological treatises with self-conscious didactic
intent are rare in the middle ages, or for that matter in classical antiquity.
To the modern mind, this combination of technical achievement and extreme reticence is most difficult to understand. It is usually accounted for
by some variation on the theme of scholars and craftsmen.[2]

Scholar-craftsman arguments take as their starting point a number of
gaps between the mental world of the learned and that of the skilled manual worker. Craftsmen are, for most of the period under discussion, illiterate, and thus unlikely to compose great tomes on their arts. Craftsmen
learn their skills as apprentices, in their masters' workshops, by word-of-
mouth and by example. They have little or no need for written material to
instruct them or assist them in their daily work. Scholars, on the other
hand are equally the prisoners of their educations, which were generally
rhetorical, not practical, and through which they absorbed attitudes of disdain for the world of physical work. Even if they could have overcome
their biases, the argument continues, it is questionable whether they were
equipped by their educations to make intelligent comments on the messy
and complex operations that the craftsmen performed. The purpose of
philosophy was to understand the cosmos, not to bake bread or make pots.
Stated in this way, the argument is, of course, a caricature of the attitudes
and abilities of both scholars and craftsmen, but the general line of argument has some merit nevertheless. We do have but a few treatises on technology, and fewer still can be shown to be from the pens of craftsmen.
We do have examples of technological comments that reveal the ignorance

of the commentator, and other examples of text which have become deeply corrupted because the scribes who copied them knew nothing of the subjects they discussed.[3] Whatever misgivings one may have about the scholar-craftsman argument, it must, I think, be accepted as a general description of conditions and limits that shaped medieval technological writings. The problem is to modify the argument so that it can incorporate the documents we do have, and as we shall see, the fourteenth-century works allow us to do just that.

GIOVANNI DE'DONDI

Let us look first at the work of Giovanni de'Dondi (1318-1389).[4] He was a physician and the son of a physician, Jacopo de'Dondi; he taught medicine at the University of Padua from 1350 or 1352 and later at the University of Pavia. He played minor roles in the political life of his day and was rather famous for his learning. From 1348 to 1364 he built in his spare time and apparently with his own hands a very complex astronomical clock which became one of the marvels of his age (FIGURE 1). At any rate, we would call Giovanni's creation a clock, but he used the word "astrarium" to describe it. (The closest analogy in modern English would be "planetarium.") By means of seven large dials it displayed the motions of the sun and the moon as well as those of the five planets then known; in addition the machine had a 24-hour dial, indications of the dates of the fixed and moveable feasts, the nodes (points of intersection of the solar and lunar orbits), and the times of sunrise and sunset in Padua. The complex gearing necessary to achieve these motions was driven by a verge-and-balance wheel escapement powered by a hanging weight. Made of brass and bronze, the completed instrument contained 297 parts, stood 52 inches (1.32 m) high and was about 30 inches (76 cm) in diameter. The *astrarium* survived until 1529 or 1530, after which all traces of it disappear from the historical record. Modern reconstructions may be seen at the Smithsonian Institution in Washington, D.C. and at the Museo Nazionale della Scienze e della Tecnica in Milan.

The reason we have been able to make reconstructions in this century of a machine that disappeared more than 400 years ago is that Giovanni left behind a detailed descriptive treatise on his work, *Tractatus astrarii*, also known in some copies as *Opus planetarium;* the text survives in 11 manuscripts, only one of which can be shown to date from Giovanni's lifetime. Giovanni describes his motives for wanting to build such a device in the preface: his wish to bring some measure of common appreciation to

FIGURE 1. Model of the astrarium of Giovanni de'Dondi, showing several dials plus the hour indicator, gearing and weight drive. With permission of the Smithsonian Institution, Washington, D.C.

the noble, but difficult, art of astronomy and his wish to demonstrate a correct, i.e. pristine and unmodified, Aristotelian-Ptolemaic version of theories of planetary motion. He relates his initial inspiration to his reading of the *Theorica planetarum* of Campanus of Novara, a work of the thirteenth century, and Campanus's description of an equatorium. (The equatorium was an unmotorized computational instrument roughly comparable to the astrolabe for solving problems in planetary astronomy; *Theorica planetarum* was one of the most significant Latin works on the subject.) The rest of the *Tractatus* is devoted to detailed descriptions of constructional matters relating to the astrarium itself, concentrating on gear trains, their calculation and arrangement. The work is illustrated by numerous diagrams, again mainly of gear trains (FIGURE 2). These are difficult to interpret by themselves without recourse to the text; clearly they were meant to assist a reader to understand some of the difficult points of construction.

Both Giovanni's machine and the treatise in which he describes it can be related to a very old tradition of complex geared devices and writings about them. Indeed, before about A.D. 1400 instrument treatises make up the single most numerous category of technical treatises. We have vague accounts of elaborate planetaria that Archimedes is supposed to have constructed, and there is physical evidence in the form of the so-called "Antikythera Machine" that somewhat simpler geared astronomical computation machines were in fact built in the first century B.C.[5] Though we have no written tradition that describes anything so complex as the "Antikythera Machines," we do have a body of descriptive treatises under the names of Philon of Byzantium (fl. late third century B.C.) and Heron of Alexandra (first century of our era) on such topics as surveying instruments, clepsydrae (water clocks), pneumatic devices (mainly toys), automata, and catapults. Significantly, both Philon and Heron are associated with the famous "Museum of Alexandria," that unique "research institute" of antiquity. Neither the tradition of writing such treatises nor the skills to make sophisticated instruments survived in Europe past the collapse of Roman power, but both are visible in Islam from at least the eighth century onwards. We know of elaborate water-clocks with associated automata, we find both Heron and Philon in Arabic versions at an early date, and we have a line of indigenous Arabic treatises stretching from the Banū Musà in the ninth century to al-Jazarī and Riḍwān in the early thirteenth.[6] The tradition reappears in the West in the eleventh century with translations from Arabic works on astronomical instruments, and there-

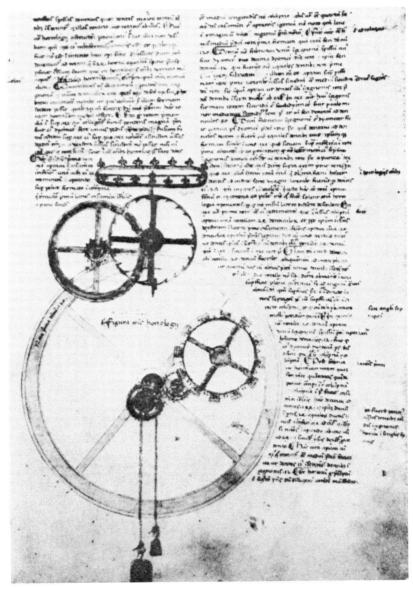

FIGURE 2. Schematic drawing of the astrarium's movement from the oldest extant manuscript, Cod. D. 39 of the Biblioteca Capitolare Vescovile, Padua. With permission.

after we find a fairly continuous development of such works. The name of Richard of Wallingford, whose *Tractatus horologii astronomici* (c. 1330) gives us the earliest extant description of a weight-driven clock, should be mentioned; the paper by J. D. North later in this volume provides some additional detail.[7]

There is ,therefore, a solid and reasonably continuous tradition of clockwork instruments and descriptive treatises running back from Giovanni de'Dondi to before the time of Christ. Making some allowances for individual variations and modes of expression, these instrument treatises share many common characteristics. They all seem to have been written by men who had practical experience with the devices they discuss; there does not seem to be a dilettante in the lot. They are all intensely descriptive, heavily oriented toward the exposition of details of design and construction. Their ostensible purpose is to permit a reader to build the machine in question, but they almost always leave the impression that the reader had better know something in advance about the subject under discussion, for they make no concessions to his possible ignorance of major matters. Giovanni, for example, explicitly makes this point when he glosses over the all-important matter of the escapement mechanism, noting only that his is like that in common use. He further advises anyone who does not understand such simple matters not to press on to the greater difficulties that lie ahead. Since we are in fact greatly interested in the early development of escapements, we find Giovanni's attitude lamentable; alas, Richard of Wallingford seems hardly more informative about the escapement of his *horologium*. The presumptive conclusion from such episodes as these is that we are dealing with a literature by specialists and for specialists. Something similar is manifest in the treatment accorded illustrations in these works. We rarely get a picture of what a machine actually looked like; we have only diagrams of the details, and these serve almost exclusively to elucidate difficult aspects of the text. Again, the specialist would know what the overall machine should look like, but he might need visual help with difficult details. Finally, and again with individual exceptions, the instrument treatises are in one respect rather unimaginative. That is to say, the locus of imagination, of creativity, is in the machines and their design, not in the texts that describe them. There are no leaps of thought, no sense of trial-and-error, no attempts to describe possible variations on the established theme. The preferred style is flat, somewhat laconic, almost disembodied, and purely descriptive.

Now these qualities may seem entirely admirable in a technological work, and indeed, they are, but they are not the only qualities such writings can show, as we shall see when we consider Guido da Vigevano. First, however, let us glance backwards at the scholar-craftsman argument mentioned earlier.

What does the existence of a well-established tradition of instrument treatises mean for the general thesis? It would seem that in the specialized business of designing and describing complex instruments, the gap between scholars and craftsmen was narrower than elsewhere in technology. Obviously the ability to read and work with numbers was a *sine qua non* of complex instrument design and construction. Hence a craftsman working with such devices is more likely to have been literate enough to compose a treatise, as for example al-Jazarī was. Scholars, on the other hand, would seem to have been on safe grounds in showing an interest in the science of astronomy and devices closely related to it. Recall that both Giovanni and his father Jacopo were physicians and thus professionally related to astronomers and astrologers. The relentlessly descriptive and uncompromising quality of most such texts suggests that the authors thought to address only like-minded men; there is no question of reaching for the untutored. The intellectual labor of designing complex instruments, and especially the mathematical demands of such activity, made it into a type of High Technology (to borrow Derek Price's term from another context) which was more attractive to scholars than the ruder business of mills, or mines, or siegecraft. Lastly, and almost paradoxically, though such instruments were extraordinarily complex and demanding, they were within the intellectual grasp of men properly trained in (let us say) Ptolemaic astronomy in a way that the still greater subtlety of, for example, metallurgical chemistry or the fluid dynamics of water-wheels could never have been. In other words, instrument design fitted within the compass of ancient and medieval natural philosophy in a way that no other branch of technology did. For all these reasons we may accept the manifest existence of the tradition of instrument treatises without discarding the scholar-craftsman argument.

GUIDO DA VIGEVANO

Let us turn now to Guido da Vigevano,[8] who, like Giovanni de'Dondi, was a physician. His career had taken him to France where in the 1330s he found himself court physician to Jeanne of Burgundy, Queen to Philip VI, the first of the Valois line. The new king had twice pledged himself to the reconquest of the Holy Land, vows which we know were empty gestures, but which were taken with great seriousness at the French court, at least for a time. Guido seized on the opportunity presented by the crusading fervor to present in ca. 1335 a treatise to his monarch on new and better means for combating the infidel. He called it "Treasury of the King of France for the recovery of the Holy Land beyond the sea, and of the health of his body, and of the prolongation of his life, together with a safeguard against poisons." True to his calling, Guido devoted the first half of his work to medical advice about protecting one's health while in

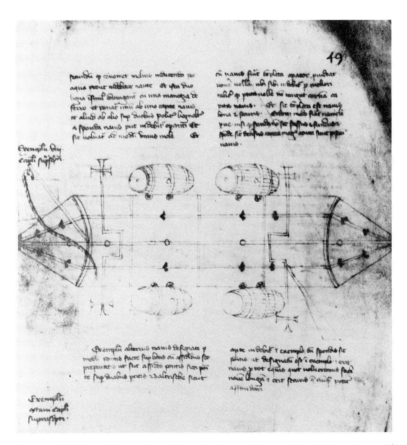

FIGURE 3. Guido da Vigevano's paddle-wheel boat with crankshaft drive. The paddle-wheels are indicated by the cross-shaped extensions at each end of the crankshafts. With permission of the Bibliothèque Nationale, Paris.

the field. More than mere professional pride may have been at work here, for disease had been at least as effective as Muslim arms in stopping earlier crusades. Moreover, deliberate poisoning of the politically powerful was somewhat more common in the fourteenth century than is customary in polite society today, and Guido's own remedy against aconite poisoning, which involved a soup made of slugs that fed on aconite leaves, was a matter of special pride to the physician.[9]

In the latter half of his manuscript Guido devotes his attention to siege engines with which to assault the paynim's strongholds. His proposals contain a number of technological novelties, among them:

(1) Machines made up of pre-fabricated elements, all portable and with some measure of interchangeable or multi-purpose parts.

(2) Paddle-wheel boats propelled by crankshafts (FIGURE 3).

(3) A fighting-wagon or "tank" propelled by crankshafts and gears, and with a steerable front axle (FIGURE 4).

(4) A similar "tank" propelled by a windmill mounted on its top (FIGURE 5).

(5) Extensive structural use of iron.

There is little evidence that anything Guido proposed was ever attempted in practice. Structural iron, pre-fabricated parts, paddle-wheel boats, and tanks all lay in the future. But if Guido was merely an armchair engineer, he was remarkably well informed about the practices of his day. Many of the technical words he employs are unique to his treatise in their Latin forms, but they are clearly and closely related to Italian words we find in vernacular texts from a later date. Obviously Guido had been listening and observing in workshops for some time before he put pen to parchment. Even his most outlandish suggestion, the windmill-powered armored vehicle, shows that he understood well the craft basis from which his proposal grew, i.e. the business of the millwright. He was also apparently capable of independent invention, for his crankshafts are the first of their type anywhere, and they do not reappear for nearly a century (and then only as the carpenter's brace-and-bit). We cannot be certain that French carpenters of the 1330s did not have the brace-and-bit, but even if Guido borrowed from such a source, his adaptation of the crankshaft to propulsive purposes is itself a significant act of inventive intelligence.

If we seek some tradition to which Guido might be related, we have difficulties that did not arise in respect to the instrument treatises. There is no strong or well-defined body of works to which the *Texaurus* seems closely related. There are, to be sure, noteworthy books on military affairs

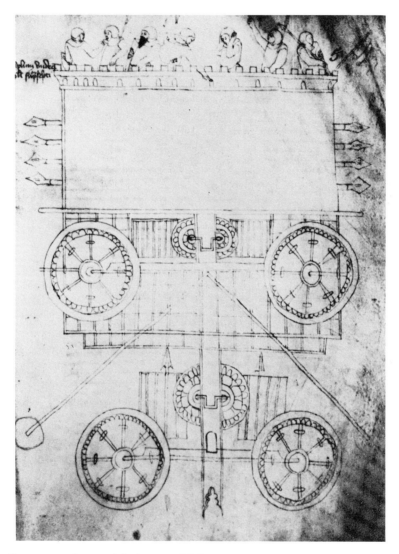

FIGURE 4. Guido's crankshaft-powered fighting wagon or "tank". The drawing is deliberately distorted to display the mechanism. With permission of the Bibliothèque Nationale, Paris.

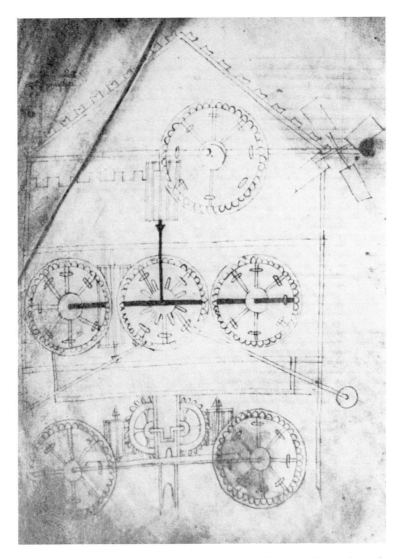

FIGURE 5. Guido's similar fighting wagon equipped with windmill propulsion drive.
With permission of the Bibliothèque Nationale, Paris.

from antiquity: one thinks immediately of Vegetius, whose work survives throughout the middle ages. On the Byzantine side we have the *Tactica* of Emperor Leo VI (886-911). Both of these are concerned with military machines only peripherally, however. The Alexandrian writers, Philon and Heron, concern themselves with catapults, but their treatment of such weapons is firmly within the limits of the instrument treatise tradition. Apollodorus of Damascus, a contemporary of Hadrian, wrote on siege-craft, and his work was expanded upon in the eleventh century by a Byzantine author known as Heron the Younger. I am unaware, however, that either Apollodorus or Heron were known in fourteenth-century France, either in direct copies or through Latin imitators. Only that strange anonymous work of the late fourth century, *De rebus bellicis*, which pro-poses a number of imaginative, if impractical, military devices to help save a dying empire could have been a potential source of inspiration to Guido.[10] It has also been recently suggested that two authors closer to Guido's own day might have influenced him: Aegidius Columna, author of *De regimine principium* (c. 1285) and Marino Sanudo (Torsellus the Elder) who wrote his *Liber secretorum fidelium crucis super Terrae Sanctae recuperatione et conservatione* between 1306 and 1313. Both works con-tain remarks on siege engines and fortifications, but even the author of this suggestion admits that neither work contains the kind of detailed technological information that Guido presents.[11]

   Guido's relationship to tradition is therefore somewhat ambiguous, but his novel traits stand out quite clearly. He is not describing machines that he or anyone else has actually constructed; rather he is putting forth a set of proposals in specific, detailed fashion about machines that could be made in the service of a valued goal. These proposals are practical or realistic only to the degree that they are rooted in the living technology of Guido's own day; otherwise they are imaginative fantasies, contempla-tions of the barely possible, and, like most armchair musings, a mixture of the admirable (interchangeable components) and the ridiculous (wind-mill-powered tanks). This is not the place to enter into a discussion of the role of technological fantasies in Western thought, but it is significant that with Guido the locus of imagination begins to shift from the world of objects, such as astronomical clocks, to the realm of text. In the follow-ing century, composition of imaginative technological treatises would be carried to unprecedented lengths.

   Guido also heralds a new style of technological writing in his use of words and pictures. His drawings, unlike those in the instrument treatises,

are not diagrams, but naturalistic illustrations (or as close an approximation as was possible in the trecento to developed Renaissance naturalism),
and they figure prominently in the organization of the manuscript (see
FIGURES 3–5). We can learn from Guido's figures what his machines looked
like, often, alas, at the expense of some of their interesting technical details. In the fifteenth century, pictures would rise to a central position in
many technological treatises, subordinating and in some cases suppressing
verbal descriptions. Guido's words were, of course, Latin, but based as I
have indicated on the vernacular of the workshop; again, in the fifteenth
century vernaculars would replace Latin in technological writings, possibly as a consequence of greater literacy among craftsmen in their native
tongues. Finally, unlike Giovanni and the instrument writers, Guido
intended his work for an audience that did not consist of technical specialists, the king and his court. The fifteenth-century treatises on technology
likewise reach for an audience that was broader than just those professionally skilled in the crafts discussed. To be sure, we do not find works
aimed at a very general audience until printed technological treatises
appear in the sixteenth century, but the line of development toward a
more public form of discourse about technology can be traced forward
from Guido's time without any break.

   In short, Guido da Vigevano's *Texaurus Regis Francie* is less a representative of an older tradition of technological writings than a harbinger
of new forms of such works that appear in force only in the fifteenth
century.[12] That Guido was known to the later writers is manifest by his
patent influence on Roberto Valturio's *De re militari*, one of the most
widely read of the new treatises and the first to reach print. Looking at him
as a precursor, we can use Guido to suggest a second modification of the
scholar-craftsman argument mentioned at the outset. The medieval reticence on matters technological came to an abrupt end after 1400 when a
much larger number of new treatises in a new style began to appear. This
development had its own dynamic and rationale, which we cannot discuss
here, but we should note that these new works were written and illustrated
by men from both sides of the sociological divide, by scholars and physicians, such as Giovanni da Fontana and Conrad Kyeser (the latter being
extremely influential in Germany), by craftsmen such as the anonymous
artillery man who composed *Feuerwerkbuch* (the first didactic manual
for gunners), by men such as the Sienese notary Mariano Taccola, and
by artist-engineers such as Francesco di Giorgio Martini and, the greatest
of them all, Leonardo da Vinci. These new writings were concerned with

some of the old subjects, instruments and siege engines, but they include much else besides: civil engineering, architectural matters, machine design, pumps, hydraulics, mills—the list is practically endless. Although these new treatises do not mark the end of social and intellectual distinctions between scholars and craftsmen, they do represent a more open and discursive phase in the history of our sources. The apparent existence of this open audience, combined with the fact that such works were at times patronized and collected by the wealthy and high-born as elaborate encyclopedias of technology, suggests a fundamental realignment of attitudes about technical matters. The exact outlines of these changes are still not clear to us, but the fourteenth century seems to have been a transitional period of considerable importance in bringing them about.

CONCLUSION

To conclude by summarizing: I have tried to introduce some of the major types and features of medieval writings about technology by holding up two works from the fourteenth century as examples. Giovanni de'Dondi's *Tractatus astrarii* represents an older, more stable form of technological treatise concerned with the design and construction of relatively precise and complex instruments. This is a very well-represented species, and I have tried to account for the frequency with which such works appear (in apparent contradiction to the scholar-craftsman argument) by pointing to the intellectual roots such activities had in natural philosophy, astronomy, and mathematics. I have tried to suggest that Giovanni's work represents the medieval version of High Technology and that this was a suitable subject for scholars to write upon. Guido da Vigevano's *Texaurus Regis Francie*, by contrast, represents a less cultivated form of expression concerned with less exalted subjects. Guido's treatment of his material takes the form of imaginative proposals directed at an unspecialized audience. I have indicated that his work is possibly less significant in itself than in its role as forerunner of a great flood of fifteenth-century treatises on all manner of technological subjects. These imaginative and discursive works which Guido heralds are, I have argued, a new type of technological writing and they suggest new intellectual attitudes and possibly a new social arrangement with respect to technology. We do not yet have a good grasp on all these changes, but it does seem that the fourteenth century, Machaut's century, is a turning point along the road that leads to Leonardo da Vinci, the printed works on technology of the sixteenth and seventeenth centuries, and ultimately to the birth of modern technology in the eighteenth century.

NOTES AND REFERENCES

1. White, Lynn, Jr. *Medieval Technology and Social Change.* Oxford: Oxford University Press, 1962.

2. White, Lynn, Jr. "Medieval Engineering and the Sociology of Knowledge." *Pacific Historical Review,* 44 (1975), pp. 1-21. This article presents a more balanced discussion.

3. See C. S. Smith and J. G. Hawthorne, "*Mappae clavicula:* a Little Key to the World of Medieval Techniques." *Transactions of the American Philosophical Society,* N.S. Vol. 64, pt. 4 (1974), pp. 14-20 for an example.

4. Bedini, S. A. and F. R. Maddison. "Mechanical Universe: The Astrarium of Giovanni de'Dondi." *Transactions of the American Philosophical Society,* N.S. Vol. 56, pt. 5, (1966).

5. Price, D. J. de S. "On the Origin of Clockwork, Perpetual Motion Devices and the Compass." *Bulletin 218: Contributions from the Museum of History and Technology.* Washington, D.C., Smithsonian Institution (1959), pp. 82-112. Also Price, D. J. de S. "Gears from the Greeks: the Antikythera Mechanism—a Calendar Computer from *ca.* 80 B.C." *Transactions of the American Philosophical Society,* N.S. Vol. 64 (1974). p. 7.

6. Price,[5] pp. 96-102. Also *The Book of Knowledge of Ingenious Mechanical Devices* by Ibn al-Razzaz al-Jazari. Ed. and Trans. D. R. Hill, Dordrecht and Boston: D. Reidel Publishing Company, 1974.

7. North, J. D., ed. and trans. *Richard of Wallingford: An Edition of His Writings.* 3 Vols. Oxford: Clarendon Press, 1976.

8. Hall, A. R. "Guido's *Texaurus,* 1335." In *On Pre-Modern Technology and Science.* Eds. B. S. Hall and D. C. West. Malibu, Ca.: Undena Publications, 1976, pp. 11-52.

9. White, Lynn, Jr. "Medical Astrologers and Late Medieval Technology." *Viator,* 6 (1975), p. 300.

10. Thompson, E. A. *A Roman Reformer and Inventor.* Oxford: Clarendon Press, 1952.

11. Hall, A. R.,[8] p. 12.

12. Hall, B. S. "*Der meister sol auch kennen schreiben und lesen:* Writings about Technology, ca. 1400-ca. 1600 and Their Cultural Implications." In *Early Technologies.* Ed., D. Schmandt-Besserat, Malibu, Ca.: Undena Publications, 1978.

DURING GUILLAUME DE MACHAUT's 77-year life span, manuscript illumination in France underwent many important changes both in style and in subject matter. When Machaut entered the Luxembourg family's service, the height of fashion in court circles was the mannered, precious, punctilious style of Jean Pucelle, and so it remained until the end of the century. However, during the reigns of Phillip VI and John II a different, more vigorous and realistic style appeared, which blossomed under Charles V. Several artists and groups of artists sharing the same aesthetic creed were urged by common necessity to visually translate philosophical concepts and practical subjects into manuscript decorations. Several manuscripts of Machaut's works exemplify this important turning point in French illumination.   MPC

French Illumination in the Time of Guillaume de Machaut

MARCEL THOMAS

Bibliothèque Nationale
Paris, France

H AVING rashly agreed to prepare a short paper on French illumination in the time of Guillaume de Machaut, it did not take me long to realize that I had fallen into the trap that tempts anyone who decides to deal briefly with a vast amount of facts loosely connected together within a long period. One always runs the risk of telescoping them into one another, thereby giving a quite misleading impression as to their real spacing in time. The further back in the past the period under discussion lies, the stronger is this tendency to foreshorten history.

When discussing dates, prehistorians allow themselves, in many instances, a margin of approximation running into thousands of years; medievalists very often do not feel disgraced if they are unable to decide whether an undated manuscript belongs to the tenth rather than to the eleventh century. Similarly, living in the twentieth century, we could be tempted to consider the lifespan of a man who lived 600 years ago as so short that no very significant changes could have taken place in his cultural environment between his youth and death.

No mistake could be more fatal in dealing with French illumination in the fourteenth century. The art of illumination evolved so quickly and so deeply during Guillaume de Machaut's life that it is difficult to point out in a short paper the numerous and significant changes that he witnessed. I think that it can safely be assumed that Machaut was interested in illumination since an author is seldom indifferent to the material form his works assume.

To understand why French illumination underwent such deep and rapid transformations during Machaut's life, one should remember that during the years 1300-1377 France was ruled by seven successive kings. Had Machaut lived until 1380, he might have witnessed an eighth coronation. So many changes in government (including the replacement of the direct

0077-8923/78/0314-0145 $01.75/2 © 1978, NYAS

Capetian line by the Valois branch of the royal family) could not help but be reflected in the social, cultural, and, above all, artistic life. French kings and the royal family were great patrons of the arts—and it is in the nature of art patrons to encourage new fashions and trends. Wishing to compete with fellow amateurs, each patron supports the creation of works of art, which he somehow considers to be partly of his own creation.

Machaut's active life as an author and composer was spent mainly under the first three Valois kings (Philip VI, John II, and Charles V), since the first of his poetical compositions to be dated with certainty is from 1324, and his oldest known major literary work, the *Dit dou Lion*, dates from 1342. Although Machaut was not yet an active writer, we should not omit a description of the trends in manuscript illumination during the first 20 years of the century. If Machaut had become interested in illumination in his teens, what would he have seen? What was considered beautiful at the time—or more to the point, perhaps, what was fashionable?

When Machaut was born, Philip IV, the Fair, the grandson of St. Louis had been on the throne for 15 years. It was during his reign that noticeable stylistic changes first appeared in illuminated manuscripts. These changes were in fact so striking that while it is somewhat difficult to date with accuracy manuscripts within the second half of the thirteenth century, the task becomes much easier after 1290.

The superiority of Parisian illuminators over provincial or foreign ones was already undisputed, and an ever-growing number of important and beautiful manuscripts were made accessible to lay amateurs—primarily royalty and nobility. These manuscripts were produced by copyists and illuminators working independently and supporting themselves by their labor. This trend was in sharp contrast to previous centuries when books were written and decorated almost exclusively by clerics for clerics.

Partly as a result of these sociological changes, art gradually became less abstract, more humanized and realistic. Representation of living persons tended to become more supple and more expressive; gestures and attitudes more natural. As a reaction to the thirteenth-century predilection for purplish reds and deep blues, lighter colors—pale green, rose, mauve, grey—came into fashion.

If as a young man of about 17 Machaut could have seen the beautiful *Life of St. Denis* presented in 1317 to Philip V, he would have found there a very good example of Parisian taste (FIGURE 1). Possibly, he would have been as amused as we are today by the charming scenes of everyday

FIGURE 1. "Life of St. Denis." (Paris, Bibliothèque nationale, MS fr. 2091, f. 111)

life in Paris that the artists had included in the lower part of their large paintings. This touch of realism, derived from marginal drolleries originating during the previous century in England and northern France, already indicates the path that fourteenth-century illumination will follow.

Hardly 10 years later, another change became obvious—one that is still more important from the art historian's point of view: we may, without exaggeration call it the Pucellian revolution. This change is of such significance that it would take far more space than is available here even to attempt to define it and point out every innovation (many of which derived from Italian formulas) that it introduced into French illumination. Therefore, I will only briefly discuss some of its major accomplishments.

Jean Pucelle is one of the very few fourteenth-century illuminators whose name has survived; what is more, owing to a recent discovery, we now know the exact date of his death—1333. We also know with almost absolute certainty that he took a decisive part in the decoration of three very important and still-extant manuscripts: the Bylling Bible, the Belleville Breviary (both in the Paris Bibliothèque nationale, see FIGURE 2), and Jeanne d'Evreux's Prayer Book, now in the Cloisters Museum, New York.

The obvious stylistic differences between these manuscripts, despite the fact that all three can be dated to around 1327, have been the subject of much discussion: so far, no completely satisfactory explanation has been found. As this is not the place to delve deeply into this perplexing subject, I will merely point out that a relatively large number of manuscripts, with dates both before and after Pucelle's death, may be attributed partly to him, partly to his workshop, to his pupils, or to his followers. I fully realize how vague this statement sounds, but the question is most delicate. Since we now know the date of Pucelle's death with certainty, a complete re-evaluation of works previously attributed to him has been undertaken but is still far from complete.

With very few exceptions, Pucellian manuscripts are of a liturgical, or at least religious nature: prayer books, Bibles, breviaries, and so on. This means that the artists in charge of the decoration could reuse the same iconographical scenes and types. Therefore, once a pictorial arrangement came to be considered harmonious and pleasant, it tended to become fixed. This fixity, this tendency to repetition, can be demonstrated by the fact that a breviary belonging to Charles V and decorated around 1370 faithfully reproduced images already existing in the Belleville Breviary (see FIGURE 3), one of the three manuscripts actually bearing Pucelle's name, and therefore, of necessity done before 1333.

FIGURE 2. "Belleville Breviary." (Paris, Bibliothèque nationale, MS lat. 10483, f. 24vº)

FIGURE 3. "Charles V's Breviary." (Paris, Bibliothèque nationale, MS lat. 1052, f. 36v⁰)

Although no firm conclusion can be drawn from this fact, it should be pointed out that a large number of these manuscripts were commissioned by ladies—more specifically, by ladies of very high and even royal rank, such as Jeanne d'Evreux, Jeanne de Navarre, Blanche de Savoie, and Bonne de Luxembourg. As Kathleen Morand aptly said in her excellent (although now somewhat outdated) book about Pucelle[2], "The patronage of the royal court did much to encourage miniature painting and was responsible for the appearance of a distinctive Parisian elegance and refinement." That so many patrons of Pucellian manuscripts were women probably contributed to the exquisite subtleness of his drawing—a subtleness that the occasional use of the "grisaille" technique (which does not allow any weakness in contour or modeling) does much to enhance. On that count, no better example could be found than Jeanne d'Evreux's Prayer Book.

Emergence of a new style is already noticeable in some important manuscripts decorated about 10 years after Pucelle's death. Thanks to their famous possessors, we can establish a fairly precise chronology for a few outstanding ones. They are of a liturgical or para-liturgical nature, and their illuminators continued to adhere to the Pucellian tradition between 1330 and 1365. Good examples of this trend are Yolande of Flanders' Prayer Book, copied around 1340[3] or Charles V's Breviary, copied around 1365[4] (see FIGURE 3).

On the other hand, until very recently, we knew much less about the non-liturgical manuscripts produced during the same period. This lack of knowledge was all the more unfortunate since it is in these manuscripts that the new trends first appeared. These trends continued to mature and emerged in full bloom during the reign of Charles V (1364-1380), obviously with his approval and encouragement.

Fortunately, in a remarkable paper published a few years ago,[5] my colleague, François Avril, drew the attention of scholars to a copy of a *Bible moralisée* kept at the Bibliothèque nationale.[6] This manuscript is particularly important since, because of the large number of paintings it contains, it provides a kind of epitome of Parisian illumination at the time of John the Good's coronation in 1349. We should remember that during this very year Bonne de Luxembourg died. She had been John's queen and Machaut's patron before he linked his fate with the Navarre family.

Despite its religious character, a "moralized Bible" is not a liturgical book. It is intended primarily to familiarize laymen with Holy Scriptures' teachings without requiring too great an effort from their uneducated

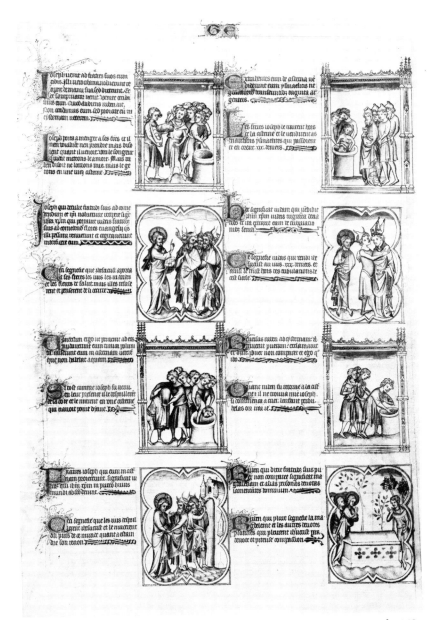

FIGURE 4. "Bible moralisée," fourteenth century. (Paris, Bibliothèque nationale, MS fr. 167, f. 11v⁰)

minds. The completion within a reasonable time of the 5000 small paintings in this Bible required the enlistment of many artists. François Avril
has numbered 15 of them, each endowed with specific mannerisms that
enable us to recognize his hand in other manuscripts of the same period;
needless to say, all remained perfectly anonymous.

A quick examination of some of these biblical and gospel scenes suf-
fices to prove that the art of illumination was at the time undergoing
a deep mutation in France (see FIGURE 4). Since this Bible's iconographical
themes were, of course, mainly religious, we should not be surprised to
recognize many instances of previous Pucellian formulas—but they have
been altered by a new mannerist trend.

Other artists in this collective venture evinced a definite leaning toward
naturalism. They abandoned the harmonious and idealized type developed by Pucelle and his followers in favor of a more individualized and
expressive representation of humanity. In so doing, they announced clearly the style that would be in favor during the 1360s and early 1370s (see
FIGURE 5).

Some illuminators of this Bible even seem to have had a definite leaning
toward ugliness. It could almost be said that they understood their work
with a modern cartoonist's spirit (see FIGURE 6). It is especially interesting
to note that one of the earliest illuminated manuscripts of Machaut's
works contains paintings done by one of these artists. The interest that
this particular illuminator evinced for contemporary fashion is also characteristic of the new realistic trend that soon came to prevail everywhere
(see FIGURE 7).

François Avril very plausibly connects this trend with the arrival in the
French capital of many artists from places that Pucelle's influence had
not reached. At this time, northern France and Flanders were the only
regions endowed with enough creative powers to give birth to such an
evolution, already visible in a manuscript of the *Voeux du Paon* belonging
to the Morgan library, which in all probability originated in Tournai.

One of Machaut's oldest manuscripts is quite characteristic of this deliberate emphasis put by the artist upon contemporary fashion. Dresses and
feminine hairstyles are carefully personalized and contrast sharply with
the draperies and gracefully flowing robes favored by Pucelle and his followers. Inicidentally, this Machaut manuscript can symbolize another phenomenon also typical of the period: the multiplication of literary and didactic works written in French, or translated into this language from
Latin.

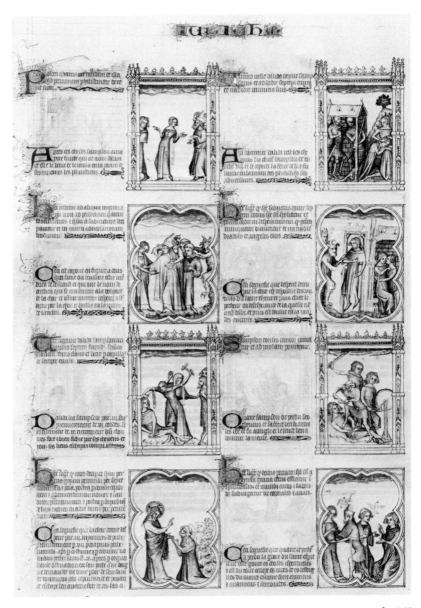

FIGURE 5. "Bible moralisée," fourteenth century. (Paris, Bibliothèque nationale, MS fr. 167, f. 59)

FIGURE 6. "Bible moralisée," fourteenth century. (Paris, Bibliothèque nationale, MS fr. 167, f. 29)

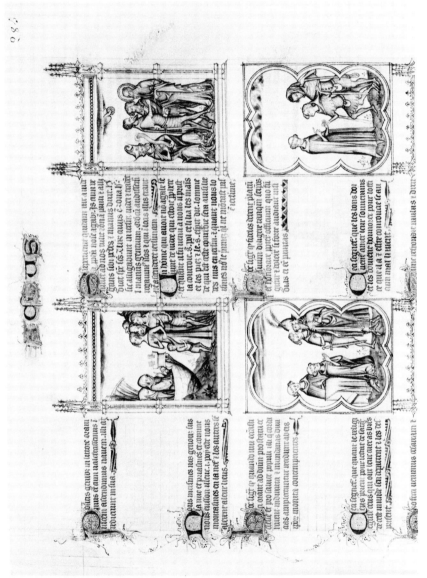

FIGURE 7. "Bible moralisée," fourteenth century. (Paris, Bibliothèque nationale, MS. fr. 167, f. 285)

That phenomenon was, of course, going to influence illumination in a very decisive way since it imposed new obligations upon the men who made their living out of it. Obviously, it was not a new idea to illustrate a text with scenes drawn from everyday life: we have already shown an early example of this tendency. Nevertheless, such subjects had generally been relegated to a modest and truly marginal level. For contemporary texts, one had to be satisfied with stereotyped and more-or-less interchangeable scenes of fighting knights or enthroned kings who could pass for biblical figures or contemporary persons as needed.

It is interesting to note that several illuminations datable to around 1360 represent Charles V as a clearly recognizable individual.[7] Although that king's "puny countenance"—to quote Christine de Pisan—was not overprepossessing, contemporary artists now felt that they had to treat it as a portrait, not as an idealized symbol.

This altogether new interest in accuracy and verisimilitude is even more tangible in a manuscript generally called *Livre du sacre*—or Coronation Book—kept in the British Library.[8] Copied and decorated immediately after Charles V's coronation, so as to record a precedent for similar ceremonies, this manuscript was conceived, as far as its illustration went, as a kind of graphic—we should almost be tempted to say photographic—report. Every part of an intricate pageant, which we know to have been carefully organized by the king himself, was truthfully translated into pictures. Even minute details, such as the exact form of the King's scepter were absolutely true to life. We are certain of this since Charles's scepter has survived; anyone seeing it in the Louvre can verify the artist's strict adherence to reality.

In much the same spirit, a state visit in 1379 of the German Emperor Charles IV was vividly evoked in paintings included in a famous manuscript of the *Great Chronicles of France* (see FIGURE 8). From the contemporary accounts, we know that the illuminator faithfully represented what everybody could have seen in the streets of Paris.[9]

Other reasons, to which I have already alluded, induced illuminators to display more inventiveness than their predecessors. Obviously, a strong correlation can be found between stylistic and iconographical changes noticeable during the reign of Charles V and the great number of Latin works then being translated into French.[10] While these texts in their original form were normally unillustrated, their translators or their translators' patrons clearly wanted something else.

For example, let us touch for a moment upon the work of a very pro-

FIGURE 8. "Grandes Chroniques de France." (Paris, Bibliothèque nationale, MS fr. 2813, f. 473 v⁰)

lific artist, active throughout Charles's reign, who, for reasons impossible to explain here, is now generally called "pseudo-Bondol." Assisted by several aides, he illuminated a large number of very important and beautiful manuscripts—many of which went into the King's famous Library, then housed in one of the Louvre towers.

When the text he had to illustrate was of a very common nature—such as a Bible—he did not display much inventiveness and kept very close to a well-established iconographical tradition. This tendency is particularly obvious in a beautiful but little-known French Bible now kept in Hamburg.[11] Its small paintings adhere to the Pucellian canon; although their colors are different, their style keeps close to tradition. The Italian *loggias*, the Sienese way of representing mountains as piles of low plateaux set upon one another are from that point of view very typical throughout the volume.

Some years later, the same master was commissioned to illuminate two splendid and little-known manuscripts containing three major treatises by Aristotle translated into French by Nicole Oresme from the Latin version some time before 1375. In coming to grips with the new task of illustrating abstract ideas, as opposed to events, pseudo-Bondol shows that he is capable of originality. One interesting new trend to be found in this manuscript is an effort to introduce real landscape as a background for open-air scenes.

The same artist included another interesting landscape in a beautiful copy of Machaut's works, finished a short time before the author's death (see FIGURE 9). Here, a real blue sky replaces the old conventional checkered or foliated background. It is undoubtedly an important step toward realism and "space conquest." Although the exact date of this manuscript is not known, it is highly probable that it was copied during Machaut's lifetime. Since it contains the Prologue, which did not appear in early manuscripts, it is of a very late date; but nothing prevents us from believing that Machaut could have seen it with his own eyes, and—who knows—somehow supervised its organization and even decoration.

No survey, however brief, of French illumination during Machaut's lifetime would be complete without a mention of several other masters whose contributions were far from negligible.

It would be especially unfair not to grant a brief line to the charming "Master of Charles VI Coronation"; he is referred to by this name because one of his most important works was the representation of the new king's crowning in the copy of the *Great Chronicle*, that I have already alluded to (see FIGURE 10). Though ill-at-ease with the difficult problem

FIGURE 9. Guillaume de Machaut's Works (Paris, Bibliothèque nationale, MS fr. 1584, f. D)

of representing a multistory building with an interior spiral staircase, he amply redeems his gaucherie by his naturalistic treatment of minor characters. Look, for example, at the usher on the ground floor who is keeping the crowd in its place while another argues with a late guest or potential gate-crasher. As for the disgusted-looking man on the extreme left, nobody will ever know for certain whether he has had his feet trodden upon

FIGURE 10. "Grandes chroniques de France." (Paris, Bibliothèque nationale, MS fr. 2813, f. 3v⁰)

or whether, if we dare risk a modern colloquialism, he has put his foot in it!

At the same time, we should not forget that outside Paris less skilled artists produced manuscripts that are interesting, but whose execution betrays simultaneously a more durable attachment to outmoded formulas, a will to imitate their more gifted Parisian colleagues, and an obvious gaucherie in so doing (see FIGURE 11).

To conform to an accepted "règle du genre" I will now attempt to draw some conclusions to finish off this very sketchy "exposé." The most indisputable fact about French illumination in the fourteenth century is that it evolved with unprecedented rapidity. During Machaut's lifetime, at least four different styles appeared, developed, and disappeared, one after the other.

Another incontrovertible fact is that Parisian illumination maintained throughout the century the supremacy previously acknowledged by Dante. That supremacy continued to assert itself during the next century and was at its height when the discovery of printing and etching gradually put an end to the gracious and noble art of illumination.

As time went by, craftsmen who had so beautifully mastered its technique tended more and more to become artists in the modern sense of the term; each one tried successfully to add a personal touch to pre-existing formulas and in many cases indulged in what we would call "artistic research" if we were discussing contemporary works.

A glance at FIGURE 12 will be enough to demonstrate how quickly artistic tastes were changing at the time—as of course they continued to at the same pace later. This figure is taken from one of the first (and most beautiful) prayer books that belonged to John, Duke of Berry, a younger brother of Charles V. This manuscript was begun around 1380, and once more we see a new style emerging. It is hardly necessary to underline the seductive magnificence of the drawing, composition and color of this so-called "international style." The large painting occupying the upper part of the picture is sufficient witness to that.

However, if one looks closely at the small "bas de page" scene that provides a kind of counterpoint to the large picture above, one notices a rather pathetic fact that has in itself much symbolic value. These "bas de page" paintings in the first part of the book were entrusted to none other than the painter known as "pseudo-Bondol," to whom so many much more important commissions had been granted by the late king himself a very short time before.

FIGURE 11. Prayer Book, Metz Use. (Paris, Bibliothèque nationale, MS lat. 1403, f. 40)

Are we forbidden to be a little sentimental and try to imagine the out-moded master's feelings when he saw himself reduced to such a menial rank? And what would have they been if he had known that a few years later his work was to be retouched by another illuminator, and that his small clusters of umbrella-like trees—a trademark for his generation—were to change form to suit another generation's tastes?

We are thus reminded that while a work of art keeps its permanent value

FIGURE 12. "Très belles heures de Notre Dame." (Paris, Bibliothèque nationale, MS n.a.l. 3093, f. 56)

for centuries, the kind of appreciation it gets may vary greatly within a very short time. Fortunately, art historians can be relied upon to see that, in due course, quality should prevail over fashion.

REFERENCES

1. Fr. Baron, "Enlumineurs, peintres et sculpteurs parisiens des XIVe et XVe siècles d'après les archives de l'hôpital Saint Jacques-aux-Pélerins," ds. *Bull. archéol.*, N.S. t. 6 (1970), pp. 79-115.
2. K. Morand, *Jean Pucelle*. Oxford, 1962.
3. London, British Library, Yates Thompson, MS 27.
4. Paris, Bibliothèque nationale, MS latin 1052.
5. Fr. Avril, "Un chef d'oeuvre de l'enluminure sous le règne de Jean le Bon: la Bible moralisée, manuscrit français 167 de la Bibliothèque nationale," ds. *Monuments Piot*, t. 58 (1973), pp. 91-125.
6. Paris, Bibliothèque nationale, MS fr. 167.
7. C. R. Sherman, *The portraits of Charles V of France* (New York, 1969).
8. London, British Library, MS Tib. B. VIII.
9. M. Thomas, "La visite de l'Empereur Charles IV en France d'après l'exemplaire des "Grandes Chroniques" exécuté pour le roi Charles V," ds. *VIth International Congress of Bibliophiles . . . 1969. Lectures* (Vienna, 1971), pp. 85-98.
10. L. Delisle, *Recherches sur la librairie de Charles V* (Paris, 1907) et Bibliothèque nationale, *La Librairie de Charles V* (Paris, 1968).
11. Hamburg, Kunsthalle, Kupfertischkabinett, MS fr. 1. See catalogue mentioned above, n° 208.

PAPER became popular for book production in fourteenth-century France. Though paper was invented in China near the beginning of the Christian era, knowledge of the manufacturing process was acquired by the Arabs in the eighth century and brought by them to Spain in the twelfth. Paper mills were established in Italy in the thirteenth century, and in France and Germany in the fourteenth. Several notable French mills of this period existed in the provinces of Champagne, Ile-de-France, and Auvergne. The Auvergnat village of Ambert is the location of the Richard-de-Bas mill, where paper is still being made today, using fourteenth-century methods. Early paper was expensive, although less so than parchment. There are few paper books in Europe dating from the first half of the fourteenth century, but in the second half paper manuscripts became numerous, especially by the 1380s. During the course of the fifteenth century paper replaced parchment as the most common book material.   MPC

Paper Manufacturing and Early Books

SUSAN O. THOMPSON

School of Library Service
Columbia University
New York, New York 10025

FOUR great inventions that spread through Europe at the beginning of the Renaissance had a large share in creating the modern world. Paper and printing paved the way for the religious reformation and made possible popular education. Gunpowder leveled the feudal system and created citizen armies. The compass discovered America and made the world instead of Europe the theater of history. In these inventions and others as well, China claims to have had a conspicuous part."[1] Thus begins the introduction to Thomas F. Carter's standard work, *The Invention of Printing in China and its Spread Westward*. My own subject, the manufacturing of paper and early books, must have as its introduction "The Invention of Paper in China and its Spread Westward."

The fifth century annals of the Han dynasty tell us that in the year A.D. 105 paper was invented by the eunuch Ts'ai Lun because earlier writing materials were unsatisfactory, silk being too expensive and bamboo too heavy. Ts'ai Lun's paper was said to be made from tree bark, hemp, rags, and fish nets. Archaeologists have long known scraps of paper composed of such materials that date from the period of Ts'ai Lun. Then, in 1957, paper fragments were found in a tomb thought to be not later than the Wu-ti period, 140-87 B.C., indicating that the event recorded in the Han history may simply have been official recognition of a process that had been developing for some time.[2]

Used widely in China for many purposes, paper spread to Korea and Japan, where its most notable use was for the million charms, printed with blocks and placed in wooden stupas, that the Japanese Empress Shotoku ordered to be made from 764 to 770. Authorities at the Library of Congress have suggested that this may have been the first example of mass production.[3] It was in this same century that the Far East relinquished its monopoly on papermaking. In 751, Chinese prisoners of war taught the

0077-8923/78/0314-0167 $01.75/2 © 1978, NYAS

art to their Arab captors in Samarkand, giving to the Arabs a monopoly in the West which they held for five centuries. Paper itself had long been known in central Asia, for Chinese paper had been traded along the old silk routes, and now the Arabs built up their own trade in paper. From 795 Harun al-Rashid made Baghdad a center of this trade, and other cities, such as Damascus, also exported paper.

Carter describes the replacement of papyrus by paper in the Arab world. The Erzherzog Rainer collection in Vienna consists of about 12,500 Arabic documents from Egypt. From the second century of the Hegira (719–815) there are 36 dated documents, all on papyrus. From the next century there are 96 on papyrus and 24 on paper. From the fourth century, 9 are on papyrus and 77 on paper.[4]

When the Moors conquered Spain they took papermaking with them. A manuscript at the monastery of Santo Domingo of Silos, possibly from the tenth century, is partly on paper that may be of Spanish manufacture.[5] In any case, by the twelfth century the Spanish paper industry was so flourishing that its products were exported. Paper also entered Europe from Asia through Constantinople and from Africa through Sicily.

Papermaking had begun in Italy by the thirteenth century, knowledge of it probably having come from the Arabs in Sicily. There is a documentary reference to papermaking in Fabriano in northern Italy in 1276 (the actual work having begun perhaps in 1268), the first known location in Christian Europe and the site of fine papermaking ever since.

The other two countries where paper was made in the fourteenth century are France and Germany. The latter may have had mills in the early part of the century, but there is documentation for the last decade. Ulman Stromer of Nuremberg built a mill in 1390/91. His diary tells how he brought workmen from Lombardy who reneged on their contract to keep the precious secrets of the craft to themselves. Stromer had them sent to jail for a few days. The representation of Stromer's mill in Hartman Schedel's Nuremberg Chronicle, printed and published by Anton Koberger in 1493, is the first picture we have of a European paper mill.

The question of France is much more vexed. Did papermaking come from Spain, or from Italy, or from the Near East via the Mediterranean? Legends have grown up concerning the last possibility. One story has it that a group of French crusaders, kept prisoner by the Saracens in the Levant for several years, worked in Arab paper mills, then returned to France with the new craft. Another version fixes on one man, Jean Montgolfier, as having brought back papermaking from the Crusades to the town of Vidalon in 1157.[6]

There were certainly paper mills in France in the fourteenth century. In 1926 Henri Alibaux listed 14 mills possibly established by 1400.[7] The list has been augmented by Anne Basanoff in her 1965 book, *Itinerario della Carta*. She gives a map locating 25 possible fourteenth century French mills and even refers to the fourteenth century as "the French period."[8]

Some areas are better than others for providing the necessary raw materials, abundant pure water, and marketing outlets. Ile de France, Champagne, and Auvergne have all been papermaking centers. Auvergne is the area where the Crusader legends abound and which claims priority in French papermaking. What is perhaps the first paper mill in France, the Moulin Richard de Bas, situated in the Auvergnat town of Ambert, is still making paper by fourteenth century methods.

European papermaking differed from its precursors in the mechanization of the process and in the application of water power. Jean Gimpel, in *The Medieval Machine* (the English translation of *La Révolution Industrielle du Moyen Age*), points out that the Chinese and Arabs used only human and animal force. Gimpel goes on to say: "This is convincing evidence of how technologically minded the Europeans of that era were. Paper had traveled nearly halfway around the world, but no culture or civilization on its route had tried to mechanize its manufacture."[9] The ability to transform a wheel's motion by levers into a reciprocal movement existed in antiquity, but the desire to make use of it, given the availability of slave labor, was, according to Gimpel, not present. But in the Middle Ages mills were used for many purposes.

By the fourteenth century there was a special need in the area of bookmaking for increased output of a material cheaper than parchment. The growth of the universities was only one of the more prominent factors in increasing the need for books in the late medieval period. The time was ripe for printing with movable type, the appearance of which in the mid-fifteenth century has been described as overdue. The four elements needed by Gutenberg for his invention—individual metal letters, a screw press, oil-based ink, and a flexible, abundant material to receive the printing—all existed before his time. Paper indeed had become familiar in libraries by the end of the fourteenth century.

Professor Paul O. Kristeller, one of our greatest manuscript bibliographers, has told me that he does not recall handling a European book on paper earlier than 1300. He describes such books as scarce during the first half of the ensuing century, but numerous during the second half. My own researches in manuscript catalogues encourage me to be even more spe-

cific. I believe the 1380s to be the decade when there begins to be a noticeable number of books on paper. For example, of the five fourteenth century manuscripts on paper in the 1936 catalogue of the University Library
in Erlangen, Germany, the earliest dated one is 1386, and it is a collection
of humanist texts while the other four are religious books of one kind or
another.[10] These five come from a total collection of 237 so-called "old
manuscripts," dating through 1500. As to the proportion of paper to vellum, André Blum has stated that study of the inventories of Charles V's
library bears out Barrois's assertion that in Charles's Louvre collection
paper to parchment was as one to 28.[11]

The 1968 catalogue of dated manuscripts in Belgian libraries lists 96
manuscripts from the period 819 to 1400.[12] Sixteen of these are on paper,
two of them combining parchment and paper. The oldest one, once in the
Louvain University Library, was Pierre Riga's *Aurora*, dated 1346, but it
was destroyed in 1914. The oldest surviving one is the 1358 Golden Legend of Jacques de Voragine, belonging to the Hôpital St. Jean in Bruges.
There are none from the 1360s, three from the 1370s, eight from the
1380s, two from the 1390s, and one from 1400. Roughly half are secular
texts.

The Klagenfurt, Austria, Studienbibliothek, which inherited books from
a Benedictine abbey, has a total of 181 old paper manuscripts—including
32 from the fourteenth century—many of which are undated.[13] The 12
that have dates start with a collection of religious texts from 1356, another
such collection from 1369, then jump to the 1380s with four books and the
1390s with six books.

Lyon, a city rich in books even before it became the second greatest center of French printing (after Paris), has in its public library at least seven
paper books earlier than the fifteenth century.[14] The titles include works
by St. Thomas Aquinas and Nicholas of Hanapis, a collection of prayers
from the abbey of Clairvaux, a collection of model letters formed by clerks
in the pontifical chancery, a collection of fragments of thirteenth and
fourteenth century paper taken from the registers of notaries in Marseille,
and two *terriers*, lists of taxes, obligations, and rights for villages in certain
priories. One of these lists dates from 1319, the other, from the Dau-
phiné, is even older, dating from 1292 and 1319.

These documents are reminiscent of Febvre and Martin's account in
L'Apparition du Livre of key dates in the replacement of parchment by
paper.[15] The Marseille notaries used paper as early as 1248, as did the
commissioners of Languedoc, while official records of Poitiers and Toulouse from the mid-thirteenth century were also on paper.

Another approach to studying the use of paper in the fourteenth century is to look at works of individual authors. Armando Petrucci has described around 30 documents that bear Petrarch's handwriting. Two of these are on paper.[16] Latin ms. 3196 in the Vatican Library consists of notes composed by Petrarch from 4 November 1336 to 12 February 1374. The other one is in the Laurentian Library in Florence (LIII 35) and is a collection of letters to Moggio de' Moggi.

The use of paper in the late fourteenth century is said to have increased the tendency toward cursiveness on the part of scribes. In the fifteenth century there was certainly large scale production in workshops of popular titles, usually in cursive script on paper, illustrated with pen drawings. An early example is Rudolf von Ems's *Weltchronik* from 1402, now in the Spencer Collection at the New York Public Library. Another Spencer Collection manuscript, Ulrich von Richenthal's *Chronicle of the Council of Constance*, from around 1450, is one of the best known of this type. Hugo von Trimberg's *Der Renner* dates from the late fifteenth century and is now at the Morgan Library (M.763). The Morgan also has in the Glazier Collection (G.23) a Flemish fifteenth century paper manuscript, Guido delle Colonne's *Histoire de la Destruction de Troye*. In the Library of Congress Rosenwald Collection are two Italian fifteenth century manuscripts (13 and 14) of Roberto Valturio's *De Re Militari*, one written on paper and vellum in gothic script and one written on paper in humanistic script.

During the course of the fifteenth century, paper replaced vellum as the most common book material. The majority of manuscripts continued to be written on vellum, but they were far outnumbered by printed books. It has been estimated that perhaps 40,000 editions were printed in the incunabula period. Taking 500 copies as an average size edition, this means that about 20,000,000 books came into being from type in Europe in the last half of the fifteenth century.[17] Some of these are on vellum, of course. Some copies of the 42-line Bible are on vellum as are all copies of the 1457 Mainz Psalter.

Why did paper take so long to establish itself as the major book material? In the first place, early paper, although less costly than vellum, was still very expensive. Leo Mucha Mladen has told me that there was a tradition in his native city of Prague to the effect that the first paper book to appear there was a blank codex, probably imported from Italy in 1310, which cost as much as a house. As manufacture developed in Europe, thanks to the mechanization already referred to, paper became less dear. Blum notes that public authorities sometimes exempted papermakers from taxes, as in 1354 when John the Good granted the University of Paris the right to tax-free mills in Essonnes.[18] Moreover, the substitution of linen for wool in underwear made rags much more plentiful.[19] But paper remained the one most expensive element in bookmaking, right down to the nineteenth century when wood pulp was successfully utilized.

Another reason for paper's tardy acceptance was the feeling that it was not as permanent as parchment. Roger II of Sicily in 1115 had renewed on parchment a paper charter given by his father to the abbey of Santa Maria di Terreti in Calabria 25 years earlier. In 1145 Roger did the same for four charters on paper given to the monastery of St. Philip of Fragala from 1102 to 1124. His nephew, Frederick II, in 1231 ordained in the constitutions for the Two Sicilies that all public documents be written on parchment "so that they may bear testimony to future times and not risk destruction through age."[20] As late as 1494 the bibliographer Johann Tritheim said in his book In Praise of Scribes, "If writing is inscribed on parchment, it will last a millennium. But if on paper, how long will it last? Two hundred years would be a lot."[21] Actually, early paper has excellent qualities of permanence and durability.

This negative feeling about paper may also be part of the natural conservatism to be found in bookmaking down through the ages. People have been reluctant to change the appearance of graphic communication. The best known example of this is the attempt on the part of the first printers to make their work look like manuscripts. They even copied in metal type the variations in scribal hands. There is said to be twenty-odd different lowercase e's in the Gutenberg Bible.

Another psychological factor pointed to by many scholars is the inevitable connection by the Christian mind of paper and the Semitic world from which it came. Arabs and Jews made paper in Spain and other places for centuries before Europeans.

Finally, technical considerations may have been important. Arab paper, often called charta bombycina in the old documents, was thought to be made of cotton until microscopic examination in the nineteenth century showed it to be made of rags. It had been believed until then that Europeans invented rag paper; now it is thought that the difference in the papers comes from the mode of manufacture, not from the raw materials. In any case, European papers were more satisfactory for bookmaking than the paper imported from Arab lands. The Arabs did not beat the pulp or size the paper for writing quality in the same way as Europeans, nor did they use the same type of mould.

With Fabriano paper, chainlines appear, as well as watermarks, the earliest of which is dated 1282. These intriguing designs are made by raised wires on the mould that cause the pulp to be thinner and can thus be seen when the sheet is held to the light. They have been used to denote such things as the size of the sheet or the place of manufacture. The codification of watermarks prepared by C. M. Briquet, which has been updated by the work of Allan Stevenson, who developed a radiographic method of study,[22] makes these marks extremely useful for dating old paper.

The paleographer Edward Maunde Thompson has summed up early paper in these words:

> European paper of the 14th century may generally be recognized by its firm texture; its stoutness and the large size of its wires. The water-marks are usually simple in design; and, being the result of the impress of thick wires, they are therefore strongly marked. In the course of the 15th century the texture gradually becomes finer and the water-marks more elaborate.[23]

Alexander Nicolai has made the following summary of the history of early European paper. The first period extends from the end of the eleventh century to the middle of the thirteenth when unwatermarked Arab paper was imported through Spain and Italy. The second period comprises the last half of the thirteenth century when the Italians began to make paper with chainlines but still largely unwatermarked. The third period, from 1300 to 1450, saw Italian paper with watermarks inundating Europe. The fourth period, beginning in the mid-fourteenth century and overlapping with the third, saw French paper mingling with the Italian. By the sixteenth century French paper had replaced Italian in the export trade, with the number of French watermarks exploding after 1470.[24]

My last word on fourteenth century paper is that we do not know enough about it. In preparing this paper, I consulted the paper historians Allen Hazen, Leonard Schlosser, and Jack Robinson, and I think they would agree with me. The question of paper versus parchment in pre-Gutenberg Europe is an interesting one, worthy of further study.

REFERENCES

1. Carter, T. F. *The Invention of Printing in China and its Spread Westward.* 2d ed rev. by L. C. Goodrich. New York: Ronald Press, 1955, p. ix.
2. Thompson, L. S. "Paper." *Encyclopedia of Library and Information Science.* Vol. 21: 336. New York: Dekker, 1977.
3. *Papermaking: Art and Craft.* Washington: Library of Congress, 1968, p. 15.
4. Carter,[1] pp. 135-136.
5. *Papermaking: Art and Craft,*[3] p. 19.

6. Hunter, Dard. *Papermaking: The History and Technique of an Ancient Craft.* 2d ed. New York: Knopf, 1957, p. 473.

7. Alibaux, Henri. *Les Premières Papeteries Françaises.* Paris: Les Arts et le Livre, 1926. Fold-out map following p. 212.

8. Basanoff, Anne. *Itinerario della Carta dall'Oriente all'Occidente e sua Diffusione in Europa.* Documenti sulle Arti del Libro IV. Milan: Edizioni il Polifilo, 1965. Fold-out map following p. 84.

9. Gimpel, Jean. *The Medieval Machine: The Industrial Revolution of the Middle Ages.* New York: Holt, Rinehart and Winston, 1976, p. 14.

10. *Katalog der Handschriften der Universitätsbibliothek Erlangen.* Vol. 2. Erlangen: Universitätsbibliothek, 1936.

11. Blum, André. *On the Origin of Paper.* New York: Bowker, 1934, p. 50.

12. *Manuscrits Datés Conservés en Belgique.* Vol. 1. Brussels: E. Story-Scientia, 1968.

13. *Handschriftenverzeichnisse Österreichischer Bibliotheken.* Vol. 1. Vienna: Österreichischen Staatsdruckerei, 1927.

14. *Catalogue Générale des Manuscrits des Bibliothèques Publiques de France. Départements.* Paris: Plon. Nourrit, 1900. Vol. 30.

15. Febvre, Lucien and H. J. Martin. *L'Apparition du Livre.* Paris: Albin Michel, 1958, p. 30.

16. Petrucci, Armando. "La Scrittura di Francesco Petrarca." In *Studi e Testi.* Vol. 248: 116, 119. Rome: Biblioteca Apostolica Vaticana, 1967.

17. Febvre and Martin,[15] p. 377.

18. Blum,[11] p. 51.

19. Febvre and Martin,[15] p. 29.

20. Clapp, V. W. "The Story of Permanent/Durable Book-Paper, 1115-1970." In *Scholarly Publishing.* Vol. 2: 108. Toronto: University of Toronto Press, 1971.

21. Clapp,[20] p. 109.

22. Briquet, C. M. *Les Filigranes: Dictionnaire Historique des Marques du Papier dès leur Apparition vers 1282 jusqu'en 1600.* 2 vols. Ed. Allan Stevenson. Amsterdam: The Paper Publications Society, 1968.

23. Thompson, E. M. "Paper." *Encyclopaedia Britannica.* 11th ed., vol. 20: 727. Cambridge: Cambridge University Press, 1911.

24. Nicolai, Alexandre. *Histoire des Moulins à Papier du Sud-Ouest de la France 1300-1800.* Bordeaux: G. Delmas, 1935, p. xviii.

A SUPERB COMIC NARRATOR in Medieval literature often depreciates himself, his capacities as a lover, and mocks mercilessly the conventions of love itself. Guillaume de Machaut created such a literary type, developing his blundering, inept narrator-lover from artistic conventions latent in the *Romance of the Rose*. Perhaps for Machaut this character is partially autobiographical, prone to cowardice, snobbery, sloth, misogyny, and pedantry. He is superbly comic, however, an excellent contrast to courtly ladies and gentlemen. Machaut's great triumph as realist and as narrative poet may be the development of this "fool" as a major protagonist of serious belles-lettres. Machaut may have profoundly influenced Geoffrey Chaucer. MPC

The Poet at the Fountain: Machaut as Narrative Poet

WILLIAM CALIN

Department of Romance Languages
University of Oregon
Eugene, Oregon 97403

L ONG ago Chaucerians used to claim that their author had invented the motif of the naïve, blundering, comic Narrator. Now scholars agree that the amusing "Geffrey" of the *Book of the Duchess, Parlement of Fowls, House of Fame,* and *Legend of Good Women,* is a comic type first exploited to its maximum potential in the works of Guillaume de Machaut. Imitated later, this humorous figure appears in works by Froissart, Christine de Pisan, Alain Chartier, Pierre de Nesson, and others, as well as by John Gower and his best friend. Chaucer. Machaut, however, was a developer not an innovator, working in a tradition latent in the *Romance of the Rose.*

The "I" of Guillaume de Lorris's section of *Le Roman de la Rose* is an active participant, a reliable commentator with whom we sympathize because the story is filtered uniquely through his consciousness; he is, except for the God of Love, the only intelligence capable of establishing norms by which other characters are judged. Indeed, on one level, we are expected to empathize with him and to associate him to some extent with the author.

Obviously, we must also make a distinction between Guillaume de Lorris, the Author, and the I-Narrator, a fictional, indeed allegorical character, a "poetic" or universal I, an Everyman (or at least Every Lover). We must also distinguish between the *erlebendes Ich* and the *erzählendes Ich,* that is, between the Lover who experiences certain events in a dream set in the past, and the Narrator who tells his dream story years later in the present. The significance of the dream will be disclosed at the end of the poem by the Narrator (2061-74), who maintains an existence independent from the dream-vision, the story of *Le Roman de la Rose;* yet the dream-events continue on into his waking life. For example, he tells us that he was 20 years old when he had the dream (21) but that it occurred five years before he records it on parchment (46). The dream is a true

0077-8923/78/0314-0177 $01.75/2 © 1978, NYAS

visio since its events are realized in the future. The Narrator laments his suffering caused by the Fountain of Narcissus (1606-12), Cupid's arrow (1694-6), or kissing the rose (3473-5). Not only this, he even creates a little frame-story around the dream-vision. Allegedly he writes the poem for his beloved lady (may she receive the book kindly!), she who is worthy to be loved and given the name of Rose (39-44). He will continue the story in the hope that it will please his beloved. May she reward him for it as only she can, when she so chooses! (3487-92).

The frame allows *Le Roman de la Rose* to be viewed as the Narrator's attempt toward the seduction of his lady, and it can thus be assimilated to the sub-genre of courtly lyric called *requête d'amour* (appeal for love). The Narrator's love, hitherto unrequited, prompts him to write the *Roman*, "ou l'art d'Amors est tote enclose" (38). The only other time Guillaume employs the term *art* is in conjunction with the word *engin* (tool) (497) and concerns the tricks or artifice the Narrator may use to enter the garden of the rose. *Engin* is also associated with the nets laid at the Mirror Perilous (1589-90) by the God of Love who recommends *Doux Parler* (sweet talk) as a remedy or consolation for love. *Le Roman* is a form of "doux parler," a seduction as well as a confession, sublimation, and consolation for the fictional narrator.

At the end of the story the distinction between *erlebendes Ich* and *erzählendes Ich* presumably breaks down; the lover-dreamer becomes the lover-narrator in an historical and narrative present. If the dream is to predict future events, so that the Narrator's ulterior love-affair with his lady has followed every detail of the dream, how can Guillaume de Lorris have the Narrator pluck the Rose in his dream without shocking the lady's (and the reader's) sensibilities in the "frame-story" and compromising her, thus violating every taboo of *fin' amor*? I believe neither that the author Guillaume de Lorris necessarily loved a particular lady in the late 1220s, nor that, because the outcome of Guillaume's affair was in doubt when he composed the *Roman*, he could not bring his poem to a conclusion. I *do* believe that the fictional Narrator finds himself in the same position as the author, who is forbidden by decorum to plot the end of his own dream since it is set in a seduction "frame."

It is possible that the majority of scholars are correct when they assume Guillaume de Lorris died around the year 1230, with his masterpiece incomplete. It is also possible, however, that Guillaume lived on, but, even though he intended to finish the *Roman*, even though in the poem he twice referred to the ending, he discovered *en route* that he could proceed no further since to recount the plucking of Rose would have destroyed that delicate, evanescent balance between sensuality and sublimation that gives the tone to his work. It would have required him to tell us outright whether the couple do or do not go to bed and what they do therein, a choice he could not or would not make; recounting the rose-plucking would have defied literary decorum and upset the very principles of *fin' amor* that he had taken such pains to exalt, thereby rendering the tale absurd. Does this mean that Guillaume's *Roman*, in its extant form, is an inferior work of art? Not in the least! Rather, consummate master that he was, Lorris may have realized that to end his poem would ruin it, and he preferred to stop in time. Better perhaps a perfect, incomplete masterpiece than a fine poem with a silly ending. Better a magnificent, unequaled Unfinished Symphony that compels the public to dream and question in turn for answers that have never been given and never will be.

✎§ §✎

In Jean de Meun's portion of the romance, the God of Love informs his troops that the lover whom they have come to support is Guillaume de Lorris, a loyal servant of Eros, who will begin *Le Roman de la Rose*, but will leave it incomplete to be taken up more than forty years later by Jean de Meun:

> Puis vendra Johans Chopinel,
> au cuer jolif, au cors inel,
> qui nestra seur Laire a Meün,
> qui a saoul et a geün
> me servira toute sa vie,
> sanz avarice et sanz envie,
> et sera si tres sages hon
> qu'il n'avra cure de Reson,
> qui mes oignemenz het et blasme . . .
> Cist avra le romanz si chier
> qu'il le voudra tout parfenir,
> se tens et leus l'en peut venir,
> car quant Guillaumes cessera,
> Jehans le continuera,
> enprés sa mort, que je ne mante,
> anz trespassez plus de .XL., . . .

(10535-10544, 10554-10560)

The God of Love urges that Lucina preside over this great poet's birth, and Jupiter over his upbringing. It is neither possible that Guillaume de Lorris dreamed his own death and that the book he had not yet conceived of writing would be continued by someone else; nor is it likely that Jean could have discovered the final outcome of Guillaume's dream forty years after the latter's death. In an extraordinarily "modern" use of point of view, with comic *brio*, the God of Love is fully conscious that he and his army are characters in a book, and that the events they "live" occur years before the author's birth, so that Jean undermines the privileged voice of Guillaume's I-Narrator and establishes distance between himself, the Author, and that Narrator. Jean also willfully undermines the illusion of reality, the temporary suspension of disbelief, which is the hallmark of traditional mimetic fiction.

The Narrator is still an active agent in the story, and his participation contributes drama and immediacy, as in all first-person narratives; but since he lacks the support of the Author's authority, we are neither expected to identify with him nor to espouse his views. He is reliable enough as a Narrator (*erzählendes Ich*), but not at all as a character (*erlebendes Ich*). He tells the story accurately but does not control our judgment nor mould our beliefs. We see beyond him (as a character) and gain insights into the story and the doctrinal *bellum intestinum* that he consciously does not. A host of other characters deliver long speeches that are filtered through the Narrator's consciousness, and which, although recounted to us as they were purportedly told to him, are so long that we forget the filtering process and imagine that the "Jealous Husband," "False seeming" or "Genius" address us directly. In the Narrator's absence (e.g. during la Vieille's lecture to Bel Accueil, Nature's confession to Genius), the presence of the original *erzählendes Ich* is so unobtrusive that we can fall into the illusion that he is or has become an objective, omniscient third person *Erzähler*.

Because narrative voice and point of view vary continually, Jean de Meun's *Roman* can be considered an excellent early example of polymodality: a story told by an I-Narrator, who is at various times hero, witness, or a quasi-omniscient outsider (mock-author). The *Roman* is also delegated to a series of secondary I-Narrators, who sometimes focalize it through their consciousness. Yet, the poem contains so much dialogue that no real internal focalization is possible. In fact, these insertions, these delegated voices provide such immediacy and drama that the Narrator's own point of view can appear objective, indeed omniscient, in contrast. Furthermore, since each of these other narrators, these stock characters, speaks in his own voice, and with equal vehemence and authority, we cannot assume that Jean de Meun necessarily agrees with any one over the others. Indeed, if by chance we have yielded to the rhetoric, he creates distance by undermining them once their speeches are over. As in real life, no commentator leads us by the hand through Le Roman de la Rose, although the reader eventually knows more than any of the characters, including the Narrator (as Lover). Jean fails neither to establish norms nor to affirm his own beliefs, but he does so indirectly, obliquely, by showing his "villains" to be even more absurd or ridiculous than his "heroes," which inevitably results in a certain amount of confusion. It is up to the reader to judge each character in turn, as he involuntarily, unconsciously, naïvely reveals his own shortcomings, blatantly holds forth logical inconsistencies, and sophistically misinterprets the very classical auctores he cites so badly. It is the reader's job to analyze facts and motivations, cause and effect, mind and rhetoric, whether he wants to or not. These shifts in perspective, the presence of multiple points of view, and the fact that each character speaks in his own person, expressing his own opinions, make it difficult for us to identify Jean de Meun's stand on any given issue. The "ideological content" of Le Roman de la Rose is a more complex, subtle matter than most people realize.

✑❧ ☙

Now a few words about Machaut. His first and esthetically least successful tale, Le Dit dou Vergier, is based directly upon the precedent of Guillaume de Lorris; however, since Machaut also learned from Jean de Meun, his Narrator and narrative technique differ from Lorris's. The Narrator is depicted as a young, innocent boy who has just fallen in love and seeks instruction. Being a child, he possesses the purity and enthusiasm required of a perfect lover, but he also suffers from physical weakness and intellectual immaturity and is revealed to be timid, foolish, and ignorant. At first timorous when he beholds the God of Love, the narrator shies from the latter's torch, lest it be thrown at him, and dares neither to advance nor retreat. Still afraid, yet driven by insatiable curiosity, he advances toward the dream-figures slowly, ". . . le petit pas / Tout couvertement . . ." (208-9) and finally asks the God of Love to reveal his identity and that of his court as well as for an explanation of his blindness, his torch, and arrow.

The God replies in two parallel discourses, but before the second speech and, therefore, before all the questions have been answered, the Narrator repeatedly interrupts, begging tearfully for intercession in his personal affairs. The God replies kindly but firmly that further queries must wait till later; he will continue his lecture in proper order. When, finally, his tears metamorphosed into joy, the dreamer wakes up quite suddenly to discover that the dream figures have disappeared, he is "esperdus" and "en moult grant effroy" (1212, 1213). Viewed from a Bergsonian perspective, the Narrator's curiosity, fear, and egocentric personal concerns become fixations. He reacts mechanically to his environment, shows excessive fear when none is called for, and later demonstrates both foolhardiness and bad manners. This individual's emotional needs are thwarted by the exigencies of social decorum. He is neither all good nor all bad, but rather simply inept and unable to adapt to the ways of the court.

At the end of the poem, the Narrator has acquired knowledge which, together with the good qualities he already possesses, should enable him to succeed in his love-quest. We use the term *should* advisedly. Machaut's first long poem manifests considerable ambiguity. Differing from Guillaum de Lorris, the Narrator cannot be identified with the archetypal Lover, of whom the God of Love speaks, even though the later is, in a sense, a projection of the former. The Lover meets a lady, desires her, declares his passion, and is rewarded all in the course of the God's speech. Such is not the case with the Narrator. He fell in love before the poem began; what will happen to him after line 1293 is left open to conjecture. The God of Love promises success to all *vrais amis*; he will help the Narrator provided the latter is faithful and discreet. We have only one indication of the Narrator's loyalty: before ever meeting the God and again at the poem's end he proclaims in his own voice as Narrator how faithful he is. Concerning his discretion, we are not informed at all. True, he does refer to the lady in the vaguest possible terms; but, on the other hand, he also tells the story of *Le Dit dou Vergier*. Within the grove the Narrator is instructed in the traits required of a good lover, but whether he possesses such traits or is successful in his pursuit remain unknown. The *dit* ends on a note of expectancy.

The blurring of focus in *Le Dit dou Vergier* between the Narrator conceived as a participant in the action (a hero telling his own story) and the same person portrayed as a neutral recorder of speeches delivered by the God of Love is resolved by Machaut in his second tale, *Le Jugement dou Roy de Behaingne*, by splitting his protagonist in two. Here the Narrator overhears a "debate" between a Knight and a Lady as to which is the more miserable in love, and helps adjudicate between them. The Knight appears as the subjective, passionate, emotionally involved lover; characteristics of the fumbling, eavesdropping onlooker are delegated to the Narrator-witness. However, tthe Narrator declares that he also is a lover (11-12, 2067-79) and speaks of his passion for a lady, which, though unrequited, brings him joy. This mutual bond of frustrated passion makes the Narrator especially qualified to console his friend, the Knight; in a sense the Knight is the Narrator's alter ego or surrogate. The parallelism corresponds to the one in *Le Vergier*. We are, however, told of the Narrator's sentimental problems only at the very beginning and end of the story; elsewhere he acts in antithesis to the Knight: uninvolved, passive, seemingly independent, in contrast to the Knight's spontaneous commitment to passion. He remains a spectator. It is the Knight who manifests the traits of the good courtly lover that we would normally expect to find in the Narrator. The Narrator has projected onto the love-sick Knight and then, in a burst of wish-fulfillment, finds a solution to the Knight's problems by introducing him to King John of Bohemia.

Machaut's originality lies in the creation of an I-Narrator who witnesses the story but does not himself play a leading role in it. He maintains the illusion of personal involvement and authenticity present in all good first-person narratives (*Le Roman de la Rose, La Vie de Marianne, A la recherche du temps perdu*). Thus, for example, the Narrator intervenes after the Lady's first speech to assure us that he personally saw her fall down as if dead (206-8). Yet, by depicting the Narrator as an onlooker, Machaut also creates distance and a greater sense of objectivity, the hallmark of good third-person narratives. The Narrator's obtrusive intervention and the fact that his story is is filtered through the understanding of others reminds us that *Le Jugement dou Roy de Behaingne* is fiction and not confession or reportage, that the Knight and the Narrator exist for us as characters in a work of imagination.

Machaut conceived the Narrator as a witness in order to solve a problem that often obtrudes in *Ich-erzählungen*: how can a nonomniscient narrator be expected to know all the events he recounts? Machaut skillfully creates a plausible witnessing-scene. The Narrator convincingly tells us how he lay down in the garden to listen to a bird sing, hiding lest the bird take fright. Hence he was in a position to observe the Knight and Lady without damaging his own moral character, and they talk freely, unaware of his presence. The Narrator, discreet court poet that he is, remains hidden because he thinks that they are lovers come to a rendezvous (53-5).

Finally, splitting the protagonist gives Machaut an opportunity to emphasize humorous traits in the Narrator while maintaining the Knight as a figure of tragic disappointed love. Machaut's narrator-witness is placed in the humiliating situation of an eavesdropper overhearing secrets of the heart as he lies half-buried in the leaves and grass. Once the Knight and Lady have told their stories, he would like to make his presence known but dreads embarrassing the litigants, revealing that he had been spying on them. Even when the Narrator decides to act, his problem is resolved by an outside agent. The Lady's little dog resolves the dilemma by leaping at the intruder, barking furiously and biting at his robe. Pleased at this accident which gives him a conversational opening, the Narrator brings the animal back to her. This creature recalls, of course, Petit-cru and Husdent in the *Tristan* romances as well as the Châtelaine de Vergi's *chienet afetié*. In humorous contrast to these other animals, Machaut's dog barks gleefully, tears people's robes, leaves his mistress's side, and offers no consolation to her grief. Although he apparently plays no role in her amours, he does serve as a mediator, not for her lover but for a perfect stranger, an eavesdropper who will, incidentally, bring about her condemnation. The archetypal friendly beast who leads the hero into the Other World is transformed into the merest trifle of a domestic pet and the romance hero into a bumbling busybody, good only to advise others.

In conclusion, I shall discuss only one other of Machaut's poems, *Le Voir Dit*, in which an elderly poet-narrator engages in a fascinatingly ambiguous love-affair with a young girl he chooses to call *Toute-belle*.

Machaut tells his story in the first person, through a narrator who is also the protagonist and a lover. Except for the letters and poems ascribed to Toute-belle, *Le Voir Dit* is told from the Narrator's viewpoint; he is the central focus. Although an I-narrator will often elicit from the reader sympathy and a heightened emotional reaction, he cannot create the illusion of omniscience we find in most third-person narratives. Aware of this, Machaut has the Narrator explain that he was informed of certain events by Toute-belle's confidante or by the girl herself.

For the first time in the history of French fiction the Narrator's limited perspective has an important function in the plot. If we believe his truth-claim, accept his norms, and allow his point of view and ours to coincide, we must then agree with his version of the story. However, the Narrator is not necessarily reliable nor are we obliged to accept without question his interpretation of the events he recounts. We have the right to disagree with him. We know the Narrator's interpretation of events but not that of Guillaume de Machaut the poet, for whom the Lover is a literary character just as is Toute-belle. This varying of focus is the key to the tale's structure. Illusion is taken for reality and vice versa. Truth can be revealed through illusion (a dream); or perhaps a lie is told in seemingly truthful terms and given the authenticity of a dream-vision. Narrative omniscience is totally out of place in a story that reveals the Narrator-hero's lack of omniscience. Ironically, in *Le Voir Dit*, The True Story, neither the protagonist nor the reader ever succeeds in unravelling the *Voir Dit* mystery.

As in the case of Jean de Meun, knowing no more than the Narrator, we perceive his weakness and vacillation. We do not see the reality behind Toute-belle's mask (her portrait, letters, and dream appearances), but we recognize that it is a mask and realize that the Narrator is incapable of distinguishing between it and reality. Regardless of the true state of affairs, the Narrator demonstrates a crushing lack of trust. His tragedy lies not in the Other but in himself, and the ultimate truth of The True Story concerns not his external relations to another (over which he agonizes) but his inner self, of which he is almost totally oblivious. For a master of *fin' amor*, a specialist in the ways of the heart, he is a most unaware, unlucid, and inauthentic individual. In this sense surely the reader discovers a "truth" the Narrator never dreamed of and arrives at a point of knowledge far beyond the Narrator's.

The Toute-belle perceived by the Narrator inevitably differs from the real one, whom neither her nor the public ever gets to know. She is his inspiration, his Muse, but as such takes on a universal, not a particular, aura. He conceives of her as the *domna* of tradition, not a living fourteenth-century girl less than 20 years old. He writes his best poetry when they are separated, perhaps unconsciously seeks obstacles to keep them apart, in order to live up to the *amor de lonh* (love-from-a-distance) convention and because the reality of Toute-belle's presence cannot but interfere with his idealized picture of her and silence him. Significantly, in the second half of *Le Voir Dit* her portrait comes to replace the real girl. Just as Toute-belle is dehumanized in the relationship, so too in the Narrator's mind is she metamorphosed into an object (the portrait) and a phantom (who comes to him when he dreams), onto whom he projects fantasies at will. Perhaps he loves only himself or the idea of love and projects his self onto the Other, finds himself reflected in her, is eager to be loved in order to proclaim his personal value, to write and then be reflected in his reader. In any case, the "I" and the "Thou" as present entities, potentially capable of an authentic, total relationship, are here separated and as a result relate to each other as "I-It," as subject to object.

For the elderly poet-lover of *Le Voir Dit,* the Lover is a created character, a narrated self who exists in words not historical fact; he comprises not only Guillaume de Machaut, the author, but his own other half, the *erzählendes Ich* who composes their story. Sending portions of the book as they are completed to Toute-belle, the Narrator attempts to mold Toute-belle's (and the reader's) interpretation of the events he has just lived through. Curiously enough, the same is true for the Narrator and Toute-belle as poets and correspondents: their various lyrics and prose epistles are written specifically for their implied readers and for the public at large, to create a calculated effect on others. The entire book and all it contains assume the existence of implied readers called upon to witness the fictional selves of the purported authors (Toute-belle, the Narrator as lover, the Narrator as narrator); each seeking to impose his own vision of the self on the Other and on the public, they are therefore inevitably guilty of conscious or unconscious bad faith.

Machaut's greatest triumph as a realist, and as a narrative poet, may well be the new literary type he made his own: the inept, blundering narrator, who is also an inept, blundering lover. This pseudoautobiographical character is prone to cowardice, sloth, snobbery, misogyny, and pedantry. Guilty of excess, unable to cope with everyday social life, obsessed by his failings, he acts in a delightfully comic manner, in contrast to the elegant gentlemen and ladies of the court. For the first time in French literature the fool has become a protagonist of serious *belles lettres;* Machaut's development of the *Roman de la Rose* tradition was to have a profound influence upon his most gifted successors, Froissart and Chaucer.

Sometimes in Machaut a lover recounts his experiences directly; sometimes they are told by a witness-narrator. As with the great eighteenth-century novelists, Machaut's narrator may participate actively in the story or withdraw from it; he can be reliable or unreliable. Guillaume pioneered in the development of a more sophisticated narrative technique and gave certain themes (point-of-view, illusion-versus-reality) a complexity seldom to be found in early fiction.

Machaut's triumph is of great significance for our esthetic appreciation of late-medieval narrative, and for the reevaluation of literary history. The question of narrative technique is of concern to modern criticism: witness the pioneering work by Wayne Booth, Franz Stanzel, Françoise Van Rossum-Guyon, and Gérard Genette. The examples of Jean de Meun, Juan Ruiz, Machaut, Chaucer, and others, prove that a highly sophisticated, literary use of point of view and narrative voice is not the invention of Robbe-Grillet, nor of Faulkner, nor Proust—and not even of Marivaux, Prévost, Fielding, Sterne, or Diderot. Fiction itself is a long-evolving mode with its first masterpieces embodied in the Middle Ages.

THOUGH far more medieval art has been lost than has survived the vagaries of the centuries—war, fire, neglect, and destruction—vestiges persist because of happy accident or conscious acts of conservation. The remarkably "complete" preservation of Machaut's works is due, in part, to his artistic pride and his active role in having his works copied and set in order. In the Prologue that opens two of his most complete manuscripts, Machaut exalts the role of the artist and discusses the aims and theories governing his works. Passages of his *Voir Dit* show the author working at the production, and reproduction in the form of copies, of lyrics and books, while collecting his works in a master copy of his own. In both his music and his poetry Machaut's keen sense of genre, and of style-separation are reflected in the arrangement of his manuscripts. The index to one manuscript claims to follow the order in which Machaut wished his works preserved, giving prominence to his musical compositions. In an age in which few artists took such pains to assure their artistic immortality, Machaut's actions and achievements were exceptional—and for us, spectacularly fortunate.   MPC

Machaut's Self-Awareness as Author and Producer

SARAH JANE MANLEY WILLIAMS

DePauw University
Greencastle, Indiana 46135

About Machaut's famous contemporary, the Italian humanist, Francesco Petrarch, it has been said "we know far more about his experiences in life than we know about the experiences of any human being who had lived before his time." (p. v)[1] Although we may not know nearly as much about the life and experiences of Guillaume de Machaut, we can be sure that more of his music has come down to us than of any earlier composer. Why, we may wonder, has his music been so extensively preserved, when many works by the equally famous musician, Phillipe de Vitry, have been lost? Why Machaut's work when not until several centuries later, in the age of music printing, is it usual for works by a single composer to be collected instead of being merely anthologized? There are several factors operating in Machaut's case. One of them is surely the patronage of the wealthy and powerful noblemen to whom many of Machaut's poems were dedicated, men famous for their culture and learning as well as for their political skill and military might. Another factor is Machaut's stature as *both* a poet and a composer—and until very recently, he was probably better known as a poet. But it is the result in large part of Machaut's own pride and self-awareness as author of both poetry and music, and his consequent concern for their production and physical transmission, that his works have been preserved with such completeness.

Guillaume de Machaut's own conception of his artistic role emerges most clearly in the later part of his career: in the Prologue[2] to his complete works, in his long, fictionalized autobiographical poem, the *Voir Dit*,[3] and in the heading and index that introduce one of the most complete manuscripts of his works. The Prologue shows most strongly Machaut's sense of self-importance, and his exaltation of the artist's role. From the *Voir Dit* we can form a picture of the author's day-to-day working methods, and of his concern with the copying, production, and transmission of

0077-8923/78/0314-0189 $01.75/2 © 1978, NYAS

his works. The index, said to indicate the order he wished his works to have, typifies his awareness of genre, and his keen sense of the separation of styles.

The two fullest and most complete of the Machaut manuscripts open with a Prologue setting forth the author's aims in the form of his conversation with the personified figures of Nature and Love, who present their allegorical offspring. (The famous miniatures showing the author greeting these wondrous, superhuman beings have often been reproduced.) Nature's first speech, in the form of a ballade, commands the poet (line 5) to compose "pleasant new tales of love," ("nouviaus dis amoureus plaisans" —the word "dit" is used by Machaut of both his long and shorter poems) and for this purpose she presents him with her three children, Sense, Rhetoric, and Music. Guillaume answers Nature in a second ballade, whereupon Love, who has overheard the exchange, presents the poet, in another ballade, with *his* three children, Sweet Thought, Pleasure, and Hope; and again the poet responds. These four ballades make up the first part of the Prologue. A section in octosyllabic couplets (Machaut's usual narrative medium) follows, in which Machaut, meditating on these gifts, enumerates the forms in which he will compose, elaborates a moral and psychological theory of art, and concludes with a final affirmation of his resolve to fulfill the tasks set him by Nature and Love.

Within this allegorical framework, Machaut exalts the claims of the artist. As a creator, the author is compared to the great goddess Nature herself. She announces (lines 1-5):

> I, Nature, by whom all things are formed
> That are on land or sea
> Come here to you, Guillaume, whom I have formed
> Apart; and for this end, that you may form
> Pleasant new tales of love.

("Je, Nature, par qui tout est fourmé/ Quanqu'a ça jus et seur terre et en mer,/ Vien ci a toy, Guillaume, qui fourmé/ T'ay a part, pour faire par toy fourmer/ Nouviaus dis amoureus plaisans.") The theme of creation is underlined by the repetition of rhymes built on *fourmer*, to form. Machaut proudly suggests an analogy between Nature and the poet whom Nature has set apart from others by endowing him with powers like her own. His works will therefore, she assures him, be more famous than those of others, blameless and loved by everyone; and she urges him to make enough, small, medium, and large, with the help of her offspring:

Sense, who will make him clever, Rhetoric, who will not fail to instruct him in matters of meter and rhyme, and Music, who will give him songs, as many as he likes, various and pleasing (lines 10-14).

Several comments need to be made here. The figure of Nature personifies the creative force, rather than the created, *natura naturans* as distinct from *natura naturata*, in the scholastic terminology. (There is a large literature dealing with the medieval conception of Nature.[4-9] The musical implications of Nature's activity are pointed out by de Bruyne,[9] Vol. 2, p. 276, ff.) Nature as a creative force had played an increasingly important role in the metaphysical conception of the cosmos from the twelfth century on, through the influence of the school of Chartres. She was thought of as an extension of the creative power of God (*Dei vicaria*) but, at the same time, tended to assume a somewhat independent position. The further comparison of Nature's activity to that of the artist was familiar to the philosophers. Nature, as patroness and exemplar for the secular artist, expects his works to be both abundant and new, their novelty being associated with their formal qualities. In the underlying meaning of the allegory, Machaut owes his powers to natural gifts (the gifts of Nature) rather than to theoretical mastery of skills.

When the gifts of Nature and Love are put to work (Part 2, lines 12–17), they will aid the author in his production of "tales and songs, double hoquets, pleasant lais, motets, rondeaux and virelais, complaints, ballades, in honor and praise of all ladies." The list mentions every genre cultivated by Machaut, with the single important exception of the mass, which does not fit in with the secular aims stressed in the Prologue. He vows to devote feeling and understanding to this task, one which is conducive to virtue and opposed to vice, for no task could be better suited to making a man noble, cheerful, lively, frank, elegant, and polite (lines 21–29). This Guillaume can bear witness to from his own experience, for when he is in this state, his only thought is the making of an appropriate poem or song (lines 36–40).

The poet, moreover, should be joyous. Even if his subject is sad, his manner must be gay, for a heart full of sadness cannot sing gaily. As a lover, the poet is fortified by Pleasure, Sweet Thought, and Memory, who recalls to him the image of his beloved. The behavior of the melancholy man ("li tristes"), on the other hand, is to be censured, nor could *he* possibly create anything so prettily. Finally, the very nature of Music requires the artist-lover to be joyful. "Music is a science," says Machaut (lines 85–88), "which asks that one laugh, and sing, and dance. It does not care for melancholy, nor for the man who is melancholy." ("Et Musique est une science/ qui vuet qu'on rie et chante et dance./ Cure n'a de merencolie/ Ne d'homme qui merencolie.") Wherever Music is, she makes men rejoice.

This mention of Music leads naturally to a further discussion of her powers. All musical instruments are formed according to her laws, and her works are more perfectly proportioned than any others (lines 99–100). ("Tous ses fais plus a point mesure/ Que ne fait nulle autre mesure.") In courtly society she presides over the dance, while in the divine office, what could be more noble than to praise God and His sweet Mother in song? Not only men, but saints, angels, and archangels praise God in song because He has set them in glory, and they see Him face to face. David and Orpheus, the great traditional embodiments of the power of music, are the final examples in this section: David, who could appease the wrath of God by his playing and singing; Orpheus, at whose playing and singing the great trees bowed down, and the rivers turned back to listen. The prominence given to Music in this second part of the Prologue is very striking. After 60 lines devoted to her praise, a mere 20 are allotted to Rhetoric, Sense, and the three children of Love; and only Rhetoric, whose ingenious rhyme types are enumerated, receives more than a brief notice.

Some ideas of the Prologue may seem puzzling to modern readers unless we understand something of their background, particularly the insistence that the poet must be joyous. Machaut's tone, especially in his lyrics, is more often melancholy. The apparent contradiction is rooted in the familiar paradox of "sweet suffering." Love's arrows may wound, but, as Machaut says (lines 2831–2832)[10] in another poem, "Sweet is the wound, and pleasant the sting." Since the rewards of the servant of Love will be in proportion to the torments he endures, he should even welcome the prolongation of his suffering (Jeanroy,[11] p. 98). (Jeanroy notes the similarity of courtly and Christian thought on this point.) Already in the writing of Guillaume d'Acquitaine, the first troubadour, the exaltation of love is called *joi*. (Jeanroy,[12] p. ix) This joy, furthermore, is naturally expressed in *song*. It is externalized in song because the essential nature of joy was traditionally thought to be *musical* or *harmonious*, and is explained (de Bruyne,[9] Vol. 2, p. 147) as the result of the encounter of two harmonies. (To Saint Bonaventura, for instance, in more technical terms, "Every spontaneous and unreflective delight presupposes as an objective condition the union of the subject with a delightful object which is good and beautiful, a perfect harmony between these two terms. . . . Moreover, every union of a subject with an object is founded on the love of the subject for the object; without love, there is no delight. . . . All joy leads to song. When a man loves a woman, he composes verses and songs." [de Bruyne,[9] Vol. 3, pp. 195, 196] This harmonious nature of love is symbolized in French fourteenth-century art by representations of lovers together with an instrumentalist, or lovers making music together [von Marle,[13] p. 466].) If joy is musical, melancholy is just the opposite, a disorder or pathological state that is fundamentally *unmusical* (21–23).[14] (Medieval medical treatises prescribe music as a cure for melancholy.)

Machaut's emphasis on the virtues with which the artist-lover is endowed, and on the ennobling power of love, suggests another link between love and music. In the ancient classification of music into *mundana*, *humana*, and *instrumentalis*, "human music" included the moral and psychological harmony of the soul, the physiological harmony of the body, and the harmony between body and soul (as in the classification of Hugo of St. Victor [Pietzsch,[15] p. 76]). Since pleasure is defined as the encounter of two harmonies, it follows that the capacity for musical enjoyment will depend on the moral and psychological equilibrium of the listener. In music the soul must recognize its own harmony, and immorality precludes the enjoyment of music.

The Prologue is remarkable not only for Machaut's views of the artist's role, but for its very nature as a prologue, an introduction to the author's complete works, envisaged as a unified whole that is governed by constant aims and principles. As its editor noted, in form as well as content it is a summary of the poet's work. To quote Hoepffner[2] (pp. liv, lv):

> The ballades represent his lyric poetry, the section in octosyllabic couplets, his narrative and didactic poetry; there are personifications borrowed from the *Roman de la Rose* and "examples" taken from the Bible or from writers of antiquity. . . and it is the poet himself, named outright, who occupies the foreground of the scene and who informs us of his ideas and personal feelings. And these are exactly the three principal elements of Machaut's poetry.

The clearest picture of Guillaume de Machaut involved with the copying of his works emerges from the *Voir Dit*, as I tried to show in an earlier article in *Romania*, to which I would refer you for a more detailed discussion.[16] That fascinating "True Story," as he called it, repeatedly insisting on its "truth," tells of his love affair with a young noblewoman who especially admired his music. The *Voir Dit* includes, inserted at various points in the rhymed narrative, what amounts to a small anthology of lyrics—over five dozen—said to have been written and exchanged in the course of the affair. And most informative, as evidence about the transmission and preservation of Machaut's works, are the prose letters inserted in the *Voir Dit*, making up an extensive correspondence. (I cannot seriously entertain William Calin's suggestion: "It is possible that Machaut first composed the letters, both the Narrator's and *Toute-Belle's*, and later fitted them into a frame [p. 170]."[17] A series of the early letters is written into the narrative in the wrong order. The suggestion that a poet usually so orderly and so proud of his art would have invented the letters and then disarranged them, creating such confusion for his readers, runs counter to common sense. I have argued elsewhere that traces of the lady's work and her independent opinions can be found in the *Voir Dit*.[18])

In sending lyrics to his lady, Machaut requests that she not circulate copies of what he sends her, for he is thinking of making music for them (Letter VI, p. 54.)[3] The cautionary request suggests that the practice of copying lyrics may have been common. Machaut would like to preserve for a little while that novelty he praises so highly in the Prologue. There are other passages in the letters to suggest that exchanges of poems and songs were not uncommon among the aristocratic society of the time, those few with the learning and leisure to enjoy music and poetry. Later in the correspondence Machaut writes to his lady that several great lords who have learned of their love have ordered him to send "some of your and my things," especially the versified fan letter which first declared her love; and he reports that he has complied with the order (Letter XXV, p. 189).[3] It was probably individual copies of songs, like those which Machaut tells of sending to his lady in the *Voir Dit* (sometimes with comments describing circumstances of their composition) that found their way, often anonymously, into the musical anthologies of the period. But circumstances would not favor their preservation. Only one isolated lyrical work has been found, an anonymous copy of the *Lay Mortel*, with its music, which turned up in a binding (p. 14*).[19]

In an early letter (p. 21)[3] Machaut asks his lady to let him know, "which of my books you would like: I will have it copied." A little later he has his latest work, *Morpheus* (also known as the *Fonteinne amoureuse*) copied for her, and sends it, along with a composition, the ballade on the refrain, *Le grant desir que j'ay de vous veoir* (pp. 53, 69).[3] He also reports on the various stages in the writing of the *Voir Dit* itself. Its composition goes by fits and starts. When his work is going well, he writes a hundred verses a day (p. 202).[3] But at other times the activities of his noble patrons distract him from the task; or he becomes discouraged by not hearing from his lady, and stops writing because of lack of material (pp. 262, 342).[3] As he is working on the book, he makes several professional estimates of its finished length. It will be three times as long as *Morpheus* he predicts (p. 241).[3] As he nears the end he gives a more closely calculated estimate (p. 363): "As for your book, it will be finished, if God pleases and I am able, within a fortnight. And it will take up approximately twelve quires with forty lines. When it is finished, I will have it copied, and then I will send it to you." In one manuscript of the *Voir Dit*, which corresponds to this description in being ruled with forty lines to the page (the text is arranged in double columns), the work runs slightly short of filling eleven quires, although it *is* three times as long as *Morpheus*.

We happen to know, through a ballade that Eustache Deschamps addressed to his master, Machaut (pp. 248-49),[20] of the presentation of another copy of the *Voir Dit* to Louis de Male, Count of Flanders. But again, as with the lyrics and songs, the passage of time has not been favorable to the preservation of isolated copies of narrative poems, although a copy of one, the *Dit dou Lion*, has been identified (p. 7*).[19]

Most important of the *Voir Dit* passages are those that refer to Machaut's own copy of his complete works. "I would have brought my book to divert you," he writes to his lady (p. 69), "where all the things I ever wrote are; but it is in more than twenty pieces, for I am having it done for one of my lords, and so I am having it notated, and for that, it is better for it to be in pieces." It has been suggested (p. 294)[21] that this book might be identified with the Vogüe manuscript. Indeed the *copy* Machaut is concerned with here might well be identified with that manuscript, which can be supposed to represent Machaut's repertory at the time of the composition of the *Voir Dit*. Later he refers again to this book, his master copy of his own works, as he sends his love the partly completed *Voir Dit*, warning her (p. 259)[3] to take very good care of it, since he has no copy, and he would be distressed if it were lost, and were not "in the book where I put all my things."

While Machaut writes in the *Voir Dit* of having a large manuscript copied for one of his lords, a manuscript possibly like the Vogüe manuscript, another more complete manuscript, Bibliothèque nationale français 1584, contains, on several unfolioed pages that precede its main body, the Prologue already discussed, and an index with the heading: "This is the order which G. de Machaut wishes his book to have." ("Vesci l'orde-nance que .g. de Machau wet quil ait en son livre"—to quote directly from the manuscript.) The main difference in overall order between the Vogüe and the 1584 is that the former begins with the lyric poems with-out music, the collection known as the *Louange des Dames* (although this title comes from only one manuscript); while the latter begins with the narrative poems. Of the seven principal Machaut manuscripts the Vogüe is one of three to begin with this large collection of lyrics, while the 1584 copy, reflecting Machaut's final judgment, if we take seriously the rubric preceding the index, is one of four to present the narrative poems in first place. (There is a short form of the Prologue, consisting of only the four ballades, found in some manuscripts which begin with the collection of lyrics. See the table, p. 43*.[19])

As an author, in his composition of both poetry and music, Machaut had a keen sense of genre, and of the separation of the styles which he com-manded in an impressive variety. There are differences both of scale ("small, medium, and large," as the Prologue suggests) and of style among them. His monophonic works, to illustrate from his music, are very differ-ent from his isorhythmic motets, which, in turn, differ from the shorter, though equally complex, polyphonic ballades and rondeaux. The manu-scripts reflect this stylistic differentiation in grouping similar kinds of works together. And just as the Prologue seems to pay more attention to the musical than to the poetic side of Machaut's production, the index to the 1584 manuscript groups the several hundred lyrics of the *Louange* under the heading, "Les balades ou il n'a point de chant" (the ballades to which there is no music), but lists separately by incipits each of the *lais, motes, balades notes, rondeaulz* and *virelais* in the music sections. Preced-ing the very last entries of the index, there is a final notice: "These things which follow you will find in the *Remede de Fortune*," with a listing then of each of the seven musical compositions included in the *Remede*. (My point is somewhat weakened, I confess, by the fact that an index to an-other Machaut manuscript does have individual listings for the lyrics of the *Louange* as well [p. 12*].[19])

The heightened self-consciousness of the fourteenth-century artist is very evident in Machaut. He customarily proclaims his authorship by weaving anagrams into his longer poems.[22] Introducing himself as an ac-tor in his own works, he creates a myth of the artist. In the Prologue he exalts his accomplishments as an author and articulates the aims that govern his complete works. On a more practical level, he can be glimpsed keeping track of his output in the book that contains all his works, ordering copies made of single works—and of whole manuscripts— and concerning himself with the order in which his complete works should be arranged. It is to this pride and this care that we owe, finally, the very preservation of his music.

REFERENCES

1. Wilkins, E. H. 1961. Life of Petrarch. Chicago.
2. Oeuvres de Guillaume de Machaut. 1908. Ernest Hoepffner, ed. Vol. 1. Paris.
3. Le Livre du Voir-Dit de Guillaume de Machaut. 1875. Paulin Paris, ed. Paris.
4. Gelzer, H. 1917. Nature. Zum Einfluss der Scholastik auf den altfranzösischen Roman. Halle.
5. Berndt, E. 1923. Dame Nature in der englischen Literatur bis herab zu Shakespeare. Leipzig.
6. Parent, J. M. 1938. La doctrine de création dans l'École de Chartres. Paris.
7. Curtius, E. R. 1948. Europaïsche Literatur und Lateinisches Mittelalter. Bern.
8. Lewis, C. S. 1936. The Allegory of Love. Oxford.
9. Bruyne, Edgar de 1946. Études d'Esthétique Médiévale. 3 vols. Bruges.
10. Remede de Fortune. 1911. In Oeuvres de Guillaume de Machaut. E. Hoepffner, Ed. Vol. 2: 1-157. Paris.
11. Jeanroy, A. 1934. La poésie Lyrique des Troubadours. 2 vols. Paris.
12. Jeanroy, A., ed. 1927. Les Chansons de Guillaume IX duc d'Aquitaine. Paris.
13. Marle, Raimond von 1931. Iconographie de l'Art Profane au Moyen Age et à la Renaissance et la Décoration des Demeures. Vol. 1. The Hague.
14. Panofsky, E. 1923. Dürer's 'Melencolia. I'. Leipzig-Berlin.
15. Pietzsch, G. 1929. Die Klassifikation der Musik von Boetius bis Ugolino von Orvieto. Halle.
16. Williams, S. J. 1969. An Author's Role in Fourteenth Century Book Production: Guillaume de Machaut's 'livre ou je met toutes mes choses.' Romania, 90: 433-54.
17. Calin, W. 1974. A Poet at the Fountain: Essays on the Narrative Verse of Guillaume de Machaut. Lexington.
18. Williams, S. J. 1977. The Lady, the Lyrics, and the Letters. Early Music, 5: 462-468.
19. Ludwig, Friedrich, ed. 1928. Guillaume de Machaut: Musikalische Werke. Vol. 2. Leipzig.
20. Oeuvres Complètes de Eustache Deschamps. 1878. Le Marquis de Queux de Saint-Hilaire & Gaston Raynaud, eds. Vol. 1. Paris.
21. Reaney, G. 1956. Review in Music and Letters, Vol. 37, pp. 294-298.
22. Hoepffner, E. 1906. Anagramme und Rätselgedichte bei Guillaume de Machaut. Zeitschrift für romanische Philologie, 30, pp. 401-413.

As POET AND MUSICIAN Machaut searched for a complete harmony of all components of his works. With its space and its time exactly measured, the lyric poem was allied to mathematical laws of the universe. Through the magic of numbers, Machaut tried to control the "impetus of love" as well as the "movement of Fortune." Machaut's imaginary universe is first characterized by its scientific definition according to the medieval conceptions of order. But beauty, the principle of love, aroused a more concrete, visual imagination: light and colors focus the sight on flowers, animals, and human endeavors. The lyric text proposed comparisons; but poetry relying on allegory expressed the relation between the "outside" and the "inside" worlds. The long poems, transgressing the mathematical proportions of the lyric, are based on a more figurative imagination in which harmony becomes analogy. In Machaut's narrations and descriptions, the "eye" of the narrator and the "ego" of the lover finally find place in the middle of an imaginary universe. In this music and poetry are noteworthy similarities to *mimesis*, imitation of reality, in painting of the time.   MPC

The Imaginary Universe of
Guillaume de Machaut

DANIEL POIRION

Institut de Littérature Française
University of Paris-Sorbonne
Paris, France

W HEN Machaut began to compose his polyphonic songs, much medieval music aimed rather to intellectual than to sensual effect. Mathematical laws of the universe, with their arithmetical and geometrical
figures, governed composition. Through musical rhythms human beings
might participate in the metaphysical reality of the world. Musical knowledge gives the poet a magic power. This is why Orpheus is mentioned
with David by Machaut in his poetic Prologue, injecting a kind of magic
into an Aristotelian conception of the universe, derived from a Pythagorean myth: "those are visible miracles that Music does."[1]

As for the text of the poems, we must observe that the formal structure provides the language with a parallel to the system of the universe.
The ballad looks like a trefoil, a triangular shape; the rondeau has a small
circle inscribed in a larger one; the virelai has three circles bound together.
So the progression of the text is analogous to various movements in the
universe, including the revolution of planets and stars. "My end is my
beginning and my beginning my end," says a rondeau, giving an important key of both meaning and form to Machaut's poetry, while we recognize a geometrical and a cosmological interpretation of life.[2] But the most
typical form is the motet, which distributes the text among different levels,
or we could say different spheres, the story of the lover circling about a
religious prayer. For example, over the *tenor* "Speravi" a first monologue
exposes the decision to endure the suffering of love until death, while the
most elaborate text deplores the fading away of hope.[3] The polyphony reflects the cosmos' hierarchy, with its homocentric spheres. Man himself is
subject to similar movements, like a mechanical clock (which fourteenth-
century technology helped to build in some medieval cities).

The same relationship between macrocosm and microcosm is illustrated
in the astrological calendar of the *Très Riches Heures du Duc de Berry*.

[199]

0077-8923/78/0314-0199 $01.75/2 © 1978, NYAS

We can perceive analogies in forms and figures between the astrological signs and certain realistic details in the foreground. In "January" the man pouring out a drink recalls Aquarius, while the curve of a little vessel on the table recalls the shell of Capricorn. In "February" the legs of the woodcutter are similar to the arrangement of the fishes. In "March" the horns of the oxen and the plow stocks, inspired by the Ram, announce the Bull. In "April" the feathers of a lady's hat have the same curve as the Bull's horns. May's couples at the party look like Twins; in "June" there is geometrical relation between Cancer and the scythes. In "July" a man's hood and his sickles have the same curve as the sign of Cancer. In "August" the fringes of a dress look like the palm branches of the Virgin; in "September" the swaying of the Virgin's hips recalls the gait of a country-woman. While the "October" scarecrow has the menacing spike of the Scorpion, the Balance is suggested by the gesture of the sowing man and by the stone on the harrow. So the art of painting reflects the universal harmony.

But we may go farther. In the well-known "astrological man" page of the same manuscript, a naked man is represented near another one, the medical figure of *homo venarum*; consequently we understand that each part of the human body is related to a zodiacal sign, and that the day to choose for a blood-letting is decided upon by the moon's phases as indicated on the calendar. Not surprisingly, then, the poet will cure the lover of his pain as a physician would attend to his patient. Through the magic of numbers the lover should control himself by conforming his desire to the cosmic order. The projection of poetry into a space and a time defined by music effectively realizes what the text announces: the transmutation of pain into joy, and the sublimation of erotic desire. So the poet must speak and act at an auspicious time.

Machaut's poetic language is well suited to this ambiguous function. We find, for example, many enumerations of substantives, designating the components of the feeling: "pains, sorrows, tears, sighs and moans."[4] There are also enumerations of coordinates referring to some special event: "A curse on the hour, and the time, and the day, the week, the place, the month, the year."[5] Numbers haunt the imagination, for instance in the hyperbole: "No more possible than to reckon the stars,"[6] or in the enigma in which letters have their numeral values: "And do you know what her name is? Thirteen, two times five, one together and eight and nine," which makes JEHANE.[7] Thus the imaginary universe is first characterized by a specific definition.[8] Lyric poetry proposes a cosmic and astrological vision of existence.

But the abstract idea of beauty, the principle of love, leads to a more concrete vision. With the theme of vision or "the look," of the eye through which desire penetrates the heart, the vision of physical nature reorganizes the poetry. The sun is seen with all its illuminating splendor. We perceive colors,[9] flowers,[10] bodies,[11] appearing in different series of emblems, comparisons, and allegories. Machaut exploits the resources of rhetoric bringing together different fields of nature, which the music would leave aside. Some kind of correspondence between various colors is to be found while contemplating the universe. Concrete analogy takes the place of the abstract harmony. The poetic language gets closer to the art of painting. The metaphor carries the vision from one part of the world to another, from one level of reality to another. The metaphor of water transfers the idea of pleasure or fecundity from the mineral realm to the vegetal, the animal, and the psychological one: "You are softly keeping my heart in the river of all pleasures, and in the pure spring of sweetness whose stream make me feel as happy as possible on earth."[12] The metaphor of fire unfolds the alchemical notion of refinement to explain the transmutation of pain into joy, thanks to the heat of desire: "This is why I agree to die with that pain which sets on fire my desiring heart, to please to the God of love, and to honor my lady."[13] Behind a metaphor that evokes the Phoenix we recognize the image of an alchemist.

But this rhetoric is also based on a scientific conception of the *mind*. The visual image is, etched (*emprainte*) in memory: "For the very sweet imprint of her image is in my heart so firmly impressed that the mark will stay forever."[14] The poetic analysis of the thought corresponds to the old theory of the three cells in the brain: they are (from the front to the back) *sensatio vel imaginatio, cogitatio,* and *memoria.* Inside the mind a small image of the lady exists, and, while separated from her, the lover will be able to turn his adoration and his desire toward that icon: "For day and night her colored face and her fine beauty which I desire so much through memory are manifested to me."[15] In the *Voir Dit* an artistic image, the portrait of the lady whom the narrator had not yet seen, plays the role of that icon, materializing it.[16] The poetic imagination is perhaps bordering on the iconographic style that will lead to a realistic portrait. But it is true that literature does not yet cultivate the art of portrait with all the details necessary to identify a person. The image of a woman is still alluded to but not described. In some cases, however, we find more concrete characteristics, such as when the poet deals with a mythological or an allegorical figure. The *segnefiance*, the interpretation of the emblems will follow the description. Machaut gives a portrait of the god of love, who himself explains "all the signification of the burning fire-brand, the iron-shod dart, and the beautiful wings."[17] Picturesque imagination, but without authentic *mimesis*: we enter the realm of allegory.

The allegory, whose poetic model is for Machaut *Le Roman de la Rose*, transforms and develops the traditional process of comparison. Personifications are associated with metaphors in order to communicate the philosophical message; dialogues comment upon the *narratio-descriptio*. We can follow that transformation when we consider the image of a plant: "root of pain,"[18] "love's plant,"[19] are just metaphors for lyric poetry. But in a long poem such as *Le Jugement dou Roy de Navarre*, using the same metaphor, an allegorical representation explains the change in the attitude and the feeling of a young lady who married a richer man than her young lover; she is like a growing and blooming tree:" this pretty damsel who was this scholar's friend was the charming graft, as a sweet young girl, planted in the large orchard of Love. There she had been brought up, spread with branches, covered by foliage, and finely adorned by flowers."[20] The gardener is glad to announce to the rich man that he may now enjoy the shade of the beautiful tree. And that is supposed to comfort the deserted lover because it is a law of nature. Such is the principle of the allegory: human behavior and feeling are explained by an analogy with the organization and the rules of nature. It is therefore important to observe and to describe the outer world in order to explain the inner world. If the circular movement of the cosmos and all the mathematical laws may be sufficient for interpreting human destiny in general, and even the perturbations of Fortune, the poet needs other tools to analyze the secrets of love, the events of life, the singularity of individuals, and to expose the comforting message of wisdom. But the exploration of a familiar world, of everyday life, of the surrounding nature does not tend yet to a pure *mimesis*, to the pleasure of committing a beautiful landscape to memory, to the delight of working with objects. It is only a detour to arrive at the real aim of art and literature, to the fundamental interest in writing: the expression of feelings, opinions, and ideas. The poet uses things only to formulate thoughts. The allegorical imagination is a means, not an end.

In that kind of allegorical composition animals play a significant role. This is a standard function of the bestiary: real or fantastic animals have a moral, a religious, or a philosophical significance that varies with the context but reflects the influence of primitive and childish imagination. This unconscious origin may explain that a more profound, deeply rooted sense occasionally appears in the allegorical text. For example, in *Le Dit de l'Alerion,* Machaut has the strange idea of using the metaphors of falconry to demonstrate the differences among four love stories, and their lessons. The behavior of the birds, the allerion, hawk, eagle, and gyrfalcon serves as a metaphor of love. In Machaut's poem, the hawk eats other little birds.[21] And by insisting on the parallel experience of love, Machaut helps us rediscover the sexual meaning of that symbolism.[22] As a matter of fact the Celtic imagination, as seen in the French *lais* and novels, was haunted by feminine creatures at the same time seductive, provocative, and possessive. This erotic obsession is confirmed by the reference to a strange habit of the hawk, which holds a little bird to warm up its feet: "Let us see how a lady might be compared to a sparrow-hawk which seizes a little bird in the evening to keep its feet warm."[23] He gives a courteous explanation, but the sexual suggestion is there! This is one of the interests of that imaginary universe: it may be related to some kind of unconscious message, a function the allegory now shares with the myth. Describing the world the poet is analyzing himself.

This personal perspective certainly is intentional when the poet, in *Le Dit de l'Alerion,* pretends to be using his own experience. The poet describes the havoc of the plague in 1349, at the beginning of *Le Jugement dou Roy de Navarre,* with an allusion to an attack of melancholy he himself was suffering. It is an autumnal disease, but this time the poet is confronted with a dramatic event, not with the cycle of the seasons and the corresponding humors. He gives a vivid picture, somewhat apocalyptic and visionary, of the epidemic. After numerous cataclysms the plague arrives:" the air became dull, dirty, black, and obscure, ugly and stinking, murky and purulent, so that everything in it was putrefied, and people began thinking their bodies were also putrefied, literally losing their color. For they were ill-treated, pale and sick; they had boils and abscesses which made them die."[24] Then Machaut confides with reader, recounting what he was doing during that time, and speaking of the precautions he took to avoid the contagion, cloistered in his house, until the rejoicing of people let him know that the air was pure again, and that he could go out safely for a walk:

> And when I saw people merry-making happily as if they had not lost anything, I stopped being frantic and regained my nature, opening my eyes and turning my face toward the air, which was so pure and sweet that it called me to leave my prison where I had spent the whole season.[25]

We discover then, at the beginning of this poem, two levels of the imagination: one is fantastic and visionary, the other is sincere and subjective. The poetic picture has a perspective.

This perspective accounts for all the originality and the modernity of the imaginary universe as conceived by Machaut. The scenes of life are no longer perceived and interpreted from God's point of view, at least not without referring to the human situation of the poet which defines the appearance and the meaning of things. We understand why Machaut explains, at the beginning of each long poem, where he was, how he saw and heard the characters he stages. For example, in *Le Jugement dou Roy de Behaingne* he hides himself in a bush: this invisible presence gives his story a kind of authenticity within the fiction of the poem. But the role of the Ego and its perspective are emphasized by his masterpiece, *Le Livre du Voir Dit*. The entire story is presented from the lover's point of view, and this lover is the poet, Machaut himself. This is not the abstract "self" of the courtly songs, but the concrete adventure of an old man in love with a young lady. Courtly songs are added to this story, which personalize the traditional experiences of the lover. The circumstances are accurately explained, and even if we are not convinced that all the details are autobiographical, it has, at least, as an artistic work, a definite perspective. The imaginary universe shows finally an evolution toward a phenomenological type of vision.

This evolution seems to fit with some trends of the new philosophy as illustrated by William of Occam, but also with a new discovery in the art of painting. A few years later, in *Les Très Riches Heures du Duc de Berry*, the Limbourg brothers, representing the month of October in their calendar, place the Louvre castle in a perspective which gives the impression that we are there, near the sowing man. All the objective relationships between forms and figures, which we have seen in the former paintings of the calendar, have a personal touch in this picture. The allegorical stage of the other scenes, surrounding the more precise architecture of various castles belonging to the duke of Berry, becomes a realistic landscape. Machaut's poetry signals that evolution. At the end of the fourteenth century the human imagination already foresees the modern world: a nature that human vision will try to explore, and that the human mind will define in the perspective of its own curiosity.

NOTES AND REFERENCES

1. *Oeuvres de Guillaume de Machaut*, Publiées par Ernest Hoepffner. Paris. 1908. Vol. 1, p. 10, lines 145-146 ("ce sont miracles apertes Que Musique fait").

2. Guillaume de Machaut. *Poésies Lyriques*, publiées par V. Chichmaref. Champion. Paris. 1909. Vol. 2, p. 575, XV, 1-2 ("Ma fin est mon commencement Et mon commencement ma fin").

3. "De Bon Espoir, de Tres Dous Souvenir Et de Tres Dous Penser contre Desir M'a bonne Amour maintes fois secouru." And the second part: "Puis qu'en la douce rousee D'umblesse ne vuet florir Pitez . . .", Chichmaref.[2] Vol. 2, p. 489-490.

4. "Peinnes, dolours, larmes, soupirs et pleins, Gries desconfors et paours de morir Sont en mon corps en lieu de cuer remeins." Chichmaref.[2] Vol. 1, p. 139, CXLVII, 1.

5. "Je maudi l'eure et le temps et le jour, La semainne, le lieu, le mois, l'annee Et les .ij. yeus dont je vi la douçour.". Chichmaref.[2] Vol. 1, p. 192, CCXIII, 1-3.

6. "Ne qu'on porroit les etoiles nombrer, Quant on les voit luire plus clerement, Et les goutes de pluie et de la mer, Et l'areinne seur quoy elle s'estent, Et compasser le tour dou firmament." Chichmaref.[2] Vol. 1, p. 209, CCXXXII, 1.

7. "Et scez tu comment on l'apelle? .XIII., .V. double, .J. avec lie, Et. .VIII. et .IX." Chichmaref.[2] Vol. 1, p. 191, CCXIII, 15-17.

8. *Le Jugement dou roy de Navarre.* Hoepffner.[1] Vol. 1, pp. 177-78; he then mentions the four elements and the four qualities: "Bien savoit la cause des choses Qui sont ou firmament encloses, Pourquoy li solaus en ardure Se tient, et la lune en froidure, Des estoiles et des planettes Et des douze les signes les mettes, Pourquoy Dieus par nature assemble Humeur, sec, froit et chaut ensamble, Et pourquoy li quatre elements Furent ordené tellement Qu'ades se tient en bas la terre, Et l'iaue pres de li se serre, Li feus se trait haut a toute heure, et li airs en moien demeure." (Pages 179-180, lines 1243-1256).

9. "Qui de couleurs saroit a droit jugier Et dire la droite signefiance, On deveroit le fin asur prisier Dessus toutes; je n'en fais pas doubtance. Car jaune, c'est fausseté, Blanc est joie, vert est nouvelleté, Vermeil ardeur, noir deuil; mais ne doubt mie Que fin azur loyauté signefie." Chichmaref.[2] Vol. 1, p. 235, CCLXXII, 1-8.

10. "Tout ensement que la rose a l'espine Se differe d'odeur et de biauté." Chichmaref.[2] Vol. 1, p. 23, VIII, 1-2.

11. "Tres douce dame debonnaire, Que nulz ne porroit trop loer De gent corps ne de dous viaire." Chichmaref.[2] Vol. 1, p. 129, CXXXV, 1-3.

12. "Car pris Tenez mon cuer, sans pensee vilainne, Tres doucement en flun de tous delis Et de douceur en la droite fonteinne Dont li ruissiaus toute joie mondeinne Avoir me fait." Chichmaref.[2] Vol. 1, p. 157, CLXXIII, 10-14. See also pages 378 and 406-414.

13. "Pour ce vueil bien morir de la dolour qui par desir mon cuer enflame, Au gre d'Amours et a l'honneur ma dame." Chichmaref.[2] Vol. 1, p. 22, VI, 12-14. Also pages 26, 46, 94, and 97.

14. "Car la tres douce imprecion De son ymagination Est en mon cuer si fort empreinte Qu'encor y est et yert l'empreinte Ne jamais ne s'en partira." *Dit dou Lyon.* Hoepffner.[1] Vol. 2, p. 166, lines 207-211.

15. "Car nuit et jour sa face coulouree Et sa fine biauté que tant desir Par souvenir m'est ades demoustree." Chichmaref.[2] Vol. 1, p. 152, CLXVII, 9-11. Also pages 29, XIV, 8-14; 38, XXIII, 1, 7.

16. "Je pris ceste ymage jolie, Qui trop bien fu entortillie Des cuevrechiés ma douce amour, Si la desliay sans demour. Et quant je la vi si tresbelle Je li mis a non: TOUTE BELLE, Et tantost li fis sacrefice Non pas de tor ne de genice, Ainsois li fis loial hommage De mains, de bouche et de courage, A genous et a jointes mains; Et vraiement ce fu du

mains; Car sa douce plaisant emprainte Fu en mon cuer si fort emprainte Que jamais ne s'en partira." *Le Livre du Voir Dit*. Société des bibliophiles. Paris, 1875, p. 63.

17. *Le Dit dou Vergier*. Hoepffner.[1] Vol. 1, p. 21: "Et la signefiance Dou brandon de feu qui ardoit Et dou dart qui ferrez estoit, Et de quoy ses eles servoient." (lines 224-227).

18. "Voy que je m'affine Pour ma dame fine Qui onques ne fine De la doleur affiner Qu'en moy s'enracine" Chichmaref.[2] Vol. 2, p. 310, IV, 49-53.

19. "Car sa grant biauté la plente D'amours plente En moy." Chichmaref.[2] Vol. 2, p. 336, VII, 209-211.

20. "Celle damoiselle jolie Qui estoit a ce clerc amie, C'estoit li ente faitissette Comme une pucelette Ou grand vergier d'Amours plantée. La pot estre si eslevee Et de branches si estendue Et de fueilles si bin vestue, De fleurs si cointement paree." Hoepffner.[1] Vol. 1, pp. 222-223, lines 2493-2502.

21. "Vi aussi que d'un oiselet Qu'il avoit pris tout nouvelet qu'un petit s'en estoit peüs." Hoepffner.[1] Vol. 2, p. 258, lines 539-541.

22. "Einsi est il d'aucun s'il aimme." Hoepffner.[1] Vol. 2, p. 282, line 645.

23. "Or veons comment ce seroit Qu'une dame resambleroit L'esprivier qui l'oiselet prent Et vers le vespre le sourprent, Pour les piez tenir en chaleur" Hoepffner.[1] Vol. 2, p. 286, lines 1349-353. This behavior of the hawk has been mentioned by Vincent de Beauvais in his *Speculum naturale*.

24. "Car l'air qui estoit nes et purs Fu ors et vils, noirs et obscurs, Lais et puans, troubles et pus, Si qu'il devint tout corrompus, Si que de sa corruption Eurent les gens opinion Que corrompu en devenoient Et que leur couleur en perdoient. Car tuit estoient mal traitié, Descoulouré et deshaitié. Boces avoient et grans clos Dont on moroit." Hoepffner.[1] Vol. 1, p. 148, lines 313-324.

25. "Et quant je vi qu'il festioient A bonne chiere et liement Et tout aussi joliement Com s'il n'eussent riens perdu Je n'os mie cuer esperdu, Eins repris tantost ma maniere Et ouvris mes yeus et ma chiere Devers l'air qui si dous estoit Et si clers qu'il m'amonestoit Que hors ississe de prison Ou j'avoie esté la saison." Hoepffner.[1] Vol. 1, p. 153, lines 476-486.

FROM *CLERC* TO *POÈTE*:
THE RELEVANCE OF THE *ROMANCE OF THE ROSE*
TO MACHAUT'S WORLD

THE POETIC *ŒUVRE* OF GUILLAUME DE
MACHAUT: THE IDENTITY OF DISCOURSE
AND THE DISCOURSE OF IDENTITY

FROISSART'S *LE JOLI BUISSON DE JONECE*:
A FAREWELL TO POETRY?

THE *Romance of the Rose* was one of the most popular poems in the Middle Ages. A love story with astonishing diversions and digressions, acerbic anti-feminism yet courtly manners, the *Romance of the Rose* depicts the winning of a young lady, the Rose, by a Lover who endures tests and trials. Though Geoffrey Chaucer translated the poem, and the fifteenth-century "feminist" writer Christine de Pisan wrote epistles against it, the 22,000 line thirteenth-century poem (begun by Guillaume de Lorris and later completed by Jean de Meun) seems banal, mystifying, or ridiculous to the modern reader forgetful that it is an *allegory*.

Allegory is the presentation of abstract ideas as persons or literary characters. Allegory requires an intellectual cultivation of taste for us, unaccustomed as we are to perceiving Wrath as a fiery, frazzle-haired, vicious, knife-wielding woman; or Knowledge as a tall, imperious scholarly lady; or Arithmetic, a well-shaped woman facile with figures.

Nevertheless we verge upon allegorization and allegory when, sorely tempted by some beautiful food or wine that we ought not to have, we say at table, "I nearly succumbed to my desire but my discretion controlled me." Such a statement describes a battle within the mind between two conflicting strong ideas, which symbolic warfare leads to decision. Mind is the battleground upon which desire and discretion fight for control over human action. To vivify such comment, we need only to give the words flesh, making "desire" a tempestuous, red-haired, red-garbed, warrior maiden who vigorously attacks with lance and shield the restrained powerful blue-gowned warrior, Lady Discretion. Desire, on her red-caparisoned charger, nearly unhorses blue Discretion. But recovering her balance, Discretion wields the decisive stroke, and winning the fray, thus takes the day. Discretion vanquishing Desire in the battle for choice, the discreet feaster forgoes his rich cake or roast pheasant. A proper medieval allegory took place in the mind at a modern banquet table.   MPC

THE *Romance of the Rose* not only marks the end of a splendid narrative tradition in Old French literature, it also initiates a new, highly influential, evaluation of poetic discourse. Scholar-critics now are beginning to grasp the implications of the *Rose's* "originality." In Machaut's World, the *Rose*, constantly recopied and later printed well into the sixteenth century, symbolized Poetry itself. In fact, much poetic activity in Western Europe during the fourteenth century constitutes a set of *responses* to the *Rose*. Guillaume's protagonist and his narrator are a single voice, and yet two: "himself" at two moments in his "life." He is at once *clerc* (the learned-poet) and *trouvère* (the lover-poet): the fusion creates a new *persona*, the Poet-Lover. This fiction required that "Loving" be a matter of "Writing." This redefinition of Poetry led directly to the invention of "the Poet" in the modern vernaculars; Poetry itself and *as such* becomes canonical, its servants and masters: the Poets. Machaut, Petrarch, Chaucer, and Froissart are later exemplars of this great *Romance of the Rose* invention.   MPC

From *Clerc* to *Poète*:
The Relevance of the
Romance of the Rose
to Machaut's World

KARL D. UITTI

Department of Romance Languages and Literatures
Princeton University
Princeton, New Jersey 08540

N O ONE needs to be reminded that the thirteenth-century Old French
Romance of the Rose, especially the lengthier second part, by Jean
de Meun, was read, commented upon, and intensely reacted to by poets,
thinkers, and moralists of all sorts during the fourteenth century; indeed,
the importance of the *Rose* (not only in France but far beyond, in, e.g.,
Italy and England) as a dynamic and influential poetic text lasted well into
the sixteenth century. (Some 18 editions and/or printings of the *Rose*
appeared between 1481 and 1537/38.) François Villon's debt to the thirteenth-century poem, we recall, is immense; his *Grand Testament* is explicit in this regard:

> Et comme le noble *Roumant*
> *De la Rose* dit et confesse
> En son premier commancement
> C'on doit jeune cueur en jeunesse,
> Quant on le voit viel en viellesse,
> Excuser, helas! il dit voir...

> (ed. Rychner-Henry, vv. 113 ff.)

Approximately 250 manuscripts of the *Rose*—almost all of them containing both parts and many of them dating from the fourteenth century—
have so far been identified; a number of these codices are sumptuously produced, lavishly decorated with beautiful miniatures and rubrics.

Obviously, reasons can be, and have been, adduced to explain the extraordinary presence of the *Rose* in what we here have called "Machaut's

0077-8923/78/0314-0209 $01.75/2 © 1978, NYAS

World." Scholars have written—some very cogently, others less so—
concerning the "influence" of the *Romance of the Rose* on, for example,
the *Seconde Rhétorique*, on Dante, on Machaut, on Chaucer, and so forth;
the *Rose* Quarrel of around 1400 has been studied (though not so thoroughly as one would wish) and note has been taken of some of the uses
to which such poets as Villon and Marot have put the *Rose*. As Félix Lecoy
has remarked, the fourteenth- and fifteenth-century defenders of the *Romance of the Rose* "se contentent de l'imiter ou de s'en inspirer, et les
innombrables songes, en particulier, qui remplissent les recueils poétiques
de notre moyen âge finissant sont, plus ou moins directement et à des
degrés divers, des répliques (au moins pour la forme) du songe du *Roman
de la Rose*."¹

I do not intend even to mention, let alone examine, a fraction of these
complex questions today. Much more spadework and detailed textual
analysis are required before any thorough and general exploration of these
matters can properly be undertaken. Basically, I intend to make one simple suggestion, then go on to show (1) why I think the suggestion is worth
while, and (2) how it might lead, if adequately followed through, to a
deeper understanding of the relevance of the *Romance of the Rose* to
Machaut's World. Here is the suggestion: The *Romance of the Rose* is,
and was perceived to be, a defense and a renewal of poetry. Jean de Meun
was the first to have—or at least to implement—the idea that Guillaume
de Lorris's poem (*ca.* 1230) contained the essential structural ingredients
of a truly poetic *summa*; that is, he perceived that the makings and the
architecture of an authentically poetic vehicle for the containing of meanings and value on an encyclopedic scale were present and amenable to further development. This *summa*, I repeat, is poetic and is consciously
viewed as such, in opposition to the logically constructed scholastic
summae that prevailed in thirteenth-century philosophy. Furthermore,
the *Rose* constitutes a transformation; in fact it celebrates the very idea of
poetic transformation and energy, carrying this idea even further than the
idea of *translatio* had been carried by traditional twelfth-century Old
French narrative. In a very real sense, the *Romance of the Rose*—the
conjoined texts of Guillaume de Lorris and Jean de Meun—stands, with
respect to its twelfth-century models, in a position analogous to the position occupied, in scholastic philosophy, by Aquinas's *Summa theologiae*
with respect to the "new" Aristotle.

I have referred to Jean de Meun's consciousness of all this—a consciousness, as we shall presently see, that is at once poetically implicit (i.e.,

built into the very stuff of the poem) and, I believe, critically explicit. These matters require—and surely deserve—extensive treatment; some hints will have to suffice here.

When one looks at Guillaume's Prologue, one notes immediately that the first 20 lines follow the classical format of traditional Old French romance narrative. The Prologue starts out with a general *sententia*, or proposition ("Aucunes genz dient qu'en songes/ n'a se fables non et mençonges"), followed in this case by a refutation based on the narrator's appeal to accepted authority (the *auctor* named Macrobius) and his own affirmation: that dreams are meaningful, that, indeed, things dreamt of often subsequently turn out to be true. The first-person narrator is thus entirely clerkly, in the twelfth-century sense; Guillaume's text is, so to speak, hitched to the brilliant tradition illustrated by such *clers lisants* as Wace, Benoît de Sainte-Maure, Chrétien de Troyes, Jean Bodel, Jean Renart, and so many others. Our expectation is, then, that the learnèd clerkly narrator will tell us a story which he will illustratively relate to the theses sketched out in vv. 1–20. However, to our surprise, in vv. 21 ff., we read: "El vintieme an de mon aage,/ el point qu'Amors prent le paage/ de jones genz, couchier m'aloie/ une nuit. . .'" ('In my twentieth year, when Love habitually takes his toll of young men, one night I was going to bed. . .'). Our protagonist is the narrator himself; the story will be about him! This is no ordinary clerkly narrator; this is someone who is relating his own experience, an experience tied into dreams and love, akin to the matter and the manner of courtly lyric poetry—the poetry, say, of a Conon de Béthune or a Thibaut de Champagne. Yet it is narrative too, written down in traditional rhyming octosyllabic couplets (not the stanzas of courtly song) and specifically called a *romanz* (v. 35), a *romanz* entitled "li Romanz de la Rose" in which, we are promised, the entire art of Love is enclosed ("ou l'art d'Amors est tote enclose" [v. 38]). Lyric and narrative are conjoined, welded together, as it were, or amalgamated, creating a new, stronger poetic alloy: "La matire est et bone et nueve" ('The subject and its treatment are good and new'). Protagonist and narrator in this *aventure* are one and the same: together they constitute something different from the sum of their parts, something I shall call the Poet, or, perhaps more accurately at this stage, the Poet-Lover. Diction and love-experience are fused, as of course they are, traditionally, in *troubadour* and *trouvère* lyric; but here the fusion occurs in the broader, more generally valid romance narrative context of the *translatio studii*.

Guillaume makes sure we get the point; in vv. 44 f. he specifies a kind of doubling-in-unity. The dream about to be recounted, we recall, *took* place when the narrator-protagonist was 20 years old. That is the *story* (or history): it is situated in the past. The present tense of the telling itself occurs some five or more years later: "Avis m'iere qu'il estoit mais,/ il a ja bien .v. anz ou mais,/ qu'en may estoie, ce sonjoie,/ el tens enmore- us. . ." ('I thought it was Maytime, a good five or more years ago, that I was in May [or: in the Maytime of life]—this I dreamt—in the time of love. . .' [vv. 44 ff.]). This temporality is extraordinary: past, present, and, to boot, the archetypal, eternal time of lyric love. The protagonist, then, is the 20-year-old lyric lover (later to be identified as Amant); the narra- tor, though the "same person," is the 25-year-old romance-type clerk: the two literary rôles (here poeticized) are identified as constituting the inseparable and intrinsic nature(s) of the Poet-Lover, of him who alone knows how to "enclose the entire art of Love." The occasion of this most remarkable poetic invention is a very joyous one: in the glorious Spring- time of lyric poetry birds burst into song, tree sprout leaves, flowers and grass bloom, young men are enjoined "a estre gais et amoreus" (v. 79). and Amant's adventure, about "to begin" (in the past), is about to be *told* (in the present).

I repeat: Clerkly romance narrator and lyric *trouvère* join forces to create something new, that is, the Poet-Lover, related of course to the *poetae* of Antiquity, but not identical to them, a new poetic *auctoritas*, rooted in vernacular French tradition and yet transforming this tradition too. Lyric authenticity guarantees romance narrative; romance narrative expands the scope of lyric.

Now then, as I stated earlier, Jean de Meun understood and accepted Guillaume's invention. Explicit proof of my assertion is to be found in vv. 10463 to 10650, i.e., at almost the exact midpoint of the conjoined texts of Guillaume and Jean. (Lecoy's edition of the two parts totals 21751 lines.) Midpoints in Old French romance, we recall, are frequently signifi- cant; they often have to do with matters of identity and naming. Thus, Lancelot's name is revealed by Guenevere at the midpoint of Chrétien's *Charrette*; it is at the midpoint that Yvain meets his lion and becomes known as the Chevalier au lion; one finds a prose summary of the *Char- rette* at the very center of the *Vulgate Lancelot* proper. Verses 10463– 10650 (which, of course, we owe to Jean de Meun) are spoken by Amors himself, who exhorts his troops to free Bel Accueil from Jealousy's castle. The speech is actually about writing (Love's servants are poets) and, quite specifically, about the writing of the *Romance of the Rose*. Amors defines what I have called the Poet-Lover: he was, and is, the collective *poetae* of Antiquity. Tibullus, Gallus, Catullus, and Ovid "bien sorent d'amors trestier" ('knew well how to treat of love'). But now they are dead and gone. (One is reminded of the *translatio studii* topos as worked out in the Prologue to *Cligés*; once *clergie* thrived in Greece and Rome, however now, Chrétien reminds us, the Ancients' "burning ember" is "extinguished," and *clergie* is established in France.) "Vez ci," exclaims Amors, "Guillaume de Lorriz" ('Here stands Guillaume de Lorris'—the first time Guillaume's name is mentioned in the entire poem!), who one day *will* "conmancier le romant/ ou seront mis tuit mi conmant" (will 'begin the romance/ in which all my commandments will be placed' [vv. 10519 f.]). Jean is identifying Guillaume in his dual rôle as Poet-Lover; meanwhile, he, Jean, has *de facto* assumed the clerkly narrator's rôle, retaining Guillaume principally as lyric protagonist. Guillaume *symbolizes* the Poet; he is subsumed by Jean essentially into the poetic process. He is, I repeat, Poet; i.e., he, at a point in time, incarnates what we would call *Poetry*, just as, in earlier times, Ovid, Catullus, and others had once done. (Incidentally, this whole passage is a conscious transformation of some famous lines from Ovid's *Tristia* [IV:10:55 ff.][2]; Ovid transformed is ominpresent in both parts of the *Rose*. As Michelle Freeman has shown in her acute analysis of the Narcissus episode in Guillaume de Lorris, Ovid is transformed through the prism of previous Old French narrative: he is incorporated into Guillaume's and Jean's sense of poetic process, into *translatio*.[3]

But this is not all. Eventually, Amors tells us, Guillaume too will die, and his tomb will be filled with balms, incense, and precious spices. Then "Jehans Chopinel" will come; he will be born at Meung, on the banks of the river Loire, and will hold "the romance so dear that he will decide to finish it perfectly," for "quant Gillaumes cessera,/ Jehans le continuera" (vv. 10557 f.). Says Amors: "I shall fit him out with my wings and sing him such melodies that as soon as he is no longer a child he will be so indoctrinated with my learning (*sciënce*) that he will sing out (*fleütera*) our words on the highways and in the schools, in the language of France, everywhere," and "no longer will those who listen to him have to die of the sweet pains of love." The book will be known to all as the "*Miroër aus Amoreus*" (v. 10621)—the '*Lovers Encyclopedia*' or '*Speculum of Lovers*.'

These lines are fraught with implications. First and foremost, of course, is the fact that Jean de Meun is never identified as a Lover; his love-service is exclusively poetic: he *teaches* Love's science in musical words (*fleütera*).

Guillaume's *romanz* is thus transformed into Jean's *miroër*. Ironically, however, in his capacity as Jean's protagonist, Guillaume the Poet-Lover is caused to take on the rôle of pure Lover at the same time that he is characterized, as we have just seen, as a link—a transition—between Ancient *poetae*, or servants of Love, and Jean himself. It is as Jean's romance-type character—as Amant—that Guillaume finally succeeds in winning the Rose. (Guillaume the *Poet*-Lover, Jean reminds us, "will cease and desist," i.e., he abandoned an enterprise—loving and writing—which Jean himself, according to Amors, will "hold so dear" that he "will continue it and bring it to a perfect end.") Yet—and what I am about to say is not so much contradictory as it is illustrative of the enormous poetic dynamism of Jean's inventive continuation—the fiction of the poem does not permit our reading Jean de Meun back into the rôle of a twelfth-century-style *clerc*. For one thing, *clercs* never—or very rarely—sing; for another, Jean retains Guillaume's first-person device: narrator and protagonist ostensibly stay one and the same. Despite the passage we have just analyzed, where, we remember, it is Love himself who is speaking, the poem, so to speak, "writes itself out" in precisely the framework within which it started. (Jean Chopinel, one day, *will be born;* he *will finish* the romance; the story time, i.e., Guillaume's twentieth year, remains intact; Jean situates himself *as poet*—not as *clerc* or *trouvère*—with respect to the story. He too will incarnate Poetry.) Thus, Jean's *attitude* toward the Poem—for it is here surely Poem, or Poetry, that is at issue, something symbolized by the *text* that we, following Guillaume de Lorris, call the *Roman de la Rose*— consequently prefigures the attitude such future poets as Dante, Machaut, Villon, and Chaucer will adopt toward the *Rose* and, above all, what it, poetically and (therefore) concretely, represents. To put the matter quite bluntly, for them the *Rose* is Poetry itself. It is also, and very importantly, *summa*-like Poetry: the *Speculum* of *Lovers* (= Poets) complete and entire at once as *text* and *as process*. In this new sense the *Rose*, thanks to Jean de Meun, becomes in turn canonical in a new, or renewed, definition of the canonical, and, therefore, through the *Rose*, the poet's function and notion of identity are also defined anew, in a fashion, I daresay, that is still very much with us today.

Let me conclude with one or two remarks of a speculative nature. It is my contention that though, in the *Rose* at least, he never quite says so, Jean de Meun saw himself—and subsequent generations would see him— as a *poeta*, or *poète*, a term habitually applied to Classical *auctores* in the preceding vernacular canon. What he represented, both following (i.e., "continuing") Guillaume de Lorris *and* transforming him, has no real an-tecedent in Old French literature. Along with, perhaps, Rutebeuf, Jean de Meun created the *idea* of poetry (in contradistinction to the idea of *non-poetic* discourse), namely, the idea of an open-ended, marvelously ambigu-ous, and yet *total* discourse (usually metrical, hence not prose; in the vernacular, therefore opposed to logical philosophy—to what Dante would later call *gramatica*). This open-ended total discourse is the province of a new kind of person whose chief duty, as he judged it (and this personal judgment is significant), was, no more and no less, the defense of this discourse. We are no longer exclusively in the domain of *clergie*, no longer confined to the *grand chant courtois*; we are in a new country and at a new time: the truth of poetry has been reaffirmed and, once again, redefined.

Such an idea of poetry surely implies a corresponding idea of poet, just as, in the twelfth century, the notion of *clergie* required its counterpart, the *clerc*. It is therefore no coincidence that the terms *poète, poeta*, and *poetria* (and so forth) applied to contemporary masters and to their art start to gain currency in Europe during the years immediately following upon the completion of Jean de Meun's continuation (and *chef d'œuvre*). To all in-tents and purposes, Jean built the road that Petrarch would travel in 1341, on his way to Rome, where, in one of the most spectacular ceremonies to take place in "Machaut's World," he would be crowned with laurel and thereby declared the "first" modern *poeta laureatus.*

NOTES AND REFERENCES

1. Guillaume de Lorris and Jean de Meun, *Le Roman de la Rose*, ed. Félix Lecoy, I. Classiques français du Moyen age (Paris: Champion, 1970). p. xxxi. Quotations from the *Romance of the Rose* itself are taken from this three-volume edition; the line refer-ences will be given, when appropriate, in the main body of my text.

2. Here is a sample from Ovid's poem; the elegiac genre is at issue in these lines, as well as Ovid's relationship to it. The structure is virtually identical to that of Amor's discourse.

> Vergilium vidi tantum: nec avara Tibulle
> tempus amicitiæ data dedere meæ.
> successor fuit hic tibi, Galle, Propertius illi;
> quartus ab his serie temporis ipse fui
> utque ego maiores, sic me coluere minores,
> notaque non tarde facta Thalia mea est.

See also my "Remarks on Old French Narrative: Courtly Love and Poetic Form," *Ro-mance Philology*, XXVI:1 (August 1972), pp. 77-93, and *op. cit*, XXVIII:2 (November 1974), pp. 190-199.

3. Thus, as Freeman points out, it does not suffice merely to note that the Fountain of

Narcissus episode (ed. Lecoy, vv. 1423-1698) "derives" from Ovid's *Metamorphoses* (III, vv. 339 ff.); its derivation is filtered through the "Blood Drops Scene" in the *Conte du Graal (Perceval)*, of Chrétien de Troyes: thematic, imagerial, and precise verbal correspondences link Guillaume's text to that of Chrétien. Ovid, Chrétien, and Guillaume are purposefully—metaphorically *(translatio)*—conjoined. See "Problems in Romance Composition: Ovid, Chrétien de Troyes, and the *Romance of the Rose.*" *Romance Philology*, XXX:1 (August 1976), pp. 158-168.

MACHAUT'S POETIC WORKS reveal a new concept of "poetic identity," deriving in large part from the innovations of the *Romance of the Rose*. A new relationship between the individual who "produces" poetry and the text which results from this activity is adumbrated by Machaut himself, a relationship much closer to the concept of *poeta*, our modern notion of poet, than to anything found in earlier French writers. This new concept of poetic identity seems related to an expansion of the Old French term *poëte* to include contemporary, vernacular masters as well as the poets of antiquity (the *auctores*) —a semantic development which took place in the fourteenth century. Two aspects of Machaut's vast *œuvre* illustrate the nature and function of his poetic identity, his self-presentation as *poeta*: first the organization of his poem the *Voir-Dit*: second, the use of "codicological resources" for poetic effects. MPC

The Poetic Œuvre of Guillaume de Machaut: The Identity of Discourse and the Discourse of Identity

KEVIN BROWNLEE

Department of Romance Languages and Literatures
Princeton University
Princeton, New Jersey 08540

To a considerable extent the modern notion of "poet," derives from fourteenth-century developments, for it was during the fourteenth century that a new relationship between the individual who "produces" poetry and the text resulting from this activity was first adumbrated. In fact, it is only in the fourteenth century that the semantic field of the very term *poëte* in French is expanded to include vernacular, contemporary poets. Previous centuries used the term only with reference to the classical *auctores*. The earliest examples of this "expanded"usage that I have been able to find come from the *Ballades* written by Eustache Deschamps in connection with the death of Guillaume de Machaut.[1] In his *Ballade* "A dame Péronne, après la mort de Machaut,"[2] Deschamps portrays himself as requesting that Péronne (the Toute-Belle of Machaut's *Voir-Dit*) accept him as *ami* in place of his dead master. The *Ballade* begins as follows:

> Après Machaut qui tant vous a amé
> Et qui estoit la fleur de toutes flours,
> Noble poete et faiseur renommé,
> Plus qu'Ovide vray remede d'amours,
> Qui m'a nourry et fait maintes douçours,
> Veuillés, lui mort, port l'onneur de celui,
> Que je soie vostre loyal ami.*

* "After Machaut who loved you so much, and who was the flower of all flowers, noble poet and renowned maker, truer remedy of Love than Ovid, who raised me and did me many kindnesses, may it please you, since he is dead, that, for his honor, I be your faithful lover." (trans. Brownlee)

[219]

0077-8923/78/0314-0219 $01.75/2 © 1978, NYAS

It is interesting to note that Deschamps seems to feel that the term *poëte* requires a gloss, *faiseur* ('maker, creator'), a term already in use to designate vernacular poets. Deschamps is, it appears, aware of the linguistic innovation involved in using the term *poëte* to refer to Machaut. In his second *Ballade* "Sur la mort de Guillaume de Machaut" (no. 124 in the SATF ed., vol. I, pp. 245-46, dated 1377), Deschamps uses the term in a context that links the classical poetic inspiration of the *auctores* with the contemporary vernacular master:

> La fons Circé et la fonteine Helie
> Dont vous estiez le ruissel et les dois,
> Ou poetes mistrent leur estudie
> Convient taire, dont je suis moult destrois.
> Las! c'est par vous qui mort gisez tous frois,
> Qui de tous chans avez esté cantique,
> Plourez, harpes et cors sarrazinois,
> La mort Machaut, le noble rethorique.†

Again, Deschamps' use of the term seems to be deliberate and to involve an intentional expansion of its field of meaning. As Jean de Meun linked Guillaume de Lorris and the Latin elegiac poets in the *Rose*, so Deschamps links *auctores* and Machaut. This connection is rendered explicit in the first text cited above when Deschamps utilizes the same generic term in order to designate *both* Ovid and Machaut (and by extension, himself— self-depicted as Machaut's successor). The idea of *auctor*, then, is, as it were, built into, fused with, the idea of *poëte* in its expanded sense. What had been implicit in Jean de Meun, I repeat, becomes explicit in Eustache Deschamps.

That Deschamps should take Machaut as an *exemplum poetae* is no accident. This is the way in which Machaut, of course, conceived of himself. The complex poetic identity he constructed may be regarded as the real innovation—even though Machaut did not explicitly apply the term *poëte* to himself. Building on the *Rose*, Machaut established a new kind of poetic identity, greatly expanding the range of the lyric voice and, thus, he authorized the new sense of the term *poëte* as applied to him by Deschamps.

---

† "It is fitting that Circe's fountain and the spring of Helicon, from which you were the stream and the canal, and at which poets study, be silent—which makes me very sad. Alas! It is on account of you who lie dead and cold, who were the canticle of all songs. Weep, harps and Saracen horns, for the death of Machaut, the noble rhetorician." (trans. Brownlee)

Before going on to consider Machaut's poetic identity on its own terms, it is first necessary, however, to examine the way in which Guillaume conceives of himself as related to the *Rose*. Indeed, this relationship may be said to be built into the "global" poetic voice which Machaut establishes for his complete *œuvre*.

Let me explain. Guillaume de Machaut conceived of his diversified artistic production as a unified and coherent body of work and presented it as such. Of central importance in this presentation is the *Prologue* found in MSS *AFG*.[3] This *Prologue* involves the explicit establishment of the poetic voice that will be speaking in all the works to follow. It is the character of this poetic voice—which I shall discuss in detail later in this paper—that allows an extremely diverse collection to be presented as a unity, as an *œuvre*. Machaut's *Prologue* situates him and his artistic endeavor within the particular poetic tradition of the *Romance of the Rose* in such a way as to embody the principles of continuation and transformation that are inherent in that tradition.

It is highly significant that Guillaume, in the *Prologue*, receives his poetic mission from *dame* Nature and "li dieus d'Amours," who act as necessary complements to each other. These are the same two figures whose combined effort in the *Rose* had made the taking of the castle, the winning of the rose and thus, in a sense, the writing of the poem possible.

Further, Guillaume's taking up of his poetic mission is depicted as his entering the "trés dous service" of "li dieus d'Amours." Again, there is a significant correspondence with (as well as departure from) the *Rose*. First, there is a structural transformation: the midpoint of the *Rose* has become the *Prologue* to Guillaume's *œuvre*. By "midpoint" I mean both the midpoint of Guillaume de Lorris' section of the poem (where Amant enters the service of Amors) and the midpoint of the conjoined *Rose* texts (where Amant reenters the service of Amors). Secondly, it is Machaut the *poet, not* Machaut the lover, who enters Amors' service; as in the case of Jean de Meun, Machaut's service to Amors consists not so much in loving well as in writing well. It is as if the 25-year-old poet-narrator in Guillaume de Lorris or the narrator-figure in Jean de Meun were to undergo directly the ceremony of entering into the service of Love that they describe Amant as undergoing. In Machaut's *Prologue* the lover-protagonist of the *Rose* is conflated with the poet-narrator and a new poetic identity explicitly emerges—that of the *poëte*.

The fact that this identity may be seen as deriving from the *Rose* provides it with a very important kind of authorization in terms of the values of the Old French literary canon. Further, this explicit link with the *Rose* establishes a poetic "point of reference" that will continue to operate throughout Machaut's diversified corpus. Time and again in his varied *œuvre*, Machaut utilizes the *Rose* as a source of poetic constructs; he does so on every level. In the *Fonteinne amoureuse*, to cite only one important example of this process of poetic transformation (itself poeticized in the *Prologue*), both the central image (the Fountain itself—supported by a pillar on which Pygmalion has sculpted the story of Narcissus) and the narrative structure (a dream sequence involving both the poet-narrator and the lover-protagonist) may be viewed as self-conscious exploitations (and transformations) of features central to the *Rose*.

That Machaut viewed the *Rose* as embodying an entire poetic tradition —in whose context he sought authorization for his own artistic endeavor —is, I think, sufficiently clear from even a preliminary consideration of the *Prologue*. But if Machaut conceives of himself as being, in a very important sense, a poetic descendant of the *Rose*, he is equally conscious of the fact that his own work represented a kind of fragmentation of the *Rose*. With Machaut, we no longer have a self-consciously encyclopedic but structurally unified single poetic work (a poetic *summa*, as it were). Rather, in his case we find a series of separate works that are *presented as though*, together, they formed a unified *œuvre*.

Utilizing the *Rose* as authorization while at the same time departing from it in a radical way, Guillaume de Machaut developed a new kind of poetic identity which is much closer to the concept of *poeta* as used, say, by a Petrarch than to any of the terms previously employed by Old French writers in speaking of themselves or of their poetic activity.[4] It is, I think, this exemplification of the concept of the *poeta* that explains why Deschamps should apply to Machaut the *term poëte* in the expanded sense we described. It is also Machaut's self-presentation as *poeta* that approaches, for the first time in medieval French literature, our modern notion of "poet."

This breakthrough into "modernity," however, was accomplished in a typically medieval fashion. Machaut, self-consciously working within the vernacular literary tradition, created his new poetic identity by conjoining the Old French tradition of the clerkly narrator figure, the first-person lyric (and lyrico-narrative) voice, and a new conception of the professional artist—in part a development of thirteenth-century scribal activity in "editing" and organizing codices.

For the purpose of illustrating the nature and function of Machaut's concept of poetic identity, I would like to focus on two aspects of

Machaut's vast *œuvre:* first, the organization of the *Voir-Dit;* then, Machaut's creative use of codicological possibilities for poetic effect.

The *Voir-Dit* recounts the story of the love affair and literary relationship of Guillaume and Toute-Belle. It is a relatively long work and one that combines three generic forms, consisting of 8437 octosyllabic lines with a large number of intercalated lyric poems as well as 46 prose letters (ed. Paulin Paris).[5] The poet claims that his story is true (hence the title) and that he is undertaking it at the request of his *dame:*

> Car celle pour qui Amours veille
> Vuet que je mete en ce Voir-Dit
> Tout ce qu'ay pour li fait et dit
> Et tout ce qu'elle a pour moy fait
> Sans riens celer qui face au fait. (p. 17)‡

The story-line may be summarized briefly as follows. The poet-narrator, old, alone, ill and melancholy, receives a declaration of love from a beautiful young *dame* who has never seen him. This declaration is in the shape of a poem and a letter. Toute-Belle's declaration of love is combined with an expression of her admiration for Guillaume's poetic stature and a request that he send her poetry and music. Further, she asks him, in effect, to take her on as an apprentice—to correct her own compositions. Thus Toute-Belle wants Guillaume to be simultaneously her lover, her teacher and her poet (i.e., the poet who celebrates her). There seems to be a suggestion here of the Abelard and Heloïse relationship, but in a very different key. Remy de Gourmont has suggestively referred to Toute-Belle as a Laura who self-consciously "created herself" by initiating a relationship with the "French Petrarch."[6]

The love affair, then, is first a literary correspondence in which lyric poems figure as "crystallizations" of various stages of the love experience. Each of these poems is "explained" by means of the circumstances and emotions that are depicted as motivating it. At the same time the lyric poems function narratively. They are the means by which the love affair is advanced. This correspondence continues for almost a year. Guillaume has a copy of the *Fonteinne amoureuse* (his most recent *dit*) made for Toute-Belle and receives her portrait ("ymage") from her. He promises

---

‡ "For she whom Love watches over wants me to put into this True Story all that I have done and said for her, and all that she has done for me, hiding nothing that might be pertinent." (trans. Brownlee)

that he will serve her better than "Lancelos ne Tristans servirent onques leurs dames" (Letter X).

Guillaume's love service, however, is entirely identified with his poetic output. Even at the beginning of the "affair" (i.e., before the lovers meet), Guillaume's consistent success as a poet seems to be contrasted with his inadequacy as a lover. He is always able to make poetry out of his love experience—which may be said to be presented *qua* poetic subject matter (*matière*) almost from the beginning of the work (cf. esp. Letter II). On the other hand, he is plagued by doubts concerning his fitness for the role of lover (from his very first letter in which he denigrates his physical appearance). In addition, illness (Letter VI) and fear (Letter X) repeatedly cause him to delay making the journey to see his beloved.

Finally the two lovers meet for the first time (p. 79) and Guillaume spends eight days near his beloved during which time Toute-Belle consistently takes the initiative and Guillaume behaves with a timidity bordering on ineptitude. They meet several times in a *vergier* where, on one occasion, they communicate by mean of lyric poems (pp. 92-93). Toute-Belle requests Guillaume to compose a poem on the spot and reads it while it is being written down.

A series of relatively brief separations and meetings ensues until finally Guillaume leaves Toute-Belle's presence in order to visit his patron, the Duke of Normandy (p. 129). Later, Guillaume returns to Toute-Belle (p. 140) and the two set out on a pilgrimage together. It is at the end of this episode (in a reversal of the *aubade* situation) that the lovers achieve some kind of physical intimacy. Toute-Belle sends for Guillaume at dawn and reveals her naked body to him. Guillaume falls to his knees and addresses a prayer to Venus in the form of a lyric poem asking for her help as one who has always served her loyally (pp. 155-57). Venus answers his prayer, appears in the bedroom and conceals the lovers in a fragrant cloud. Before the goddess departs, Guillaume composes a *chanson baladée* to celebrate the event. In an interesting imagerial reversal Toute-Belle gives him a golden key ("la clef de mon tresor"); he gives her a ring and then rides off (p. 163). This episode is, in fact, the high point of Guillaume's love affair and the last time the lovers see each other in the course of this work.

On his way home from the rendez-vous Guillaume has an *aventure*. He is captured by Esperance who complains that he 'has not yet said anything special about me in your story' (". . . tu n'as encor de moy dit/ Rien d'especial en ton dit." p. 168). He promises to compose a *lay d'Esperance* and is released (p. 170). This episode is quite important on several counts. First, it is a particularly rich example of the textual self-consciousness that characterizes the *Voir-Dit* throughout. Secondly, it calls attention to the "global" identity of the poet-narrator, which transcends any particular poem in which he might figure. (This is accomplished in part by reference to earlier *dits*—the *Remede de Fortune*, where Esperance had helped Guillaume; the *Jugement dou Roy de Navarre*, where writing poetry was depicted as an "amende."). Thirdly, this passage (related and later commented on by the narrator in the same tone he uses to describe his journey to Toute-Belle) has significant implications for the avowed "truthfulness" of the work. We are clearly meant to understand the truthfulness of a poetic artifact. In this context it is perhaps also significant that the *lay d'Esperance* occurs near the midpoint of the Paulin Paris edition.

The two lovers, I repeat, are not to meet again. Indeed, close to 60 percent of the work (in the P. Paris edition) deals with the "love affair" after this final separation. They continue their correspondence however, which concerns the composition of the *Voir-Dit* as much as it does the love experience that the *Voir-Dit* is mean to relate. Guillaume begins to doubt his beloved (pp. 213-32). He has a dream in which he sees Toute-Belle's "ymage" turn from him and dress in green. Then, still dreaming, he speaks at length to the dauphin, mingling political advice with a request for help in his love affair. The dauphin, in a learned speech full of references to the *auctores*, advises him to wake up. He does so and finds the portrait unchanged. The dream (which contained its own gloss) was thus both true and false.

The correspondence continues and the reader sees the love affair being transformed into a literary work before his eyes (cf. esp. Letters XXXIII, XXXIV, and XXXV). The "book" (i.e., the *Voir-Dit* itself) is discussed and worked on; it is sent back and forth between the lovers. Writing as process(including the physical process of making a manuscript) is thematized. Guillaume suffers acutely from his separation, and Toute-Belle sends for him (pp. 281-82), claiming that she has taken care of Dangier and Malebouche. However, not only does the projected meeting not take place, but the lovers quarrel. Guillaume feels that Toute-Belle has been indiscreet (a somewhat strange pretext in the context of this work) and hides her portrait, which comes to him in a second dream (pp. 315-30), reproaching him, using several Ovidian *exempla*. In Letter XLII (p. 342) Guillaume states that work on the *Voir-Dit* has come to a halt "puisque matière me faut." Difficulty in love seems to have caused difficulty in writing. However, the literary work is of course ultimately to be completed whereas the love affair will not be resumed. The lovers are finally reconciled, but the possibility of a physical relationship, indeed of a love experience as direct experience, seems to have receded out of reach. What remains, becoming by contrast increasingly tangible, is the poetic expression of the love experience: the *Livre du Voir-Dit* itself.

The organization of the *Voir-Dit* thus presents Guillaume as a successful poet despite his having been an unsuccessful lover (as a Jean de Meun rather than as a Guillaume de Lorris). The relationship with Toute-Belle fails as a love relationship but succeeds consistently as poetic inspiration—both in the sense of "generating" the lyric poems that are embedded in the narrative (presented as deriving from Guillaume's love of Toute-Belle) and in the sense of producing the *Voir-Dit* as a whole. In the final analysis the love experience is valorized, it seems, primarily as a generator of poetry, as poetic inspiration, even as *matière*. At the same time love is shown to be immortalized only because it has been treated in poetry.

All of this fits in both with Machaut's global poetic identity and with his self-presentation in the *Voir-Dit* as above all a professional writer, for whom the activity of composing poetry and music is closely linked with the physical aspects of the writer's craft: the transcription and circulation of manuscripts and the business of patronage.

Indeed, one of the most striking aspects of Machaut's self-portrayal as poet in the *Voir-Dit* is his explicit concern with the supervision of the arrangement and copying of "editions" of his collected works—i.e., with the making of codices.[7] This activity is thus presented as an inherent part of Machaut's poetic activity—of his poetic identity. The poet no longer simply composes verses, songs or *dits*, but, in addition, concerns himself with the transcription of his various works—with their codicological existence. What had been in the thirteenth century largely the business of scribes (who often exercised considerable editorial, even literarily significant roles[8])—i.e., the arrangement of codices—becomes with Machaut the busines of the poet himself. Indeed, the notion of organizing a codex is transferred by Machaut into the organizing of an *œuvre*. We have moved one step beyond the accomplishments of the thirteenth-century scribes—to something approaching the modern idea of the *book*.

This explicit expansion of the realm of the writer's activity—of the poet's activity *qua* poet—is of the utmost significance. For, by depicting himself as supervising the transcription and arrangement of his own complete works, Machaut has done nothing less than to raise the codex to the level of literary artifact, making it function as an inherent part of his poetic production.

This implication is confirmed by MS evidence. In each of the major MSS one notes an evident concern with the arrangement and ordering of the whole, which was almost certainly the result of Machaut's personal supervision. Codex *A* "opens with an original index on 2 unfoliated pages bearing the superscription . . . 'Vesci l'ordenance que G de Machaut wet quil ait en son livre.' "[9] That our author was indeed presenting his *œuvres complètes* as such is further confirmed by the *Prologue*. For this is not the prologue to any individual work of Machaut but to his collected, complete works, which are thus presented as a unified *œuvre*.[10]

Not only does the existence of a global prologue of this sort make even more explicit the poeticization of the codex, but it serves to inform our reading of Machaut's *œuvre* in a very important way. For it is in the *Prologue* that Machaut establishes the poetic voice that will be speaking in all the works that follow. It is the character of this poetic voice—what might be called Machaut's identity as *poeta*—that allows an extremely diverse collection to be presented as a unity, as an *œuvre*.

The paramount importance with which the *Prologue* is thus invested necessitates, I think, a detailed examination. The text is composed of four *ballades*, each preceded by a brief prose introduction. There follows a sequence of 184 rhyming octosyllabic couplets.[11] (It is important to note that the form of the *Prologue*, combining lyric and narrative verse patterns, appears as, in the words of Ernest Hœpffner, "un raccourci de toute l'œuvre du poète.") In the first *ballade* (as indeed throughout the *Prologue*) the dignity and importance of the poetic vocation are strongly affirmed. Nature invests Guillaume (who is explicitly named) with his poetic mission: "Je, Nature . . ./ Vien ci a toy, Guillaume, qui fourmé/ T'ay a part, pour faire par toy fourmer/ Nouviaus dis amoureus plaisans." ('I, Nature come here to you Guillaume, I who have formed you apart in order to have formed by you new, pleasing love songs.' vv. 1–5). Nature then gives the poet *Scens, Retorique* and *Musique*—his poetic means or technique ("la pratique"). In the second *ballade*, Guillaume accepts his poetic vocation explicitly and wholeheartedly. This vocation involves following "le bon commandement" of "dame" Nature. In the third *ballade* Amours comes to Guillaume to give him his *matere*. Service of Amours is presented as involving above all the process of poetic composition: Amours gives Guillaume ". . . grande sustance/ Dont tu porras figurer et retraire/ Moult de biaus dis." ('. . . great substance/nourishment from which you can shape and derive many beautiful poems. . . .' vv. 17–19) In the fourth *ballade*, Guillaume thanks Amours and reaffirms his "poetic commitment": ". . . ces fais que j'ay a ordener,/ Pour lesquels arriere tous mis/ Seront autres, puis qu'a ce sui commis." ('. . . this activity which I have to arrange, for the sake of which all others will put aside, since I am committed to this.' vv. 14–16)

In the narrative sequence of 184 lines that follows we have both a commentary on, and expansion of, the ceremonious taking up of the poetic vocation depicted in the *ballades*. The poet-narrator claims that no one has ever served Amours better than he will. Then he restates the equation of love service with poetic service and reaffirms his commitment to both. He lists the specific genres in which he will write (vv. 11–16) and the different kinds of rhyme he will us (vv. 151–156). We have what could be called a poeticization of poetic technique. Technical mastery becomes one of the characteristics of the poet-figure who is presented to us here. Hence the importance of "rhétorique": "Retorique versefier/ Fait l'amant et metrefier/ Et si fait faire jolis vers/ Nouviaus et de metres divers." ('Rhetoric makes the lover versify and metrify and thus it causes pretty, new verses in divers meters to be made.' vv. 147–150). Finally, poetic activity is depicted as joyous activity, as celebration (vv. 49–55, 77–86). As such it is exalted, being associated with the highest kind of celebration, the angels' music of praise before God. They "louent en chantant" (v. 118). David and Orpheus are praised by Machaut as archetypal poets analogous to angels in their "chants." With v. 138 we have a celebration of poetry carried, as it were, to the second power, a poet making poetry about a poet making poetry: "Cils poetes dont je vous chant." The reference is to Orpheus and the use of the term "poetes" is significant. The sense of the Greco-Roman *poeta* as 'maker, contriver' (cf. Deschamps' *faiseur*) as well as 'poet' in the modern sense, seems to be particularly relevant to the poetic identity that is established in Machaut's *Prologue*. The emphasis is on the act of *making* poetry (more explicitly, making verses—hence the importance of technical mastery) rather than on the experience that gives rise to the poetry (i.e., the personal love experience of the lyric poet). Thus, the concept of the *poète* allows for (implies, even) a multiplicity of experiential perspectives from which poetry can be written, a variety of poetic stances. Yet all of these can remain in the first person, since they can all be placed within the context of serving love through poetry.

What emerges, then, from the *Prologue* is a carefully constructed poetic *persona*, a complex first-person poet-narrator whose identity serves to unify the vast corpus to follow. This is the *poëte* figure: acutely aware of the dignity of his calling, highly conscious of technical expertise, glorying in the breadth and diversity of his artistic production. We have here a conflation of the clerk, the lover and the poetic craftsman into a single but multifaceted poetic voice. Poetic activity (which includes music) is conceived of as service to ". . . li dieus d'Amours, qui mes sires/ Est. . . ." (vv. 3–4.). And, it is implied, this service can take many forms. The *poëte* is no longer bound by the lyric convention which requires him to sing his own love-experience only. The service of the *poëte* is different than that of the lyric poet; though the latter is, as it were, included in the former, the "lyric stance" becomes only one of a number of possibilities open to him. This potential diversity of expression is consequently one of the defining, and new, characteristics of the vernacular *poëte*.

By establishing the figure of the *poëte*, the *Prologue* thus provides the diverse sections of Machaut's corpus with a principle of unity, built, so to speak, into the body of the work itself.

The "expanded" concept of poetic identity developed by Machaut is, of course, intimately connected with his "expanded" concept of the nature of the poetic artifact. The implications of this connection are indeed far reaching. If Machaut the *poëte* himself presides over the arrangement of his complete works into a single document in which they exist as a unity —as the *œuvre* of the *poëte*—a new kind of reading seems to be invited: that of the *œuvre* as a whole. In this case the ordering of the various individual works (as well as their structural interrelationships) may be viewed as having a potential *literary* significance. Different orderings in different codices could be viewed as different "versions" of Machaut's *œuvre*. Here again there seems to be poeticization of what had previously been scribal activity. In this connection one is reminded of the various versions (involving in part different arrangements of the individual poems) of Petrarch's *Canzoniere*.

That the arrangement of the various works in a given codex may indeed have a literary significance has already been suggested (though in a limited context) by Ernest Hœpffner in his explanation of why the *Jugement dou Roy de Navarre* follows directly on the *Jugement dou Roy de Behaigne* in all the surviving MSS, in seeming contradiction to the chronology that Hœpffner takes to be the organizing principle used by Machaut in ordering his narrative poems.

Let us suggest another example of the way in which the ordering and "structure" of the codex—considered as *œuvre*—may be said to function literarily. The *Dit dou Vergier* directly follows upon the *Prologue* (in which it is mentioned by name) in MSS *AFG*. One of Machaut's shortest narrative poems (1296 lines in the Hœpffner edition), it involves a series of very obvious structural and even verbal variations on the *Romance of the Rose*. For this reason it has in general been dismissed by critics as a rather poor "imitation" of the *Rose*. This is, I think, to miss the point. The *Dit dou Vergier*, strategically placed at the beginning of Machaut's *œuvres complètes*, serves to strengthen and to make more explicit the poetic kinship with the *Rose* that we have already seen established in the *Prologue*. Further, it may be regarded as an important "opening signal" for the reader—situating Machaut's poetic endeavor self-consciously in the Old French tradition represented by the *Rose*.

In addition, if we confine ourselves to codex *A*, an interesting correspondence seems to be established between the *Dit dou Vergier* (the first narrative poem in the MS) and the *Dit de la Rose* (the last one), in which Machaut once again performs a kind of variation in miniature on the narrative structure of Guillaume de Lorris. If the *Dit dou Vergier* may be regarded as an opening signal, does the *Dit de la Rose* serve as a closing signal for the poetry section of the codex—of this version of Machaut's *œuvre*—with a particularly explicit poetic exploitation of the *Rose* being used to link them? Between the *Dit de la Rose* and the large section of codex *A* consecrated to Machaut's musical works comes only the disembodied technical tour de force in a lyric mode: "Les Biens que ma dame me fait" (64 lines, all with the same rhyme). This may, I submit, be viewed as a kind of interlude.

In the course of this paper I have done nothing more than merely outline, in preliminary fashion, the new kind of reading of Machaut that is invited and conditioned by his identity, his self-presentation, as *poète* and by the transformation of the codex through the concept of *œuvre* that is an essential part of this identity. Much more, of course remains to be said, both in the realm of medieval poetics and in that of French literary history.

NOTES AND REFERENCES

1. Between the twelfth and fourteenth centuries the term *poète* as used in Old French vernacular literature underwent a striking semantic development. The word is used in the twelfth century only to refer to classical antiquity, to the *auctores* and their world. Of its two principal meanings in this connection, the first is that of "seer." The Französisches Etymologisches Wörterbuch (*FEW*) gives the following definition: "un lettré,

savant et sage, spéc. Grec ou Romain, considéré comme possédant des pouvoirs intellec-
tuels extraordinaires." (vol. IX, p. 122); Tobler-Lommatzsch gives: "Weiser, Seher
(vates); Priester" (vol. VIII, p. 2058) for this meaning of *poëte*. Of the eight illustrative
citations that follow, seven are from the *Roman de Troie*. In addition, and more important for our purposes, the term *poëte* as used in the twelfth century designates specifically the great poets of antiquitiy, the *auctores*. This meaning is the first one given in
the *FEW*: "dans le vocabulaire des arts libéraux. auteur canonique servant de modèle
d'expression en poésie." A look at Tobler-Lommatzsch reveals that the term was applied, in this sense, either to a specific *auctor* or used with no explicit reference, simply
as a means of citing authority. Thus on the one hand we find Wace referring to Homer
as a "poetes" (*Brut*, ed. Arnold, vol. I, v. 1452). And in the early thirteenth century
the term is used, in the *Romans de Dolopathos* (ca. 1222-25) to refer to Vergil (a character in the romance): "Onkes poetes ne fu tex,/ S'il creust k'il ne fust c'uns Dex." (eds.
Brunet and A. de Montaiglon, Paris, 1856, p. 46). On the other hand we find, in the
Müchener Brut, the term *poëte* used to authenticate the parentage of Mars: "Filz fu
Jovis, lo roi de Crete,/ Si cum nos dient li pöete." (v. 3978). This kind of usage may still
be found in *Li Livres dou Tresor* of Brunetto Latini. It is important to note that neither
Wace nor Benoît de Sainte-Maure use the term *poëte* to refer to themselves or, indeed,
to any contemporary vernacular writer of verses. The semantic field of the word is
restricted to the *auctores*. The same restriction still applies to the word when it is used
by Brunetto Latini. It is interesting to consider the prologue to the *Roman du Castelain
de Couci* in this regard, for in this late thirteenth-century narrative poem about the life
of a late twelfth-century lyric poet, the term *poëte* is not used to refer either to the
narrator or to the protagonist. Rather, we find the following terms employed: *menestrel*
and *jongleour* (v. 31)—which are used pejoratively—*faiseur* and *trouver* (vv. 46, 47)—
which are used positively. (eds. J. E. Matzke and M. Delbouille, Paris: SATF, 1936).

2. *Œuvres complètes de Eustache Deschamps*, eds. le marquis de Queux de Saint-
Hilaire and Gaston Raynaud (Paris: SATF, 1878-1903), Volume III, pp. 259-60, *Ballade*
no. 447.

3. I use the *sigla* first adopted by Hœpffner:

A— Paris, B.N., f. fr. 1584
B— Paris, B.N., f. fr. 1585
C— Paris, B.N., f. fr. 1586
E— Paris, B.N., f. fr. 9221
F–G— Paris, B.N., f. fr. 22545, 22546
M— Paris, B.N., f. fr. 843

Vg Previously owned by the Marquis de Vogüé, now at Wildenstein and Co., New
York.

4. A wide variety of terms has been applied to medieval writers—rather indiscrimi-
nately by modern critics, somewhat less so by the medieval writers themselves. We
refer, for example, to Chrétien de Troyes as a "poet," but he would not have used such
a term. Indeed, Chrétien does not have a single, all-inclusive term which serves to
designate his poetic activity. Rather, he uses different expressions to describe the particular aspects of that activity which are poetically relevant in a given work. Thus in
the prologue to *Erec et Enide* (ed. Mario Roques. Paris: Champion, 1976) we find the
declaration that "Crestiens de Troies . . . tret d'un conte d'avanture/ Une molt bele
conjointure." (vv. 9, 13-14). In *Cligés* (ed. Alexandre Micha. Paris: Champion, 1970)

the vocabulary is somewhat different: "Crestiens comance son conte/ Si con li livres nos reconte." (vv. 43-44) With *Yvain* (ed. Mario Roques. Paris: Champion, 1971) we find yet another variation: "Por ce me plest a reconter. . . ." (v. 33). While in the prologue to the *Charrette* (ed. Mario Roques. Paris: Champion, 1972), Chrétien, speaking in the first person, states: ". . . je . . . anprendrai . . . romans a feire. . . ." (v. 3, 2). Finally, in the prologue to *Perceval* (ed. William Roach. Geneva: Droz, 1959) we find "Crestïens qui entent et paine . . . A rimoïer le meillor conte/ Qui soit contez a cort roial." (vv. 62-5). The variety of the expressions Chrétien makes use of in this context is striking. So also is the fact that he does not seem to feel it necessary to provide any kind of gloss on his name with respect to professional identity.

Wace, in contradistinction to Chrétien, does provide a kind of professional gloss on his own name. We find at the beginning of the *Roman de Rou* (ed. A. J. Holden. Paris: SATF, 1970-72) the author identifying himself as "un clerc de Caen, qui out non Mestre Vace" (Part I, v. 3). Both of these terms are significant and both are related to Wace's poetic enterprise in fundamental ways. *Maistre* is more than an ecclesiastical or academic title. The etymological sense of *magister*, 'teacher,' with all the authority such an identity involves, is very much present in Wace's use of the term and in the poetic stance he adopts. *Clerc* is perhaps an even more important word with respect to Wace's self-portrayal, as the prologue to Part III of *Rou* makes evident (esp. vv. 8, 103, 141). Holden's gloss of the term as "un lettré, savant" does not go far enough. For Wace the *clerc*, by his activity of reading and writing (the two seem to be inseparable) is the indispensible vehicle for the transmission of all cultural values. Without the activity of the *clerc* mankind would have no memory, no knowledge of its own past. Wace's self-description as *clerc* thus makes use of the *translatio studii* topos to invest himself and his literary activity with tremendous importance. Yet the terms *clerc* and *maistre* as used by Wace also represent a kind of selectivity. It is only certain aspects of what we would regard as his over-all literary or poetic activity that Wace chooses to valorize by so designating himself. Further, these are both terms whose primary meaning lies outside of the literary sphere. The degree to which they are to be applied to specifically literary activity varies according to context.

The same is true for the term *jongleor*, widely used in the twelfth and thirteenth centuries. Faral's introductory description is particularly relevant in this connection:

Un jongleur est un être multiple: c'est un musicien, un poète, un saltimbanque; c'est une sorte d'intendant des plaisirs attaché à la cour des rois et des princes; c'est un vagabond qui erre sur les routes et donne des représentations dans les villages; c'est le vielleur qui, à l'étape, chante de 'geste' aux pèlerins; c'est le charlatan qui amuse la foule aux carrefours; c'est l'auteur et l'acteur des 'jeux' qui se jouent aux jours de fête, à la sortie de l'église; c'est le maître de danse qui fait 'caroler' et baller les jeunes gens; c'est le 'taboureur,' c'est le sonneur de trompe et de 'buisine' qui règle la marche des processions; c'est le conteur, le chanteur qui égaie les festins, les noces, les veillées; c'est l'écuyer qui voltige sur les chevaux; l'acrobate qui danse sur les mains, qui jongle avec des couteaux, qui traverse des cerceaux à la course, qui mange du feu, qui se renverse et se désarticule; le bateleur qui parade et qui mime; le bouffon qui niaise et dit des balourdises; le jongleur, c'est tout cela, et autre chose encore.

(Edmond Faral. *Les Jongleurs en France au moyen age.* Paris, 1910. Quoted in Raleigh Morgan, "Old French 'jogleor' and Kindred Terms." *Romance Philology*, VII [1954], p.

283). The semantic field of *jongleor* thus was extremely wide—but did not at all correspond to that of our modern "poet."

Jongleor, clerc, maistre, trovëor—all are terms used by twelfth and thirteenth century poets to refer to their own poetic activity, or, more exactly, to certain aspects of it, for none of these terms takes into account the whole gamut of activities that we associate with the modern term "poet." Further, none of these terms serves as a real link between medieval poets and the poets of antiquity, the *auctores*. If Homer, in Benoît de Sainte-Maure's phrase, was a "clers merveillos," (*Le Roman de Troie par Benoit de Sainte-Maure*, ed. L. Constans. Paris: SATF, 1904-12, v. 45) the reverse was not true. Benoît was not a *poëte*.

5. Paulin Paris, ed. *Le Livre du Voir-Dit de Guillaume de Machaut* (Paris: Société des Bibliophiles François, 1875). Quotations from the *Voir-Dit* are taken from this edition with page references given, when appropriate, in the main body of my text. For the number of lines in the P. Paris edition of the *Voir-Dit*, as well as an important discussion of the work itself, cf. Williiam Calin, *A Poet at the Fountain: Essays on the Narrative Verse of Guillaume de Machaut* (Lexington: University of Kentucky Press, 1974), p. 167 ff.

6. Remy de Gourmont. "Le Roman de Guillaume de Machaut et de Peronne d'Armentières." *Promenades littéraires.* 5ième série, Paris, 1913, pp. 7-37.

7. Cf. in this regard Sarah Jane Williams. "An Author's Role in Fourteenth-Century Book Production: Guillaume de Machaut's 'livre ou je met toutes mes choses'." *Romania*, 90 (1969), pp. 433-454. This excellent article discusses Machaut's self-depiction as professional writer in detail and concludes that "it is easier, on the whole, to understand the execution of the manuscripts we actually have if we think of them as deriving from an author's personal copy as we attempted to reconstruct it on the basis of Machaut's own statements [in the *Voir-Dit*] : unbound, both for convenience in copying and to facilitate the addition of new works in appropriate places, capable of being copied in 'pièces,' and subject to rearrangement each time a copy was made." (p. 454).

8. Cf. Elspeth Kennedy. "The Scribe as Editor." *Mélanges de langue et de littérature du Moyen Age et de la Renaissance offerts à Jean Frappier, professeur à la Sorbonne, par ses collègues, ses élèves et ses amis* (Geneva: Droz, 1970), pp. 523-31.

9. Williams,[7] p. 447.

10. In the words of Ernest Hœpffner: "L'auteur, lorsqu'il écrivait ces vers [i.e., the *Prologue*] avait sous les yeux son œuvre poétique tout entière, ou au moins à peu près terminée, et c'est sur l'ensemble de ses productions lyriques, sur ses dits, sur ses compositions musicales que porte le jugement qu'il émet dans le Prologue." Ernest Hœpffner, ed. *Œuvres de Guillaume de Machaut* (Paris: SATF, 1908-1921), Vol. I, p. liv. Quotations from the *Prologue* are taken from this edition with line references given, when appropriate, in the main body of my text.

11. It is important to note that the *Prologue* is found in its complete form only in MSS *A* and *F-G*, those which Hœpffner calls "les meilleurs manuscrits, qui sont en même temps les plus complets . . ." (I, liii). MSS *E* and *H* (=B.N., f. fr. 881) contain only the ballades of the *Prologue*, without the sequence of octosyllabic couplets.

THOUGH FROISSART is best known as a chronicler, preserving "historical" events for posterity, he was also a facile, significant, respected poet in his time, delighting in the vagaries of love. Froissart's last lyrico-narrative poem has been described as a possible farewell to the life of love. This poetic text merits comparison with Froissart's earlier and related poem, *L'Espinette amoureuse*. Differences abound in the dream sequences: the change from the "mirror motif" to the device of "the portrait"; the presence of certain mythological *exempla*; the role of the poet-figure in both compositions. Froissart's poetic work brilliantly utilizes the tradition of the *Romance of the Rose*, as this tradition was translated by Guillaume de Machaut.   MPC

Froissart's *Le Joli Buisson de Jonece*: A Farewell to Poetry?

MICHELLE A. FREEMAN

Department of French and Romance Philology
Columbia University
New York, New York 10027

I N preparing this paper I have allowed myself to trespass on the—to me—*terra* largely *incognita* of fourteenth-century French poetry. I do this because I wish to draw your attention to an overly neglected poet who happens to be situated far down the line of that most complex poetic tradition we have just heard about. Jean Froissart speaks after Guillaume de Lorris, Jean de Meun, and, of course, also after Machaut. Moreover, he endeavors to reinvent something worth saying, all the while conscious of the shadow cast by his remarkable models and predecessors. I have great sympathy for a man in such a situation, especially since I find myself in a similar spot, having to cast around for something worth saying after my colleagues Drs. Uitti and Brownlee have discussed the *Romance of the Rose* and the *œuvre* of Machaut. Although the third of three speakers in this part of our program, I do not claim to offer the last word on this awesome poetic tradition—only, perhaps, the good sense to speak in last place. But this is perhaps fitting in more than one way, because I should like you to consider a last poem—the last *dittier* and *traitiers amoureus* that Froissart composed—a poem entitled *Le Joli Buisson de Jonece* and dated 1373.[1]

The *Buisson* is a complicated, even baffling, work; it raises many problems and questions. I shall outline a few of these questions here, with a view to asking, in general, whether this composition might not (and interestingly) constitute a personal farewell to poetry on Froissart's part, consciously undertaken by him, as such.

The *Buisson* begins with what might be considered as a conventional set of opening remarks. It makes use of conventions typical of clerkly romance prologues; these, however, are filtered through the first person perspective of the one who testifies to a truth he has lived and/or wit-

0077-8923/78/0314-0235 $01.75/2 © 1978, NYAS

nessed. The speaker (Froissart does not sign this poem) recalls adventures
from out of the past: "Des aventures me souvient/ Dou temps passé"
(vv. 1, 2). Since he has both the required understanding and memory at
hand, as well as the necessary writing implements, he will, before his
death, relate how it was that he visited the Buisson de Jonece: "Que je
remonstre avant me mort/ Comment ou Buisson de Jonece/ Fui jadis, et
par quel adrece" (vv. 8–10). He further elaborates on the fact that this
activity is appropriate for the present moment; later he might well forget
and be unable to take advantage of his still-fresh memories.

The end of the poem justifies this initial argument, since the poet-narrator claims that, conveniently, upon awakening from his dream, he remembered everything perfectly, even as he had dreamed it:

> Mais noient je ne m'escondis, —
> Ne je ne puis ne ne poroie,
> Ne faire ossi ne le vorroie, —
> Que, quant je me fui esvilliés
> Et une espasse esmervilliés,
> Que je n'euïsse endroit de mi
> Plain memore, sans nul demi,
> De mon songe, tel et si fait
> Qu'en dormant je l'avoie fait (vv. 5119 ff.).[2]

This construct, aimed at capturing the reader's loyalty and faith, recasts in
a personal, lyric framework what had been a familiar convention in *translatio studii* romances of the twelfth century: namely, memory—a faculty
of mind as well as of poetry and rhetoric—is invoked here as an authority
that allows for the continuing of the past in another time while also insur-
ing believability. In so-called "historical" romances, where, traditionally,
the clerkly narrator belabors the point that the history he recounts deserves
belief, the *clerc* would often refer to some happily retrieved *written* source
—some book—which faithfully contained an eye-witness account and
which he promises to translate according to the measure of his talent and
honesty. Froissart makes claim to no book but rather, in accordance with
the *Rose* tradition and with his own personal dream-experience, he will
translate or, more appropriately, *transcribe* an objective individual memory. (The psychomachia takes on a new twist.)

Just as the Prologue to the *Rose* made a distinction between the dreamer's experience and the interpretation of that dream in distinguishing between Cicero's Scipio and Macrobius' commentary, so too does Froissart's
poet-narrator distinguish between two kinds of intelligence or perceptions

as well as between two different kinds of time in the person and world of the dreamer who returns to youth—who through an illusion is twenty again—and the older, more responsible transcriber of the dream, the 35-year-old. The notion of transcriber, or recorder, is central both to Froissart and this poem.[3] In part it represents also a departure from the *Rose*/Machaut tradition. Though perhaps echoing the *Rose*, the poet-narrator renews the image when he calls our attention directly—at first glance seemingly gratuitously—to his writing instruments: "Encre et papier et escriptore,/ Kanivet et penne taillie" ('Ink and paper and writing-desk,/ Penknife and sharpened quill' [vv. 4, 5]). He does not explicitly depict himself as he who reinterprets an experience of a previous age in order to learn anew from it, in order better to understand who he is. Nor does he wait five years to write down his dream after it has come true. Rather, in his dream, Froissart visits the past and transcribes *his* dream, not *the* dream-come-true, immediately upon awakening. The poet-narrator shows himself as someone who can serve an event that did in fact transpire; because he travelled to a spot worth mentioning, he remembers it clearly, and can accurately record it. He is voyager, witness, and recording-secretary.

Just as the *Rose* surprises its audience by *not* pursuing an expected clerkly sort of *exordium* after its first lines, so does Froissart's Prologue baffle its reader by *not* immediately shifting into the expected discourse on Spring, the dream, and the love affair. Instead, it continues to exploit several *exordium* conventions proper to a non-lyric situation. Froissart does this, however, in a mode reminiscent of the *Rose* in that he presents his *persona* as a lyric poet in a monologue-dialogue situation with a guide and teacher, i.e., with his own thoughts (Pensées, v. 102) which later are suddenly transformed into the person of Philozophie (v. 192). Philozophie tries to convince the poet to take up once again the *gens mestiers* of composing, according to his *Nature, biaus dittiers*. This astonishing set of arguments occupies some 900 lines, thereby attributing to what seems to be proper prologue material about one-sixth of the entire poem! Indeed, our poet-narrator is portrayed as someone reluctant to speak, or at least as someone loathe to focus his thoughts once again on a subject-matter proper for a straightforward *biaus dittiers*.

Philozophie suggests to the poet a number of reasons, or rationales, in order to induce him to compose once more: (1) It is, we recall, his nature to write poetry; this is what he was born for, what he was trained to do (an argument reminding one of the clerkly topos that one must not hide one's light under a bushel); (2) Philozophie tries to shame the poet: What would his previous patrons think of him now? Froissart's *persona* calls upon Froissart the poet's personal history as court poet to list a goodly number of his former patrons. (Later he will also recall and list, in chronological order, those poems he has written in the past that pertain to the same genre as the *Buisson;* these categories of thought call to mind such twelfth-century clerkly prologues as we find in Chrétien de Troyes' *Cligés* and elsewhere.[4]) Finally: (3) Our poet-narrator brings up the objection of his advanced years—he is 35, like Dante, *in mezzo del cammin di nostra vita* —and has nothing new to say. At this point Philozophie begs him to think back to his past and to those works he has just listed, commanding him to write about them:

> Et adonques me renouvelle
> Philozophie un haut penser
> Et dist: "Il te couvient penser
> Au temps passé et a tes oevres
> Et voel que sus cesti tu oevres" (vv. 460–464).[5]

The poet replies that he finds it difficult to remember, so Philozophie encourages him to think about something concrete, to draw out of his coffer that portrait of his Lady-love who abandoned him ten years previously. The poet is tempted; he agrees to see the portrait once again and when he contemplates art's life-like transformation/recording of his Lady's beautiful reality, he is transported into song—the *virelai* "Vémechi ressuscité" (vv. 563 ff.)—as well as into other lyric conventions; he thinks constantly of his love and only has eyes for the portrait.

A few interesting details help one to appreciate this crucial moment of transition: (1) The poet vows to serve not the Lady but the portrait (vv. 607 ff.)[6]; (2) His thoughts which derive from contact with the portrait are alluded to as *ymagination* (like memory, another faculty of mind), and they lead him into the dream, to the trip back to the Buisson de Jonece; and (3) He dates the night of the contemplation of the portrait and of the dream as 30 November 1373 (vv. 859-860). A touch of realism is thereby added to the illusion of the experience as well as to that of the poem. The detail represents an exact reversal—May as opposed to the month of November, six months later—of the season in which both his dream *and* the *Romance of the Rose* were both (most lyrically) situated. In addition, this November night (when he retrieves the portrait) constitutes still another reversal: the reversal of Machaut's actions on a similar November night in the *Voir*

Dit when that poet-*persona* placed a portrait of *his* Lady in a coffer. Furthermore, it is singularly remarkable that Froissart did not choose as the symbol of his inspiration and as his point of departure a mirror but a portrait. It is not the seemingly natural reflection of the *Rose* (which prevailed in Guillaume's poem) nor is it the mirror given to Froissart in his earlier work, the *Espinette amoureuse*, in which, during the course of his dream he sees his Lady and hears her speak to him comforting words of love, but rather, here it is a portrait, an accurate, unchanging, real work of art, commissioned as the result of his experience of 10 years earlier that constitutes his way into the new poem, the *Buisson de Jonece*. Froissart thus seems to be making a conscious break with the double *Rose*/Machaut tradition as well as with his own *Espinette*. Given the context of Philozophie's command to work on his previous works and given his own declaration of service to the portrait, I feel impelled to interpret the portrait as an emblem of his past work, i.e., a work that incorporated conventional lyric devices in such a way as to convey an accurate depiction of reality. Put more simply, I see the portrait in the *Buisson* as serving as a metaphor for the *Espinette amoureuse*.

The *Espinette*, we recall, ended with the poet, though still in love, utterly unrequited and, like Guillaume de Lorris, still in doubt. Consequently, there remains room for improvement and for a final decision by the Lover with respect to his Lady and to his youth. A response to the *Espinette* is thus very much in order; it demands continuation and completion (as, of course, did Part I of the *Rose*).

Now, I should like to sketch out two areas of comparison between the *Espinette* and the *Buisson* as well as to suggest possible interpretations.

In the *Espinette* Froissart describes himself as a child, alluding to the kinds of games he played, until one day he wanders into a garden and meets a young man—a stranger—who unexpectedly calls him by name. Everything is quickly explained: the young man is Mercury, he has known the child and has been in charge of him for the past ten years. In his company Mercury brings Juno, Pallas, and Venus. He prompts the youth to judge Paris' judgment of the three goddesses. The Froissart of the *Espinette* approves of Paris and opts again for Venus. Mercury departs, muttering that it is the same with all young men who fall in love. Shortly thereafter, the young man meets a young Lady who is busy reading a romance. He reads a few passages to her; they speak together for a while; the young man—reader of romances—falls in love. As things turn out, the young Lady decides to marry somebody else. This bad news causes the Lover to fall prey to a feverish illness; upon recovering, he departs—apparently for England!—in order to recuperate. These events take up the midpoint and a good part of the second half of the poem, respectively. The midpoint comprises 50 stanzas of a formal *Complainte*, or lament, in which the poet tells the story of Phebus and Daphne—this is an analogue to what might become his and his Lady's fate. This tale is followed by other lyric meditations on his most melancholy situation. The second half of the poem concerns the Lover's voyage and stay in England. While there, he has a dream in which he believes he sees his Lady in a mirror; she fulfills his desire by promising her love to him.

The second half of the poem, then, recasts the more realistic events of the first half into a dream experience which, in turn, explains and resolves them. The *Buisson de Jonece* does much the same sort of thing with respect to the *Espinette*. However, in the former poem, considerably less concern is expressed for the events themselves, for the "real" experiences or encounters—at least seemingly. Insetad, far more time is devoted to talking, notably in digressions, in dialogues with such allegorical characters as Venus, Jonece, Desirs. These conversations often pick up on, and amplify, lengthily mythological *exempla* only hinted at—merely authorized—in the *Espinette*.[7] Furthermore, the amplifications are often brought off within a great list of such *exempla*, and this bestows on the digression even greater proportions and allows it almost to stand on its own. Secondly, rather than constituting a neat *recueil* recited grandly in the very middle of the romance (along with an additional 15 stanzas recited by the dream-Lady in one speech as a response), the lyrics in the *Buisson* appear to be more gratuitous: they behave rather like comments, like ornamental grace-notes or emotional reactions on, and to, the action at moments of transition, of great feeling, or composed to kill time while the Lover waits around for a word from his Lady.

Comparing what might well be considered as one midpoint with another, one could mention the group of poems that occurs roughly at the center of the *Buisson*. These poems are sung by members of the company in the dream-garden, the Poet-Lover being only one voice *inter alia*. These differences suggest to me a fragmenting of the fully integrated situation of the *Espinette*. The implications of such a fragmentation deserve fuller explanation.

Next, let me note that one of the clerkly digressions contained in the *Buisson* is brought on when the Lover asks Jonece to explain the nature of the *buisson* itself. Jonece answers in some detail, saying he once heard a teacher of philosophy compare the firmament to a green *buisson*, or 'bush.' (Thus, the *buisson* as a datum of the reality of the dream derives from a metaphor—and a pedantic one to boot—created in the reality that exists presumably outside the dream.) Jonece declares that the leaves are the stars, that the *buisson* has seven branches, each one representing one of the seven planets (Moon, Mercury, Venus, Sun, Mars, Jupiter, and Saturn). He then goes on to explain the significance of each planet according to its influence on the life of a human being. The Moon takes care of the newborn until he reaches the age of four; Mercury then steps in for the next 10 years ('til the age of 14-15) at which point Venus takes charge for another 10-year span. This is exactly the paradigm enunciated in the initial, "autobiographical" verses of the *Espinette*. There our protagonist, we recall, was described in childhood (the tutelage of the Moon) and, subsequently, from the time at which Mercury and Venus exchanged responsibility for him. Since the *Buisson* was composed (supposedly) 10 years after the *Espinette* experience,[8] presumably the interim was governed by the Sun, and (consequently) the young man busied himself with seeking honors and recognition. (This period of time may be the era alluded to in the Prologue of the *Buisson* when the speaker offended Nature by having written for money.) If, then, we can presume that this pattern applies here, then it would seem that at the categorical age of 35, Froissart's *persona* should find himself at yet another transition point, switching from the Sun to his new guardian: Mars. Of course, Mars is identified as the god of war; the planet urges that he be given his due by others, that he be recognized and rewarded: "Et encline l'omme a acquerre,/ Soit par grant art ou par conquerre" (vv. 1666, 1667).

When the Lover finally speaks to his Lady, he does so at length and directly (albeit in a dream), couching his service to her in 100 octosyllables this time—a narrative form—instead of composing, as he did in the *Espinette* (vv. 2340 ff.), a *Complainte* of "cent clauses desparelles." The Lady's reply, however, is both curt and ambiguous. In this *Buisson* (or *Espinette* revisited) the Lover is no better off in his new dream and no more assured of fulfillment than 10 years earlier. The games in the garden continue, leading to a new game invented by Jonece. There will be a contest to see who can devise the best poem on the subject of wishing.[9] This second set of poems (sung in the garden) describes Utopian conditions that constitute nothing more than lyrically conventional modes—or codes—of behavior. In effect they do away with all such obstacles as make possible the tenuousness and doubt characteristic of new love—of a new love that is growing but unfulfilled. Unlike the set of lyric pieces sung at the mid-point, in this collection all the poems adopt exactly the same form. To my taste they are perhaps charming but not really interesting as poetry. Fur-thermore, this time the Lover does not choose to participate, or play the game. (His Lady also refuses to be included.) Instead, he accepts the rôle of secretary, writing down each poem in turn, preserving them so that they may be judged and the prize of a crown of blossoms be awarded (a French *jeux floraux* opposed to Petrarch's laurel?). This is repeated ritualistically over and over again. Once the company decides on the God of Love as the perfect judge of the contest and the group parades off to seek him out, the poet is abruptly awakened. In this last series of poems it would seem that the Lover is increasingly withdrawing from the world of the *Buisson*. His interview with the Lady leads nowhere; he no longer participates in the lyric song-writing and he becomes increasingly a spectator and, I repeat, a recorder. What counts is what goes on around him: in a way he reverts to *clergie*. This kind of increasing withdrawal prepares the reader for the awakening as well as, I believe, for the poet's adopting, after the dream, the rôle of recorder or transcriber. This was his last rôle in the world—a complex world, I remind you—of the *Buisson*, and it is also the rôle he had adopted, after the dream, at the beginning of the poem.

All the above suggests to me the following conclusions and, indeed, a hypothesis. According to Froissart's astrological model, the *Buisson de Jonece* belongs to the realm of Mars, i.e., to the planet that governs war and public recognition—in a sense, judgment and reward. Now then, judgment and reward were the last concerns of the dream; they also con-stitute the theme of an important digression in the Prologue (vv. 795-826) to which, let me stress, the poet returns in his Epilogue. In his *Lay de Nostre Dame* Froissart worries about the ultimate judgment and reward, the Last Judgment. Froissart's poet *persona* decides to leave aside those thoughts and deeds which might cost him salvation. He does not mention the portrait again; instead, he composes a beautiful and most demanding lyric poem to *the* Lady who, as *mediatrix*, always responds and remains ever-virgin as well as ever-comforting.

Furthermore, he transposes his image of the *buisson* and the fire-like desire he endured while there in its presence into those images ascribed to —or viewed as emblems of—the Virgin: the burning bush of Moses and the tree of Jesse.

Though clearly very important, these Christian overtones nevertheless do not contain the whole story. Other, more specifically poetic, resonances are significant too. Thus, not only does Froissart leave the *world* of Venus and desire, he no doubt is making his farewell to the kinds of poetry appropriate to that world (*dittier* and *traitiers amoureus*), genres within which he had composed on four occasions in the past. And in so doing, at least as far as he himself is concerned, he causes the tradition represented by the *Rose* and by Machaut to come full circle.

By commenting in the *Buisson* on his previous poem, the *Espinette*, i.e., by answering it, reversing it, and reworking it so as to arrive at a definitive response, Froissart reënacts, to all intents and purposes, the poetic example of the *Romance of the Rose*. A love-dream poem is continued and completed by a second such poem; the characters are the same in both, but the perspectives have shifted radically; in Froissart's case, of course, the two poems constitute one *œuvre* created by a single poet (not by two poets, as in the *Rose*). Secondly, what I deem to be an ambiance of fragmentation (coupled to the heavy influence of Mars) may well signify that the hitherto unified world of romance-adventure plus love-experience will, in the future, be broken down into separate and distinct components: history and pure lyric. In bidding a final *adieu* to the poetic world we have analyzed here, Froissart will remain the traveller, the witness, and the transcriber he showed himself to have been; he will continue to concern himself with power, wealth, and conflicts. But his witness and his transcriptions will deal with *others*, not himself *as such*. The roots of Froissart the *chroniqueur* lie deeply embedded in the experience of Froissart the poet; I suspect that he was very much aware of this.

One final note: We recall the ink, paper, and pen imagery with which Froissart opens the *Buisson de Jonece*. Let us not forget that throughout the history of medieval French poetry—from the *Life of Saint Alexis*, to Béroul's *Tristan*, to Jean de Meun, down, at least to François Villon—this set of images has been used to indicate the transcription of the writer's (or the character's, or the poem's) last will and testament. Froissart, we have seen, was no exception.[10]

NOTES

1. Line references given in my text to the two Froissart texts discussed here are drawn from these major editions: *L'Espinette amoureuse*, ed. Anthime Fourrier, Bibliothèque Française et Romane, Série B:2 (Paris: Klincksieck. 1963), and, by the same editor, *Le Joli Buisson de Jonece*, Textes Littéraires Français: 222 (Geneva: Droz, 1975); students of medieval European poetry owe an incalculable debt to these magnificent editorial enterprises.

2. A literal translation:

> But I do not at all deny, —
> Nor can I nor could I,
> Nor would I wish to do so, —
> That, when I had awakened
> And marvelled for awhile,
> That I had before me
> A complete remembrance, not just a partial one
> Of my dream, such and so done
> As I had created it in my sleep.

3. Froissart uses the verb *recorder* and the noun *recort* rather frequently throughout the poem; a glance at the Glossary alone will confirm this. The words take on different shades of meaning depending on the context. Fourrier glosses the verbs as 'dire, déclarer, exposer,' and the nouns as 'propos, paroles; rappel, souvenir.' With all due respect, however, I believe that at certain moments—during the Prologue, the end of the dream, and at the moment of the decision to tell the dream—the verb *recorder* takes on a combination of the ideas of 'to recall' and 'to write down.' In addition, mention is made of *registre* and *registreur* (not glossed by Fourrier) which, I think, helps to fill out the implications of *recorder*. For example, as we shall shortly see, in her encouragements of the Lover-Poet to write once more, Philozophie asks him what would have become of the memory of Gauvain, Tristan, Perceval, Galahad, Arthur, and Lancelot if we did not possess *registre* (v. 409): *registre* provides us with their deeds. She goes on to glorify this rôle of *aministreur/registreur*:

> Et ossi li aministreur
> Qui en ont esté registreur
> En font moult a recommender.
> Je te voel encor demander
> Se no fois, qui est approuvee,
> Et n'est elle faite et ouvree
> Par docteurs et euvangelistes?
> Sains Pols, sains Bernars, sains Celistes
> Et pluiseur aultre saint preudomme,
> Que li Sainte Escripture nomme,
> N'en ont il esté registreur?
> Moult ont pour nous fait li docteur
> De pourfit et de grant consel.
> Pour tant, amis, je te consel
> Et te di en nom de castoi:
> Ce que Nature a mis en toi,
> Remonstre le de toutes pars
> Et si largement le depars
> Que gré t'en puissent chil savoir
> Qui le desirent a avoir (vv. 411–430).

> 'And, in addition, the "administrators"
> Who have been "registrars"
> Do much that is to be commended.

I want to ask you once again
If our faith, which is proven,
Is it not made and worked out
By doctors and evangelists?
Saint Paul, Saint Bernard, Saint Calixtus,
And several others—holy men—
who are named in Holy Writ,
Were they not "registrars"?
Much have the doctors done for us
That is profitable and of enormous counsel.
For this reason, friend, I advise you,
And tell you in the name of each one of them,
What Nature has placed within you:
Display it everywhere
And parcel it out generously
So that those may be grateful to you
Who wish to receive it.'

Philozophie conflates here in this one argument romances and Holy Writ, fictional stories of "our ancestors" and sacred history—dreams and reality, in a sense—in order to have the protagonist imitate such writers. By so combining "history" and "fancy," she prepares the way for Froissart's *persona* to record the *fact* of his vision. We are reminded of *translatio studii* prologues; here, however, clerkliness takes on the duty and burden of the recorder, of him who records facts, a kind of translator of reality, not the (mere) translator of texts.

During the sequence of the wish-poems, as already mentioned, the Lover participates only as a recorder or *registreur*. The two ideas are conflated here in the business of writing down the text of the poems he hears sung to him. (Apparently, Froissart's poem exists *only* on the level of the written text; no musical notation was designed to accompany it.) The point is explicitly made: "Cesti souhet mis en recort/ Et registrai ensi qu'il doit" (vv. 4944, 4945). And another, still more explicit (and lengthy), example occurs when the Lover is chosen to write down the wish-poems; this example, let me add, recalls the opening lines of the Prologue in its reference to writing implements:

Lors que Plaisance eut souhedié,
A fin que mieuls soient aidié
Leur souhet et mis en recort,
Il eurent entre iauls un acort
Qu'on les escrise et les registre.
Lors me delivran le registre,
Encre et papier, che me fu vis.
Puis mis mon sens et mon avis
A l'escrire et au registrer (vv. 4671–4679)

'After Pleasure stated her wish,
So that their desire could be more effectively helped
And remembered,
They agreed among themselves
That all be written down and recorded.

> Then they gave me the "register,"
> Ink and paper, as I recall(ed).
> I then put my sense and my will
> To the writing of it and to the recording of it.'

This notion of tracing something down accurately and truthfully on paper (like an inspired Evangelist or like an eye-witness historian) strengthens the idea of seeing the portrait—a "written" likeness—as an emblem, or analogue, of the kind of poetry with which Froissart is concerned.

4. In *Cligés*, for example, a list of titles strengthens the self-portrait of the poet-narrator *persona* in terms of his professional history. The prologues to such works as Wace's *Rou* or Benoît de Sainte-Maure's *Troie* list the chronological history of the clerk's patrons and thereby, in round-about fashion, justify the fact that the poet dares to write: dares to tackle a particular subject. (And let us recall, for what it might be worth, that, like Wace and Benoît in the twelfth century, Froissart also wrote for the King and Queen of England.)

5.   'And then Philosophie to me
                Brings back to mind an important thought,
                Saying: "You must think
                About your past and about your works,
                And I want you to work on this work.'

6.   "Il n'est bericles ne topasse,
                Rubis, saphirs ne dyamans,
                Escarboucles ne aÿmans
                Qu'on dist qui areste le fer,
                Qui me peüst faire escaufer
                Ensi que mon ymage a fet.
                Or le voel servir et de fet,
                Car moult m'en vaurra li regars."

                'There is no beryl, topaz,
                Ruby, saphire, diamond,
                Carbuncle, or magnet,
                Which, they say, stops iron,
                That fires me up
                As my portrait has done.
                Now I wish to serve it, indeed,
                For the look of it will be worth much to me.'

7. For example, when the Lover finds himself aflame in the presence of Desirs so that he cannot go on to rejoin the group or to speak to his Lady, in misery he would fain seek the company of others who might have shared his plight in the past. Desirs enumerates 13 examples, beginning with Phebus and including Orpheus and Pygmalion (in love with *his ymage*!), Tibullus, Achilles, and others, and ending with Ovid, Vergil, and (!) Aristotle. (The Achilles *exemplum* is particularly interesting in that, instead of repeating the moral of the story, Desirs [or the poet-narrator—it is hard to tell who is addressing us] refers to the tale as told by the Lover-Poet after he has seen the portrait [but before he has experienced the dream], thereby underscoring the importance of the

textual, written, ordered nature of the *book* we are reading [vv. 3352-3359].) At the precise midpoint of this litany we find *the exemplum* of the *Romance of the Rose,* Narcissus. Here, however, this lyric protagonist *par excellence* is no longer viewed as a sinner but rather as a victim: the victim of an Echo who has died and left him behind. The reflection he sees in the pool is not his own, he thinks; it is the image of his Lady for whose love he burns and languishes. Once again, and to his own purposes, Froissart has turned a model text upside-down.

8. In reality—so far as we can judge—the *Espinette amoureuse* was composed three years prior to the *Buisson.* It is interesting to note that Froissart's *persona* in the *Buisson* is also 10 years older than Guillaume de Lorris' lover-turned-poet.

9. The fact that the dream sequence ends with a *recueil* of wish-poems in which the Lover-Poet does not choose to participate may also indicate a denunciation of, and break with, his past love, since, in the Prologue, he characterizes his time of youth and love as a period during which he lived on wishes (see vv. 734-746).

10. In a sense, Froissart not only reverts to *clergie,* as I have suggested, but also, in more modern fashion, to what was meant by being an *auctor:* he deals with the recording of a recently lived reality *for the first time.* Froissart bids farewell to the personally experienced emotional reality of the lyric sort in order to take on the character of the witness and the recorder of contemporary *history.* Curiously, though highly innovative (as I have endeavored to show), Froissart, in this way, illustrates the cyclical pattern of Old French literature, i.e., the pattern that begins with focus upon the historical/sacred, moving on to textual self-referentiality (and lyric), then back again. Froissart *contains* this circularity. His innovation, of course, lies in his concern for contemporary, even recent, events. As chronicler, Froissart is no mere traditional *clerc;* he merits the title of *auctor/poëte* (verse as well as prose) for his eye-witnessed truthfulness and a sense of poetic value. Let us also remember, in connection with Froissart's farewell to *biaus dittiers amoureus,* that he had been awarded an ecclesiastical benefice in September, 1373.

No ONE has convincingly identified the beautiful unnamed lady who tries to seduce the hero of the fourteenth-century English poem, *Sir Gawain and the Green Knight*. Her husband (who is the giant Green Knight) identifies himself as "Bercilak" (or "Bertilak"). Neither his wife's motivations nor his own are explained, or the bizarre actions of beheadings, sexual temptations, hunts, feasts, and festivities. Apparently the *Gawain*-poet used as background the thirteenth-century French prose (or Vulgate) *Merlin* and its later anonymous "continuation." In this work there appears a violent knight named "Bertolais" ("Bertelak" in a fifteenth-century English version) and a woman who must surely be Gawain's temptress, for their roles exactly parallel. Thus the actions of these major characters in *Sir Gawain* becomes comprehensible   MPC

Bertilak's Lady:
The French Background of
*Sir Gawain and the Green Knight**

RICHARD R. GRIFFITH

C.W. Post College
Long Island University
Brookville, New York 11548

*S*ir Gawain and the Green Knight is an acknowledged masterpiece, the
finest of the English courtly romances, worthy to stand alongside spec-
imens of its genre in any other tongue. Composed by an anonymous artist
of the Northwest Midlands, who almost certainly produced *Pearl* and at
least two other religious poems, *Sir Gawain* has received during the past
three decades an ever-increasing share of scholarly attention and critical
acclaim—perhaps represented best by the current practice of designating
its author "the *Gawain*-poet," whereas formerly he was called "the *Pearl*-
poet." Various attempts to date his works have produced nothing more
definite than a general ascription to the latter decades of the fourteenth
century, but certainly the poet must have been a contemporary of Machaut's, and it is the thesis of this study that knowledge of the French
Arthurian works on which the poem was based—works which were certainly part of Machaut's world—is essential to understanding its action
and characters.

The literary esteem which the *Gawain* now enjoys was long in coming.
It received only limited circulation during the period immediately following its composition, and thereafter languished forgotten until it was
printed by Madden in 1839. Even after reaching wider audiences, *Sir
Gawain* was regarded as of primarily antiquarian and linguistic interest.
Among the first to take it seriously as literature was George Lyman Kittredge, who found—as later readers have found—that considered as an
integrated, self-contained work of art it is badly flawed in characterization

* I am grateful to the Research Committee of C.W. Post College, Long Island University, for a grant covering the costs of typing this article.

0077-8923/78/0314-0249 $01.75/2 © 1978, NYAS

and motivation (1, pp. ix-x; 2, p. 136). A selective recapitulation throws its failings into harsh focus:

King Arthur's Christmas feast at Camelot is invaded by a green giant who challenges the assembled knights to an exchange of blows with his axe. Arthur accepts, but Gawain, seated beside the lovely Queen Guinevere, courteously requests the task and is accepted by the Green Knight after he inquires Gawain's name. Answering him truthfully, Gawain similarly requests the Knight's true name, but is put off until after the blow. Gawain neatly decapitates the outlandish visitor, but the unperturbed Knight recovers his head, which speaks to demand that Gawain present himself at the Green Chapel a year hence to receive the return blow.

The seasons pass, and Gawain sets forth to fulfill his vow. Finding himself at a strange castle in the wilds on Christmas Eve, he asks admittance, and is welcomed by the beaver-bearded lord, who assures him that the Green Chapel is nearby. Gawain agrees to stay and rest for a few days. He is presented to the lord's beautiful wife, who sits beside him at meals, and an ancient, highly respected lady who is also in residence.

After dinner, Gawain's host proposes a game: for each of the next three days he will go hunting while Gawain remains at the castle, and each evening the two of them will exchange whatever they have "won" during the day. Gawain agrees, and while the lord hunts (deer, a boar, and a fox on the successive days) his lovely wife steals into Gawain's bedroom, bolts the door, and starts hinting—more and more explicitly—that she would like the knight to make love to her. Gawain manages to stop at kisses (one, two, and three on the respective days), and even forces her to *ask* for these, although she complains he is not living up to his reputation for chivalry and that he has not learned that a courteous knight should seek them without prompting. In the evenings the lord of the castle presents Gawain with the day's trophies, and the knight dutifully bestows kisses in return, declining to reveal their source. However, on the final day Gawain's temptress offers a green sash as a love token; at first Gawain refuses the gift, but when she tells him it is magic and will preserve his life, he changes his mind. Instead of delivering it to his host that evening, Gawain conceals it and wears it to the Green Chapel the next day.

The giant appears, and Gawain offers his neck. After two feigned blows, the Green Knight nicks him slightly. He then reveals himself as the lord of the castle, and explains that his lady's actions had been a test of Gawain's chivalry, which he had failed only in the matter of the sash (represented by the small cut he received from the third stroke). He praises Gawain and invites him to return to the castle as honored guest, but Arthur's knight is humiliated, and before departing demands the true name of his antagonist. The Host/Green Knight replies that in this country he is called "Ber__lak de Hautdesert," and adds that the ancient lady staying at his castle was Gawain's aunt, Morgan le Fay, who through her sorcery had initiated this sequence of mysterious events in the hope that Queen Guinevere would die of fright when the severed head spoke. Apparently quite satisfied with this explanation, Gawain vows to wear the green girdle ever after as a mark of his shame, and returns to Arthur's court to relate his adventure.

Gawain† may be content with the Green Knight's explanation, but the reader is less easily satisfied. Morgan le Fay, King Arthur's half-sister, is indeed the standard villainess of Arthurian romance, a "given" of the genre, and her general dislike of Queen Guinevere—perhaps reflecting feminine rivalry or anger at being displaced as "first lady" of Camelot— is well established. But if her motive was simply Guinevere's death, why this lengthy and elaborate testing of her nephew Gawain's courage and chastity? And if she is the moving genius behind this whole series of events, the sole mainspring which activates the plot, why has the author given her such a minor role? Who is her confederate, Ber—lak, and what is his motivation for joining Morgan's plot against Guinevere? And—to ask a question that has never really been raised—who is that other major character, Gawain's lovely temptress, whose name is not given, and why is she willing to risk her wifely virtue in order to test a representative of the Round Table?

The Green Knight's explanatory speech sounds like the final chapter of a third-rate mystery novel, a summing up in which the murderer turns out to be somebody you've never heard of, his accomplice is not identified, and their actions arise out of a "motiveless malignity." The comparison to a detective story is not an absolute irrelevancy, for Kittredge was a voracious and inveterate reader of mystery novels, and may have felt the work's flaw in similar terms. Indeed, since an author's success depends, in any age, upon his ability to evoke and manipulate emotional responses in his audience, it may be legitimate to regard some modern *genres*—such as the Western, the spy story, the Gothic romance, the mystery, the historical novel—as specialized analogues for the emotional responses inherent in certain kinds of human/literary situations. The party-crashing green challenger, for example, is surely related to the "new-gun-in-town" who opens the action of many a Western story. Gawain's position as a lone figure in

---

† Except for the variant spellings of *Bertilak*, I have normalized the forms of such names as *Gawain* (which occasionally strays so far afield as to appear as *Wawon*), Arthur (*Artus*, *Arthure*) and Guinevere (*Gaynor*, *Gunnore*).

a foreign environment, surrounded by strangers whose backgrounds and trustworthiness he cannot assess, is familiar to any reader of espionage novels. The revelation that the Green Knight is also a character we have already met, Gawain's host at the castle, is analogous to the identification of the murderer as an unsuspected member of the "cast" in the denouement of a detective narrative—and it may be worth noting that the *Gawain*-poet uses the standard device of providing subtle clues that hint at the identification but are recognized only after the fact: for example, the Host, like the Green Knight, has a liking for a Christmas games that involve exchanges, and even proposes a challenge involving a weapon hanging on the wall (a spear, rather than the axe which was the award he gave in Arthur's hall, and which was afterwards hung on the wall at Camelot) (3, lines 981-984, 284-289, 477-478). But the reference to Morgan le Fay is slightly different, in that it depends upon knowledge external to the poem, part of the general tradition of Arthurian writings. Her magical powers are indeed accounted for by mentioning that she had been Merlin's mistress, but her enmity toward Guinevere is assumed to be so familiar as to require no expository comment. The pattern here is that of the historical novel, and Morgan needs no more development than, say, Baron von Richthofen or Napoleon in their appropriate eras. Precisely the same response is evoked by the Green Knight's revelation that, in addition to being the lord of the castle, he is "Ber__lak de Hautdesert."

In an effort to be stringently scientific, the scholar often feels forced to reject his own knowledge of how literature works, knowledge accumulated over several decades of reading the various genres of narrative, knowledge shared by fellow-scholars and confirmed by their experience. To me, as a representative reader, Ber__lak's revelation of his name seems to parallel that characteristic scene in the historical novel when the stranger whom the hero has just rescued announces "My name is Horatio Nelson (or Charles Stuart)." One is conscious that a certain silent fanfare has sounded, a reaction is expected, and—if the effect fails—there's an awareness that "There's something I'm supposed to know that I don't know," followed by a trip to the encyclopedia.

In the case of *Sir Gawain and the Green Knight*, the matter is somewhat complicated by the speaker's qualification, "Ber__lak de Hautdesert I am called *in this land*" (3, line 2445). Here the modern reader is forced to draw upon his knowledge that *surnames* in the Middle Ages were to some degree changeable, and particularly those derived from an estate or lordship. The name *Hautdesert*, meaning "high deserted place," would apply well enough to the castle Gawain had just been visiting, which stood in a deserted area *above* the valley of the Green Chapel, although a pun suggesting that its owner was *deserving* of *high* reward or position is possible (3, pp. 128-129). The negative implication of the Knight's statement is clear: he is or was known by a different name in some other area. Now, if the title *de Hautdesert* is associated with the environs of the Green Chapel and its nearby castle, it must be the given name that is significant. And significant it clearly must be, for Gawain seeks no further explanation once he hears it. Furthermore, it is part of a long tradition of folk-tale, and Arthurian legend in particular, that the revelation of identity shall be crucially important to the seeker. Round Table romance is filled with instances of disguised knights who invariably turn out to be the protagonist's brother, monarch, or sworn enemy. Artistically, of course, it would be ridiculously anticlimactic to build up to an announcement of the Green Knight's "true name" and have it prove completely unfamiliar to both hero and audience of the poem.

Yet, for the modern audience, the revelation *is* meaningless; we are as mystified as ever. An element of mystery is often at the heart of the creative process, beyond the conscious understanding of artist as well as appreciator; but this failure in effect is clearly in a different category—the artist fully expected his readers to have the information necessary for understanding, and we lack it.

The first impulse of Kittredge and his contemporaries was the predictable one: to find the name of the Green Knight somewhere in Arthurian romance. One can only guess how much fruitless labor was expended on this project. The name appears only once in the poem, and the transcriber had recorded it for Madden's edition as B-e-r-*n*-l-a-k. Thus it stood for almost a century. The names and variant forms of names attributed to knights in the surviving manuscripts dealing with the Round Table cycle are so numerous as to encourage the belief that almost any pronounceable combination of letters would appear somewhere, but nothing resembling *Bernlak* was to be found (4, p. 13).

To understand what happened next, it is necessary to recall that for the first several decades of this century France was a dominant power in world affairs and laid claim to a cultural superiority dating back to Charlemagne. Two standard histories of English literature, for example, were by Taine and Legouis, both Frenchmen. In scholarship, source study was the prevailing occupation, and its results, so far as early English letters were concerned, seemed to support the French claims. Still in the grip of

the Romantic movement, which placed its highest value on creative originality, English scholars were dismayed when such major writers as Chaucer and Malory were revealed to have lifted line after line bodily (and
complete works as well) from Continental writings, most of them French.
Editors faced with a hiatus in a manuscript or a crux in its reading could
often resolve their problems by consulting a French source. Thus, unable
to resolve the puzzle surrounding the ending of *Sir Gawain and the Green
Knight*, scholars of the English-speaking world, in an excess of humility,
postulated that this remarkable work must be based on a French original,
in which all was made clear. Rather than looking for a work containing a
knight named "Bernlak," which might reveal who he was and why he
and his wife were hostile to Arthur's court, one should look for an entire
French romance which the *Gawain*-poet had rendered into English clum-
sily, perhaps without completely understanding it. No such romance was
known, of course, but several works contained similar themes, and it was
always possible that the crucial one would turn up some day. Furthermore,
one of the consequences of the extensive editing that was being done, with
its comparison of texts and construction of manuscript trees, was an increasing awareness of how much had been lost. If something that one was
convinced had existed was no longer extant, one categorized it as "lost."
Thus the problem of *Bernlak's* identity as a key to the motivations of the
poem was put aside or subsumed under that of the "lost French original."
As recently as 1970, A. C. Spearing was agreeable to assuming that "the
poet was using a lost source," which "may one day be brought to light
out of the tangled forest of medieval French romance" (5, p. 171).

In 1923 James R. Hulbert took a magnifying lens to a facsimile of the
page containing the Green Knight's name (which appeared in Osgood's
edition of the *Pearl*). The letter between *r* and *l*, Hulbert decided was "not
in the least like the scribe's *n*," but was instead a *c-i*, making the name
Bercilak (4, p. 12). Now there weren't any *Bercilaks* in French romance,
either. But a second strand of analogue-hunting had led to Celtic sources,
specifically for the "Beheading Game" motif, and in one of these, *The
Feast of Bricriu*, the giant challenger is referred to as a *bachlach*, "a gigantic
churl." This was close enough to *Bercilak* to suggest the possibility that at
some point in the (assumed) line of transmission the Irish common noun
had become a proper name (6, pp. 59-60; 16, pp. 531-532). Unfortunately,
this explanation of the name shed no illumination whatsoever upon the
remainder of the poem; for in *The Feast* we are told that the *bachlach* was
really the magician, Curoi mac Dairi, who was fulfilling a promise made

to the hero, Cuchulainn, that he would aid him to become first among the heroes, entitling him to receive the Champion's Portion, the choicest cut of whatever meat was being served, and entitling his lady to take precedence over all others in entering the mead hall (7, p. 13). None of this, obviously, has anything to do with Gawain—except, perhaps, insofar as the *bachlach* is indeed someone whose true name, Curoi, would be recognized by the protagonist, Cuchulainn.

Most scholars, critics, editors, and translators readily accepted Hulbert's preferred reading of the name as *Bercilak*, ignoring the other reading he presented as a possibility, Ber-*ti*-lak. One who didn't was Sir Israel Gollancz, who had access to the original manuscript, not facsimiles, and who had edited all three of the other poems in it (*Pearl, Patience,* and *Purity*), had edited the facsimile edition of the entire manuscript, and had twice revised the Early English Text Society edition of *Sir Gawain*. Surely if anyone was familiar with the scribe's hand, it would be Gollancz; and he concluded that the fourth letter was a *t*, not a *c*—a reading verified from the manuscript by Norman Davis in his recent revision of the Tolkien and Gordon edition of *Sir Gawain* (3, pp. 128, 153). Perhaps feeling that a second change in spelling would make themselves appear too inclined to shift with whatever wind was blowing, few who write about the poem have adopted this most authoritative reading. And, of course, *Bertilak*—which Max Förster derives from a conjectured Celtic *Brettulakos**, "a Briton" (8, p. 115)—is much further removed than *Bercilak* from the Irish *bachlach*.

With a better reading for the name, it is possible to undertake the search again. But we need a means of narrowing the survey. *Sir Gawain* is set in Arthur's youth: much is made of his boyishness, his inability to sit still because of his young blood (3, lines 86-89). Yet the time is clearly after his marriage, and the concurrent establishment of the Round Table, for Guinevere is Queen. Lancelot is mentioned casually as one of the many men of Camelot who assemble to say farewell to Gawain, but he is tucked into the middle of the list, with no special distinction shown him (3, line 553); thus the period is narrowed to those months after Lancelot's arrival but before his unique prowess is recognized. There is also no suggestion of any *amour* between him and the Queen, and perhaps Morgan's use of the talking head to frighten Guinevere, rather than her usual device of attempting to expose the Queen as unfaithful, confirms that their liaison had not yet begun. Although several other names listed lack the honorific "Sir" of knighthood, in Lancelot's case this is probably significant, since

most accounts of the inception of his love date it from his receiving his
sword from the Queen at the dubbing ceremony. What is needed, then, is
a work dealing with this time period.

About the year 1200 a Burgundian author named Robert de Boron
undertook an elaborate account of the history of the Grail, starting with
the legend of Joseph of Aramithea and attaching it to the Arthurian stories
by way of Merlin and Percival. Only a small part of his poetic account of
Merlin survives, but two prose versions, the *Suite de Merlin* and the Vulgate *Merlin*, indicate that de Boron carried the story up to the birth of Arthur (9, pp. 251-256). Both prose accounts continue the story beyond that
point, but with many important variations. It is the Vulgate version that
concerns us. Written during the early decades of the thirteenth century,
the Vulgate cycle goes beyond de Boron by tying his final book, which
deals with Arthur's death, to the *Merlin* and *Percival* by the addition of a
book about Lancelot, his great deeds and his love affair with Guinevere.
Toward the middle of the century a different author noted a gap between
the Vulgate *Merlin*, which culminated in the founding of the Round Table,
and the *Lancelot*, which treated the birth and youth of that hero and has
little to say about Arthur's Court until Lancelot arrives there. This blank
period, which he viewed as lasting about three-and-a-half years, is filled
in by our continuator in a businesslike fashion. Specifically covered are
the young King's battles with his rebellious barons, his defeat of the pagan
Saxons, his marriage, his acquisition of the Round Table from his new
father-in-law, King Leodegrance of Cameliard, and finally his campaign
against Rome (10, pp. 322-24).

There are several indications that this sequel or continuation of the
Vulgate *Merlin* was among the important sources for the composition of
Sir Gawain and the Green Knight, beyond the fact that this is the only
major work (except for the *Suite de Merlin*) which covers the appropriate
time-period. For one thing, Gawain is the most noted hero of the *Merlin*,
far overshadowing the young Lancelot or Arthur's earlier companions,
Kay and Bedevere. Gawain's courtesy is emphasized in several places, and
there is a scene in which Arthur accepts a challenge to single combat from
King Rions and Gawain *very courteously* asks the King's permission to
substitute for him, a scene which parallels *Sir Gawain*, in which Arthur
accepts the Green Knight's challenge and Gawain, with equal courtesy,
requests the task (11, p. 417; 12, p. 628; 3, line 201).

Second, the name of Gawain's horse, Gringolet, is mentioned seven
times in the English poem; the *Merlin* gives an account of the winning of

this renowned steed from King Clarion—again serving to date the poem as after the battle with the Saxons when this event occurred—and regularly mentions his name in association with Gawain (11, p. 513); in some other accounts involving Gawain, Gringolet is his *son*.

Third, although heralds and romance writers had assigned coats of arms to all the prominent Round Table knights, Gawain usually bearing on a green field a golden eagle, in *Sir Gawain and the Green Knight* the hero's shield is red with a golden pentangle emblazoned on it. Conservative Arthurians would no more have accepted such tampering with the accepted arms than a modern Sherlockian would accept a change in the Baker Street address, unless there were some justification for such a departure; in the *Merlin* Gawain's own shield is so demolished in battle that a grateful beneficiary of his protection, the Duke of Cambenic, presents him with a new one (its bearings are not described, but the immediacy of the occasion implies that the arms would not be Gawain's own) (11, p. 370; 12, p. 554; 3, pp. 92-93).

Fourth, of the twelve members of the Round Table listed in the poem (3, lines 110-112, 551-555), all but two are prominent in the Vulgate Cycle. One of these, Errik, is familiar from Chretien's romance, *Erec et Enide*, where he is associated with Gawain and Lancelot; Errik also appears in the English *Awntyres off Arthure*. Bishop Baldwin, the second, comes originally from Celtic sources, and is Gawain's companion in a later analogue of the temptation story called *Sir Gawain and the Carl of Carlisle*; even he may be included through confusion with a Bawdin, son of King Ban's castellan, mentioned in the *Merlin* (12, p. 124), who is elsewhere named Banin and is never described as a churchman. The remaining ten, however, are all to be found in either the *Merlin* or *Lancelot*; two of them (Doddinaul le Sauvage and the Duc of Clarence, his kinsman) are prominent only in the *Merlin*, where they are close cousins of Gawain's and received knighthood alongside him (11, pp. 127, 374; 12, pp. 177, 402).

Fifth, the unusual description of Morgan le Fay as a "goddess" is found in the Vulgate *Lancelot*, and the assertion that she was Merlin's mistress and thus learned her magic arts from him is to be found in the *Merlin* continuation (3, p. 129; 12, pp. 375, 508). Sixth, in the *Lancelot* there is an account of an attempt to seduce Sir Lancelot by a young damsel associated with Morgan le Fay, which certainly resembles the events in *Sir Gawain* (4, p. 18).

Finally, the flirtatious speeches between Bertilak's lady and Gawain have numerous verbal parallels to a scene in the *Merlin* between Gawain and a fair damsel he encounters in a forest: like the temptress in *Sir Gawain and the Green Knight*, this lady upbraids Gawain for failing to give her a kiss, declaring that he has not lived up to his reputation for courtesy; she points out that they are alone in a forest, as Bertilak's lady emphasizes that they are alone in Gawain's bedroom; and she similarly lectures the knight on the necessity of remembering in the future to salute a lady when he meets her. As in the English poem, Gawain courteously begs forgiveness for his neglect and excuses himself by saying that he was preoccupied with some other concern (11, pp. 459, 463; 12, pp. 689-691, 695; 3, lines 1230, 1283-87, 1297-1303, 1481-91). The evidence that the poet who composed *Sir Gawain* was quite familiar with the Vulgate *Lancelot*, the *Merlin*, and the *Merlin*-continuation seems beyond question (5, p. 17).

Now, as Hulbert pointed out, there is in these three French works a character named *Bertolai* or *Bertolais* (4, p. 13). Norman Davis adds that the older form of the accusative of this name would be *Bertolac*, and both scholars note that in a mid-fifteenth century English translation of the *Merlin* the name appears as *Bertelak* (3, p. 128). Neither seems to appreciate fully the significance of this latter fact, which clearly demonstrates that there was circulating *in England* a French manuscript of the *Merlin* and its continuation, in which a knight named *Bertelak* plays an important role, within a couple of generations of the composition of *Sir Gawain*.

As it happens, Hulbert, like Kittredge, had his own theory about the origins of the *Sir Gawain* plot; in his view, the initial story was not Arthurian at all, but in some version prior to the *Gawain* it had been given an Arthurian background, and the Celtic name *Bertelak* had been borrowed because of some slight similarity between his behavior toward Arthur in the *Lancelot* and the Green Knight's action in testing Gawain (4, p. 15). The views of scholarship in general are best summed up by Davis, who considers *Bertilak* and *Bertolais* "apparently the same name," but asserts positively: "none of the knights bearing the name . . . can be identified with the Green Knight" (3, p. 128).

It will be noticed that something has gone slightly askew with the scholarly process, and the situation does not improve. For more recent critics, studying *Pearl* and *Patience* especially, have concluded that the *Gawain*-poet was a far more original and creative artist than he had been considered earlier—that he indeed used in these poems traditional Biblical materials and concepts like the dream-vision and the beautiful instructress, but that he, like Chaucer, reworked these into compositions of his own quite different from those he was drawing upon. Hardly any commentor still holds that the *Gawain* is merely a translation of that "lost French original," and most attention during the past few years has been given to the folklore and mythic or patristic or archetypical elements implied in the work (1, passim). Every graduate student learns the proper sequence of events in working with a work like *Sir Gawain*: first one treats the paleographic aspects, then establishes the text, then goes through the narrative level and defines the genre, then deals with the historical and biographical components and the literary background (in conjunction with the source-analogue study); and only after these basic matters are cleared up is it time to talk about meanings. Scholarship works like a combination lock: it's not simply that certain things have to be done, it's that they have to be done in a particular order, and if one decides he's gotten a number wrong, he must start all over again at that point. It is easy enough to spot this sort of problem in a single study done by an individual scholar, but difficult to imagine any such thing's happening through the work of several scholars spread over six or seven decades. Yet it is obvious that the major part of the source study was done before the correct reading of the Green Knight's name had been established, and that the nature of the work was presumed while there were still unresolved problems on the simple narrative level. Until we know who Bertilak and his lady are and why they are acting as they do, we cannot proceed with any confidence.

Now, Hulbert's assumptions about the nature of the poem led him to look for a Bercilak associated with beheading games and chastity tests, and failing to find this direct connection with the character named Bertelak in the *Merlin*, he went no further than suggesting that some predecessor of the *Gawain*-poet had borrowed the name as generally appropriate. With a different view toward the creative process of the poem's author, we may ask different questions. The first is simple and straightforward: Is there any detail in the poem which unmistakably links the Green Knight/Host of the Castle with Bertelak of the *Merlin*? Of some help is the fact that when Gawain meets the Green Knight in his normal shape, the first thing he notices about him is that he is "of hyghe eldee," perhaps rendered accurately as "in the prime of life," definitely older than the youthful Arthur and his nephew, Gawain (3, line 844); when Bertelak appears in the *Lancelot* some years later, he is described as seeming old (13, Vol. 2, p. 11); more significantly, the second characteristic Gawain notes is his host's beard, which was *bryt* and *beuer-hwed*, "light (or intense) reddish brown," whereas in the *Merlin* Bertelak is consistently nicknamed *le*

Rous, "the red" (3, line 845; 11, pp. 310-313; 12, pp. 466-70). There are further confirmatory similarities: Gawain's host is described as "fre of his speche," whereas Bertelak is "full of feire courtesie and a feire speker"; the host seemed well suited "to lede a lortschyp" of good knights, while Bertelak brought with him "grete plenty of knyghtes," and when he leaves the Court, "many a knyght hadde he hym to conueye to whom he hadd yoven many feire yeftes, for he hadde be a noble knyght and a vigerouse" (3, lines 847-849; 11, pp. 312-13, 12, pp. 469-70). It seems a good possibility that the name "Bertilak de Hautdesert" bore before he acquired this estate was "Bertelak le Rous."

The next questions are part of an interlinked series: Did the Bertelak of the *Merlin* have any reason for antipathy toward Arthur's Court, which would induce him to submit it to so severe a test of courage and courtesy? Did he have an attractive young wife whom he would be willing to employ in a situation which might result in her being sexually unfaithful to him? Who was she, and did she have on her own account a grudge against Arthur's Court which would lead her to join in her husband's plan?

To answer these, it is necessary to summarize two linked stories from the *Merlin* continuation. The story of Bertelak is hardly complex: He owes fealty to King Leodegrance of Cameliard, father of Guinevere, and is described as "a wise knight that hadde don hym [Leodegrance] goode servise; and he was come of high lynage, and hadde be a goode knyght i his tyme." However, he hated another knight, who had killed a kinsman of Bertelak's in order to defile the widow; and on the evening of Arthur's marriage Bertelak happened to meet that knight and promptly slew him. The knight's kinsmen complained to King Leodegrance, who summoned Bertelak before him; defending himself, Bertelak recites the misdeeds of the slain man, but the King says Bertelak should have brought the case before royal justice, and calls for judgment from a court which includes Arthur, Gawain, Bohors, Agravain, Ywain, and the Duke of Clarence (Galashin). They rule that he "shall be disherited of all his londe that he holdeth on youre powere, and shall forswhere the contre for euer more." He departs "in grete thought as he that cowde moche euell, how that he myght be a-venged of the kynge leodogan and the kynge Arthur that hadde hym thus for-juged" (11, pp. 310-13; 12, pp. 466-70). Obviously, the answer to the question whether Bertelak le Rous had a grudge against Arthur's Court, and against Guinevere as Leodegrance's daughter, is a resounding "yes."

Interlaced with the story of Bertelak is another story involving Arthur's new father-in-law. Years before, when Leodegrance was a younger man, he had had an affair with the wife of his steward, and on the same night that he conceived Guinevere on his own wife he also conceived a girl child on the steward's spouse. Possibly because they were conceived (and born) under the same stars, or because medieval ideas of heredity portrayed the male as planting the seed whereas the female merely provided the nurturing soil, the two girls looked exactly alike. To confuse matters further, both were named Guinevere, the legitmate one being distinguished from the "false Guinevere" only by a little birthmark in the shape of a royal crown on her loins (11, p. 149; 12. pp. 213-214). At the time of Arthur's marriage there is a plot on the part of the false Guinevere's kinsmen to substitute her for her legitimate sister, but Merlin forestalls it. To prevent further trouble, the Queen's dangerous double is banished to an abbey off in the hinterlands (11, pp. 310-313; 12, pp. 463-468). And when Bertelak is exiled he goes into seclusion by way of this abbey and persuades her to leave with him (11, pp. 311, 313; 12, pp. 468, 470). The couple reappear in the *Lancelot*, by which time she is married to Bertelak (although their union is kept secret); this time they manage to persuade Arthur that he married the wrong sister, and the imposter is installed in the Queen's stead (and Arthur's bed), while the true Guinevere is rejected—which justifies her in going to live with her lover, Lancelot (13, Vol. 2, pp. 10-47). Here, then, are the additional components required to explain the behavior of the Host and his lady in *Sir Gawain and the Green Knight*: Bertelak has a beautiful wife whose fidelity he is sufficiently unmindful of that he is willing to employ her in a scheme which involves her sleeping with someone else; she, like her husband, desires vengeance against Arthur's court; and both have been banished to the wilds beyond the borders of the kingdom, where Gawain finally locates their castle.

It seems, then, highly probable that the moment the Green Knight identifies himself as Bertelak both Gawain and we as audience are supposed to recognize that his lady can be none other than the false Guinevere. And, it will be noted, the author has played fair with us on this count as well, for—as Burrow has pointed out—the feast scenes at Hautdesert parallel those at Camelot, and Gawain is seated in the place of honor beside the Host's lady just as he sat beside Guinevere at the Christmas feast a year earlier, clearly inviting a connection between the two women (14, pp. 65-66). Furthermore, just as Gawain's first reaction to his host is to notice his red beard, so his first reaction to the lady of the castle is to compare her (favorably) with Guinevere (3, line 945).

There is a further event associated with Arthur's marriage in the *Merlin*-continuation that is related to the plot of *Sir Gawain*. When Guinevere takes over the running of Arthur's household, she soon discovers that Arthur's resident half-sister, Morgan le Fay, is having a liaison with a young squire and forces her to break off the relationship (11, lines 508-509). Thus, at the same time as Bertelak and the false Guinevere are seeking revenge against the Round Table, Morgan is given provocation for an attack against the Queen.

Earlier, it was suggested that *Sir Gawain and the Green Knight* was to some extent analagous to the modern historical novel; it is apparent that parallels are close indeed, for what the *Gawain*-poet has done is to bring together three "historical" characters out of the Arthurian chronicle and use their accepted natures and situations to motivate a new plot of his own.

Thus, the apparent ambivalence in the character of the Green Knight is resolved by knowledge of the portrayal of Bertelak in the French sources. Rather than being a simple, almost allegorical, personification of either good or evil, he is discovered to be a rounded human being. His past record of good and faithful service to King Leodegrance, and his qualities as a loyalty-inspiring leader of a comitatus, account for Bertilak's generous hospitality, his friendly disposition toward Gawain after the final test at the Green Chapel, and for his clearly sincere invitation to return to Hautdesert as honored guest. Insofar as he embodies the standard knightly virtues in himself, Bertilak (who would certainly have been among those knights Leodegrance provided as members of the Round Table had it not been for his lapse into lawlessness) can admire them in others, and can therefore feel truly glad that Gawain—whom he has come to like and respect—had successfully passed the tests and evaded the traps set for him. On the other side of his nature, the hasty violence of Leodegrance's knight, his adherence to the ancient code of the blood-feud and personal vengeance rather than the newly established code of royal law and impartial justice, account for Bertilak's actions as the Green Knight.

Similarly, the false Guinevere can be seen as a sort of female parallel to Edmund in *King Lear*, barred by illegitimacy from the privilege and honor her sister enjoys. Reacting against those rules which deny her the position that the royal blood in her nature craves—she is, after all, no less beautiful, no less schooled in courtly manners, than the true daughter of King Leodegrance—, the false Guinevere has attempted to achieve her goals by guile. Caught, and punished by banishment from those very joys of noble society she had sought to augment, this outcast is understandably bitter and willing to ally herself, in marriage and conspiracy, with Bertilak. Nothing would please her better than to expose as sham and hypocrisy the standards of sexual behavior that had led to her being denied the role she craved, an end she could achieve by luring the foremost exponent of Arthur's ideals into an adulterous violation of his knightly obligations— like that which resulted in her own conception. It becomes easy to understand how the poet could assert, with that unchallengeable authority of the omniscient point of view, both that she was Gawain's deadly enemy and that she felt love and compassion for him. For, as representative of the society which had condemned her, the knight embodied all that she sought to destroy; yet as a noble, courageous and courteous male, with that quality of "daunger" or "hard-to-getness" which challenges a woman to emotional involvement, Gawain stands for all she has wanted —and still longs for. As medieval wife under the rule of her husband, she can dutifully abet him in his testing of Gawain; but in the tradition of *fin amour* she is playing the part of a noblewoman with her knight-lover, a role which has its own attractions.

The revelation that Hautdesert's other resident, the ancient lady, is Arthur's youngest half-sister—by any logic, not much older than the youthful king—is of course a surprise. One French romance offers an explanation for her altered appearance, averring that the continued use of the black arts she learned from Merlin renders her old and ugly (3, p. 130). If the process seems excessively rapid, little ingenuity is required to postulate that she used enchantment to make herself unrecognizable to her nephew, Gawain. But in prior analyses of the poem, in which the Green Knight is assumed to be himself a supernatural being, Morgan has seemed somewhat superfluous, just as her attempt to frighten Queen Guinevere seems extraneous to the major action (16, p. 39; 5, p. 178). But once it is recognized that Bertilak is a quite ordinary and mortal knight of Leodegrance's following, it becomes clear that Morgan is esssential to provide the magic by which the easily recognized Bertilak the Red is transformed into the mysterious Green Knight.

Indeed, the components of the plot are neatly distributed among the three antagonists. Morgan le Fay is responsible for the opening, with its startling appearance of the fearsome ogre and its yet more frightening manifestation, the speaking severed head. Bertilak is behind the challenge to Arthur's knights, the test of their reputations for courage and fidelity to their words. The false Guinevere tempts the court's representative first to sin sensually and then, offering her "magic girdle," with continuance of his life in this world of the senses at the expense of his knightly ideals. Even Gawain's long and arduous journey in quest of the Green Chapel may be seen as compensatory reminder of the exiling of Bertilak and his wife into these distant hinterlands.

Acceptance that the Green Knight is merely a familiar follower of Arthur's father-in-law, disguised by magic and seeking vengeance for banishment, renders a high percentage of what has been written about *Sir Gawain and the Green Knight* irrelevant. No longer need it be debated whether his greenness represents vegetation or a decaying corpse, whether his seasonal associations carry solar or Christian significance, whether he is Jack-of-the-Green, elf, Irish churl, Christ, or the devil himself, testing civilization against the natural life or the Old Religion against the new (15, pp. 158-165; 16, passim; 5, pp. 169, 223). No longer need the beautiful temptress of Castle Hautdesert be explained as a mere handmaiden to Morgan le Fay, as Spring or Youth contrasted with Winter or Old Age, as an unimportant and unmotivated instrument of her powerful husband, or as a youthful projection of Morgan (16, pp. 534-534; 18, p. 89); like her consort, she is a "real" person, Guinevere's illegitimate half-sister, with feelings and motivations of her own. No longer need critics desperately attempt to expand Morgan's role in order to make it account for *all* events in the narrative.

It is not, of course, necessary to deny that, in the process of inventing his plot, and especially its "Beheading Game" component, the *Gawain*-poet drew upon traditional tales of Celtic origin as well as the Vulgate cycle of Arthurian romance. Minstrel-composed poems, of which exemplars survive, doubtless suggested appropriate motifs. Indeed at one point the author declares that he is employing as source an alliterative "tale" he heard in town, whereas on two other occasions he cites "bokes" which provided his background (3, lines 31, 2521, 2523). The likeliest interpretation of these statements is that he drew from oral English, ultimately Celtic-based, sources his basic tale of the giant challenger and the test of courage, while taking from French books—specifically the *Merlin*-sequel —his characters and the "chastity test."

For the modern reader, there is something vaguely unsatisfactory about explaining a masterwork like *Sir Gawain and the Green Knight* through reference to an obscure continuation of a lesser composition in another language. We are accustomed to having the lesser work dependent on the greater, the unfamiliar upon the familiar. The first answer to this objection is that, for the *Gawain*-poet, *Merlin*-sequel represented valid—or, at least, accepted—history; it was written in prose, in chronicle (not romance) form, and in a language which was accepted as carrying more authority than English—especially as regards Arthurian material. Insofar as he was composing a "historical romance" set in Arthur's youth, this was a "standard source." The second answer turns upon our often unrecognized modern assumptions. Because we are "general readers" and are in fact reading *Sir Gawain*, we assume it was written for an audience like ourselves. But a long courtly romance would not have been written for a general audience, or even for posterity, but rather for a patron and his circle of literary friends. In a society where reading matter was hard to come by, available volumes would have circulated among members of any group that shared an interest in the written word—as records of early book collections and their owners amply demonstrate. Thus, as "poet-in-residence" for the lord involved in such a literary group (the *Gawain*-poet mentions his liege-lord in *Patience*), an author would have had a well-founded estimate of the literary and "historical" background his intended readers brought to his composition, which would have been essentially the same as his own. Indeed, had he thought in terms of a larger audience, the poet who composed *Sir Gawain* would have perforce assumed that "everybody" knew the works he and his circle were familiar with.

There is thus substantial support for the thesis that *Sir Gawain and the Green Knight* was composed with the expectation that its audience would be familiar with Bertilak and the false Guinevere from the Vulgate cycle. Incidentally, a final piece of information which the poet's readers would have had concerns the ultimate punishment of Bertelak and his lady—and indeed of Arthur's entire realm. For in the *Lancelot* we learn that the false Guinevere, while posing as Arthur's Queen, died of a malady, and Bertelak refused to permit her to be buried for nearly three years. Such unnatural affection for a rotten corpse and the concomitant refusal to allow proper burial aroused the Pope to place all Britain under interdict (12, p. 466).

It seems, then, that if the work is to be understood as its author intended, future editions and translations should be prefaced with a summary of the relevant events from the French *Merlin* continuation, specifically, and the Vulgate cycle generally.

NOTES AND REFERENCES

1. Howard, Donald R. & Christian Zacher. Preface to *Critical Studies of* Sir Gawain and the Green Knight. Notre Dame, Ind.: Notre Dame University Press, 1968.

2. Kittredge, G. L. 1916. *A Study of* Gawain and the Green Knight. Cambridge, Mass.: Harvard University Press, 1916.

3. Davis, Norman, ed. *Sir Gawain and The Green Knight.* J. R. R. Tolkien and E. V. Gordon, Eds. 2nd edit. Oxford: Clarendon Press, 1968.

4. Hulbert, James R. "The Name of the Green Knight: Bercilak or Bertilak." In *Manly Anniversary Studies in Language and Literature.* Chicago: Chicago University Press, 1923.

5. Spearing, A. C. *The Gawain-Poet.* Cambridge, Eng.: Cambridge University Press, 1970.

6. Loomis, Roger S. *Celtic Myth and Arthurian Romance.* New York: Columbia University Press, 1927, pp. 59-60.

7. Brewer, Elizabeth, ed. *From Cuchulainn to Gawain: Sources and Analogues of* Sir Gawain and the Green Knight. Cambridge, Eng.: D. S. Brewer, Ltd., 1973.

8. Tolkien, J. R. R. & E. V. Gordon, eds. *Sir Gawain and the Green Knight.* Oxford: Oxford University Press, 1930.

9. Le Gentil, Pierre. "The Work of Robert de Boron." In *Arthurian Literature in the Middle Ages.* Roger S. Loomis. ed. Oxford: Clarendon Press, 251-262.

10. Micha, Alexandre. "The Vulgate Merlin." In *Arthurian Literature in the Middle Ages.* Roger S. Loomis, ed. Oxford: Clarendon Press, 1961, pp. 319-324.

11. Sommer, H. O., ed. *Estoire de Merlin.* Washington, D.C.: Carnegie Institution, 1908.

12. Wheatley, H. B., ed. *Merlin.* 1865, 1866, 1869, 1899. E.E.T.S. Vols. 10, 21, 36, 112. This translation, although dated about 1450, provides valid indication of a medieval Englishman's understanding of the Vulgate Merlin. Since this work is more accessible physically, and perhaps linguistically as well, I have cited it, along with the French work, when lengthy passages are involved.

13. Sommer, H. O., ed. *Le Livre de Lancelot del Lac.* Washington, D.C.: Carnegie Institution, 1910-12, 3 Vols.

14. Burrow, J. A. *A Reading of* Sir Gawain and the Green Knight. New York: Barnes and Noble, 1966.

15. Moorman, Charles. "Myth and Medieval Literature: *Sir Gawain and the Green Knight.*" *Mediaeval Studies.* 18 (1956), pp. 158-72. Reprinted in Sir Gawain *and* Pearl: Critical Essays. Robert J. Blanche, ed. Bloomington, Ind.: Indiana University Press, 1966, pp. 209-35.

16. Bloomfield, Morton. "*Sir Gawain and the Green Knight:* an Appraisal." *PMLA* 76 (1961), pp. 7-19. Reprinted in Howard & Zacher,[1] pp. 24-55.

17. Loomis, Laura Hibbard. "*Gawain and the Green Knight.*" In *Arthurian Literature in the Middle Ages.* Roger S. Loomis, ed. Oxford: Clarendon Press, 1961, pp. 528-40.

18. Loomis, Roger S. *Wales and the Arthurian Legend.* Cardiff: Wales University Press, 1956.

AFTER THE YEAR 1284, marking the collapse of the great Gothic, Beauvais Cathedral, it is widely held that the architects of large churches were less willing to experiment with lofty, vertical thrusts in building design. However, daring structural innovations in two extremely tall churches were planned in the first half of the fourteenth century. At St. Ouen, a single tier of flying buttresses was substituted for the double tier, generally necessary in High Gothic churches; and at Palma, in spite of its great spans, the main arcade piers are far more slender than in any other major church. Modern photoelastic modeling of the two buildings permits detailed evaluation. Neither design was a complete triumph: the clerestory-wall-buttress of St. Ouen indicated some distress from the effects of bending: and, while bending is negligible in Palma's lithe piers, the decision to maintain a double tier of almost horizontal, long flying buttresses to support its almost flat-roofed superstructure has led to some problems. Nevertheless, utilizing modern architectural-engineering techniques, the glories are verifiable for these two audacious buildings.   MPC

Innovation in High Fourteenth-Century Gothic: The Church of St. Ouen, Rouen and the Cathedral of Palma, Majorca

ROBERT MARK

Civil Engineering and Architecture
Princeton University
Princeton, New Jersey 08540

THE year 1284, which saw the collapse of the high vaults of Beauvais Cathedral, is frequently taken as a turning point in the development of Gothic design (see, for example, Harvey[1]). After that date, it is assumed that the architects of large churches were more timid and less willing to carry out the experiments in structure that produced the classical High Gothic cathedrals. The notion is controverted by the structural innovations introduced in two major buildings that were planned in the first half of the fourteenth century, one in Normandy and the other in the Balearic Islands, south of Barcelona.

ST. OUEN, ROUEN

Begun in 1318, it was not until near the end of the Hundred Years War, in the mid-fifteenth century, that the nave of the abbey-church of St. Ouen was completed. In spite of the long delay in construction, the building is quite unified and its design, particularly of its interior elevation (FIGURE 1), is generally considered to be a superb example of late Gothic. The great, light openings in the clerestory and triforium would seem to follow from the French Rayonnant developed in the Paris region in the mid-thirteenth century; however its giant scale (the keystones of its vaults are but five meters lower than those of Chartres) invites comparison with the classical High Gothic cathedrals. In these buildings, the lofty clerestory walls are supported by a double tier (and at Chartres, an additional third tier as well) or flying buttresses. The lower tier is positioned to resist the outward, lateral thrust of the high vaults above the main arcade, while

[269]

0077-8923/78/0314-0269 $01.75/2 © 1978, NYAS

FIGURE 1. St. Ouen, Rouen. Nave wall.

the upper tier resists the effects of wind forces on the upper clerestory walls and the high wooden roof. Indeed, the forces generated by high winds on these large buildings have been shown to be quite significant, approaching the magnitude of the vault thrusts.[2]

At St. Ouen, a single tier of flying buttresses has been placed at an intermediate height, between the normal positions of upper and lower tiers, to resist both the vault thrust and the wind loadings (FIGURE 2). This step certainly was daring, and that it was largely successful is attested to by the perseverance of the building over the centuries. To discern that it was not a complete triumph required the analysis discussed in the next section.

STRUCTURAL MODELING

Analysis of the long vessels of Gothic churches is facilitated by their repeating, modular bay design. The buildings can be considered as being supported by a series of parallel, transverse *frames* consisting of the principal load-bearing structural elements: piers, buttresses, lateral walls and ribbed vaults.

A second, important analysis simplification is based on the assumption that the structural forces within the masonry "frame" are distributed as they would be in an equivalent frame constructed from a perfectly *elastic*, homogeneous material. This assumption has been shown to be adequate for predicting structural behavior in tests of reinforced concrete structures subjected to service loadings even though concrete is notoriously inelastic, compositionally inhomogeneous, and subject to tensile microcracking. For the simplification to be applied to masonry, it must also be assumed that the entire frame is undergoing compressive action: that is, that all of the individual stones are pressed against adjacent stones by interior forces. Fortunately, this assumption coincides with criteria for successful masonry performance because the tensile strength of medieval mortar is almost nonexistent; hence, structural continuity cannot be maintained if any substantial amount of tensile stress is present. Even small tensile stresses can cause cracking and begin a process of local disintegration, especially on the exterior of a masonry building subject to weathering. In point of fact, our studies have indicated that compressive stresses prevail throughout Gothic buildings and that there are usually only a few highly localized regions of tension.

Prolonged, dead-weight loadings may also cause unrecoverable flow of the masonry, but wind loadings, which are of variable magnitude and

FIGURE 2. St. Ouen, Rouen. Nave buttressing.

blow from every quarter, will not. Nevertheless, if the basic support and form of a structure remain unchanged with time, the distribution of internal forces would be little altered from the initial elastic distribution.

The structural system of the St. Ouen nave was analyzed using the relatively simple experimental technique that we have applied to study a number of historic buildings, small-scale photoelastic modeling. Stress-free, epoxy models are loaded by arrays of weights representing the distributions of wind and the dead-weight forces acting on the prototype building. The model tests are performed in a controlled circulation oven where the epoxy is brought to a rubbery state (at about 140°C) and then slowly cooled, restoring the model to its room-temperature glassy state. Relatively large model deformations that took place at the higher temperature are "locked in" after cooling so that the loadings may be removed, with negligible effect. The unloaded model, now viewed through polarizing filters, displays interference patterns which, with calibration and scaling theory, can predict the force distributions in the full-scale structure. As general force distributions were sought rather than localized stress concentrations (for which the modeling method is often applied, for example in airframe analysis), no effort was made to *detail* the cross-sections of component structural elements; the action of the cross-ribbed vault, a three-dimensional structure, was simulated in the model as a planar arch. The heavily loaded foundations were also assumed to give complete fixity to the bases of the piers (i.e., no deformations are permitted at ground level).

The nave structural system of St. Ouen was modeled at 1:167 scale. The dead-weight loadings and their distribution were computed from considering the volumes of stone in the individual bay elements and taking the specific gravity of the stone as 2.4. A similar distribution of loads, applied to the model at a scale of 1:325,000, produced a photoelastic interference pattern that revealed small amounts of tensile stress in the clerestory wall buttresses (FIGURE 3; the reading N = 1.1 is tensile). (Further details on the St. Ouen modeling can be found in References 2 and 3.)[3]

Long-term wind data obtained from French Government meteorological records indicate maximum wind speed at the 43 meter elevation of the roof peak as 130 km/hr. The distribution of wind loads on the actual building was then estimated from wind tunnel test data, and a similar distribution of loadings was applied to the model at 1:160,000 scale. (The first, dead-load pattern is "erased" when a model is reheated for its second loading.) This test also indicated bending in the leeward clerestory wall buttress, which tended to reinforce the small tensile stress already present

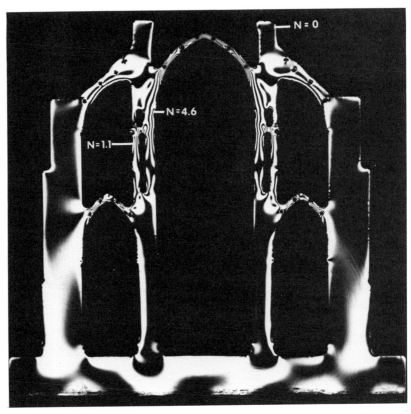

FIGURE 3. St. Ouen, Rouen. Photoelastic interference pattern in model under simulated dead-weight loading.

from the dead-weight loading. The total tensile stress from both dead-weight and bending from the extreme high wind was about four times the estimated tensile strength of medieval mortar. Hence, when a site inspection of St. Ouen revealed cracking-distress in all the clerestory wall buttresses (FIGURE 4), it was not entirely surprising.

PALMA CATHEDRAL

In spite of Palma's gigantic height (at the time of its construction, its vessel height was exceeded only by the choirs of Beauvais and Cologne), its remote site has kept it from being better known. In what is probably the most comprehensive work in English on Spanish Gothic architecture,

George Edmund Street regrets, "that I have never yet been able to visit that Island [Majorca], for so far as I can learn, it seems that the mainland owed much to it in the way of architectural development."[4]

A more modest church was planned after the Christian reconquest of the Balearic Islands in 1229. The first campaign of construction left the lower eastern apse largely completed in 1327, with the second campaign,

FIGURE 4. St. Ouen, Rouen. Clerestory wall buttress cracking-distress.

which raised the nave to its extreme height, not begun until about 1357. By the turn of the century, three high bays were standing; however construction slowed afterwards so that the entire building vessel was not complete until the sixteenth century.[5]

Unlike its northern High Gothic counterparts, Palma exhibits truly massive pier buttresses, which are utilized spatially by the creation of chapels between them, along the exterior of the church (FIGURE 5). These are evidenced on the building exterior by the pairs of chapel wall buttresses, between adjacent pier buttresses, that provide also an important visual component of the design. Heavy buttresses, flanking chapels, had been used previously in the walls of southern French fortified churches, and it is not unlikely that this design element at Palma was taken from the Cathedral of Albi (begun 1282). Low, or even nonexistent roofs are also characteristic of southern churches, which is important from a technical standpoint; the wind loads on the northern high roofs are quite significant and, as was already indicated, the effect of these forces usually required the employment of a second tier of upper flyers to resist roof wind loadings. There is hardly any equivalent loading on the structure of southern buildings; hence, in a high wind of the same magnitude, Palma is not subjected to very different *total* lateral forces than St. Ouen, even though it has a much higher interior vessel. And further, the distribution of these wind forces is far less severe, particularly on the clerestory wall, because of the missing roof component.

The interior of Palma's nave evokes Bourges, yet its structure is unlike any major High Gothic church. All the dimensions are Gargantuan: vault keystone height = 44 meters, main arcade width = 19.5 meters, bay length = 8.8 meters; but in spite of these great spans, the hexagonal, main arcade piers supporting the high clerestory walls are extremely slender (FIGURE 6). It is this aspect of Palma's design that sets it apart from all Gothic antecedents and that contributes so powerfully to creating an atmosphere of vast spaciousness, not unlike that perceived in large halls built from modern materials. The boldness of Palma's architect can be quantified; TABLE 1 compares Palma's pier dimensions and the slenderness ratio (the pier height divided by the pier width) derived from them with those of the major thirteenth-century High Gothic cathedrals. The slenderness ratio of the Palma piers is almost 50 percent greater than any previously used, and both the visual and technical difference is even more dramatic than the numbers convey. Unlike Palma, the piers of the earlier buildings are surrounded by attached shafts which were not accounted for

FIGURE 5. Palma, Majorca. View from the harbor.

FIGURE 6. Palma, Majorca. Nave piers.

TABLE 1

Gothic Cathedral Pier Dimensions (in meters) and Slenderness Ratios

Building Site	Height*	Width†	Slenderness Ratio
Chartres (nave)	8.0	1.8	4.4
Bourges (choir)	14.9	1.6	9.3
Rheims (nave)	9.6	1.6	6.0
Amiens (nave)	12.5	1.5	8.3
Beauvais (choir)	14.6	1.5	9.7
Cologne (choir)	11.9	1.3	9.2
Palma (nave)	22	1.6	13.8

* Distance from top of base to bottom of capital; i.e., straight section length of load-bearing, coursed construction.

† Diameter for round piers; distance between flats for hexagonal piers.

in TABLE 1, but which make their piers appear much heavier and in fact do provide considerable reinforcement to them.

The nave structural section of Palma was modeled at 1:144 scale. Dead weight distributions in the actual building were taken from Bellver[6] and applied to the model in a similar pattern at 1:150,000 scale as an array of point loadings. The resulting photoelastic interference pattern is shown in FIGURE 7. Palma wind velocity data over the period 1943-74 were obtained from the Spanish Ministerio del Aire and the Servicio Meteorologico Nacional as well as from the Climatic Center of the U.S. Air Force. From these data, the greatest expected wind velocity at the cathedral roof level was (also, coincidently) taken as 130 km/hr. The resulting wind pressure distribution was applied to the model at 1:21,650 scale.

The most significant result of the tests is the almost uniform pattern (of interference order 3.5) in the dead-weight-loaded main piers (FIGURE 7), which signifies almost negligible bending. The absence of bending in the piers helps to explain the stability of these very slender, main structural elements. Indeed, the maximum compressive stress level at the base of the piers was found to be 2.1 Pa (310 pounds per square inch), which is lower than the total (compression and bending) stresses found in the piers of some of the earlier, more conservative buildings.

Under wind loading alone, the maximum stresses throughout the structure are low, of the order of $0.5(10)^6$Pa in the pier. The only regions in the structure indicating problems are at the ends of the flying buttresses, where bending from the effect of winds added to the bending already

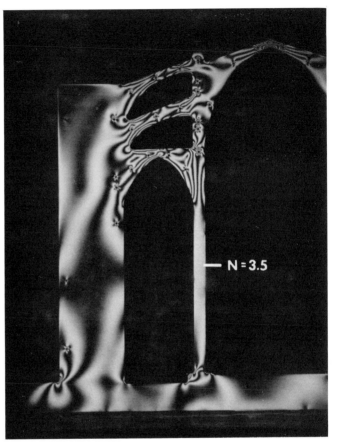

FIGURE 7. Palma, Majorca. Photoelastic interference pattern in model under simulated dead-weight loading.

present from the dead weight loading will produce tension of large enough magnitude to engender cracking. Durliat has referred to the rebuilding of the high vaults at the beginning of the eighteenth century, but gives no hint as to the cause of their failure (Durliat[5], p. 167). The analysis results and the present state of a number of flying buttresses combine to make the flyers prime candidates as possible villains. They are long and what is more critical, almost horizontal. Unless a high roof was originally planned, the upper flyers serve little purpose and this is probably the reason for their malfunction (note the propping of the upper flyers in

FIGURE 8). A single, more steeply sloped tier of flying buttresses together with parallel-sloped transverse walls above the side aisles, instead of the existing low horizontal walls, would have provided more reliable support to the vaults.

CONCLUSION

It must be assumed that both buildings achieved their designers' objectives, if not the complete perfection of their structure. At St. Ouen, the classical Gothic wall and buttress system was further refined and lightened. Palma's innovation was more radical. Even following in the southern tradition, its exterior exemplifies a greater solidity of the same Gothic elements that were lightened at St. Ouen, which in turn allowed the reduction of the piers and the unique openness of the building interior. Both

FIGURE 8. Palma, Majorca. Nave clerestory buttressing.

churches represent the highest achievement of fourteenth-century building art.

Nevertheless, there is no denying that from the point of view of structural development, the fourteenth century could not compare with the thirteenth. Nor would any period before the impact of the industrial revolution in the nineteenth century match the building activity that took place, particularly in the Ile de France region, between the mid-twelfth and mid-thirteenth centuries. Within this interval, the major technical problems of tall, skeletal stone construction were essentially solved by introducing novel structural elements such as the flying buttress. By the middle of the thirteenth century, the play with vast areas of glass and slender stone tracery seemed often to absorb the architects who now eschewed great building height. (The relative cost of these smaller buildings must not have been lost on their clients either.) Indeed, this process of "miniaturization" was underway well before the collapse at Beauvais as prime examples of two of the smaller, elegant buildings and their dates of construction indicate: Sainte-Chapelle, Paris, begun in 1241 and Saint Urbain, Troyes, begun in 1262. But all of this is not to say that structural experimentation in tall Gothic buildings had come to an end; Saint Ouen and Palma provide striking evidence that it had not.

ACKNOWLEDGMENTS

The author gratefully acknowledges the contributions to this work by several former Princeton students in the Schools of Engineering and Applied Science and in Architecture and Urban Planning, particularly A. Caine, who with the aid of J. Thompson, modelmaker, carried out the careful modeling of Palma. This research was performed as part of an interinstitutional teaching/research program between Princeton and Stevens Institute of Technology, entitled Architecture and the Scientific Revolution, sponsored by the National Endowment for the Humanities.

NOTES AND REFERENCES

1. Harvey, John. The Medieval Architect. New York: St. Martin's Press, 1972, pp. 162f.
2. Mark, R. and R. S. Jonash. "Wind Loading on Gothic Structure." Journal of the Society of Architectural Historians, XXIX, No. 3 (1970), pp. 222-230.
3. Mark, R. "The Church of St. Ouen, Rouen—A Reexamination of Gothic Structure." American Scientist, 56, No. 4 (1968), pp. 390-399.
4. Street, G. E. Gothic Architecture in Spain. Vol. 2. Bronx, N.Y.: B. Blum, 1914 (Reissued, 1969), p. 242.
5. Durliat, M. L'art dans le Royaume de Majorque. Toulouse, France: ed. privat, 1962, pp. 150-167.

6. Bellver, Juan R. "Conferencia acerca de los conceptos orgánicos, mecáicos y constructivos de la Catedral de Mallorica." *Annuario de la Asociacion de Arquitectos de Cataluna* (1912), pp. 87-140, Barcelona, Spain. Bellver worked with the Catalonian architect, Antoni Gaudi, on certain restoration to the cathedral at the beginning of this century. His work is the most comprehensive source of data on the building fabric.

MEDIEVAL CITIES displayed their civic pride and strenuously competed with one another by commissioning works of art *for* the cities (fountains, bell-towers, city halls, and universities) and *about* the cities (government itself was a subject treated "realistically" or "allegorically") In Italy, the artistic expression of the meaning of good government and ideal citizenship underwent important changes during the early fourteenth century. The development of new iconographic formulas was a succinct, clear pictorial manifestation of the changing nature of contemporary political ideas. The translation of Aristotle's *Politics* into Latin in about 1260 was a crucial event for both political art and theory. The leading Italian theorists of the time were familiar with the Aristotelian thought flourishing at the University of Paris; many, in fact, had been trained or had lectured in Paris. This exposure to Aristotelian theory led to a belief in the self-sufficiency of the State and in the "natural necessity" of political life. Several representative monuments, beginning with Nicola Pisano's *Fontana Maggiore* (1278) and concluding with Ambrogio Lorenzetti's murals of *Good and Bad Government* (1336-1339), suggest how the revival of Aristotle at the University of Paris came to be a foundation for civic political art in Italy.

MPC

French Influences on the
Early Development of
Civic Art in Italy

JONATHAN B. RIESS

Department of Art History
University of Cincinnati
Cincinnati, Ohio 45221

D URING the late thirteenth and early fourteenth centuries, political art developed in Italy. Before this time the illustration of political ideas in art had been dependent on borrowings from ecclesiastical imagery. Moreover, the representation of political ideas was, in general, most rare. By the beginning of the trecento significant numbers of public, private, and ecclesiastical monuments were being commissioned depicting a fairly uniform view of the ideal state and of good citizenship conceived as free of the need for a traditional foundation in theological justification.[1]

An important aspect of the process of creation and refinement of political art was the absorption of theories that followed from the Aristotelian revival at the University of Paris. The change in iconographic formulas is but a succinct and clear pictorial manifestation of the changing orientation of political ideas. The artistic ramifications of the re-discovery of the *Politics* in Paris is the primary theme of the present paper.[2]

The work of the leading political theorists of the time in Italy was shaped in fundamental ways by the Aristotelian culture that flourished at the University of Paris.[3] Many members of this first generation of Italian political scientists in fact either studied or lectured in Paris and thus were directly exposed to the most sophisticated political thought of the day.

The movement in Paris may be said to have begun with the translation of the *Politics* into Latin by William of Moerbeke in about 1260, although evidence of a strong and continuous interest in Aristotle can be found at the University well before this date.[4] The belief that man was by nature politically and socially motivated and that the state was natural and good, thus requiring no allusion to Original Sin to justify its foundation

0077-8923/78/0314-0285 $01.75/2 © 1978, NYAS

and to defend a largely coercive purpose, became firmly entrenched only with the dissemination of Aristotle's political and social theories: so firmly entrenched that by the first decades of the fourteenth century this nexus of principles became an important argument for the support of an independent monarch in France and for a free city-state in Italy.[5] To recognize, as one should, that there were notable anticipations of Aristotelian ideas before a new currency was granted to the *Politics* after 1260 is not to deny the profound effect that the work had, even if that effect meant only the sanction and reinforcement of an existing conviction in the state as natural and necessary.[6]

The thought of Fra Remigio de' Girolami (d. 1319), one of the earliest and among the most radical of the Italian Aristotelians, can be traced in the contemporary art.[7] This is not to suggest that Remigio himself is an immediate and direct influence on art; I mean only that his ideas help to explain and to account for unique features in the coeval political art.

Remigio's theory is clearly the product of the Parisian revival of Aristotle. Although Remigio's beliefs represent in part a modification of Aristotle's theory to bring it into conformity with the political situation in Italy, his knowledge of the *Politics* and of Aristotelian thought in Paris are the crucial formative ingredients in his writings.[8] Remigio was a close follower of St. Thomas Aquinas and may even have been his pupil. A most important resonance of his oftentimes extreme religion of the state are the polemical writings of the supporters of Philip the Fair. The kind words that Remigio has for the French monarchs, despite his strong republicanism, suggests his own awareness of an affinity with the French propagandists. Finally, the fact that Remigio lectured in Paris further corroborates the significance that the French context held for his ideas.[9]

In what is perhaps the strongest of his pro-state ideas, Remigio contended that if an individual loses his citizenship he forfeits his qualification to be a man. The worth and virtue of the individual can be assessed only in terms of his service to the state. Remigio does not deny outright the importance of religion and the church, for he is concerned with the problem of the attainment of salvation. But as man cannot perfect himself outside the state, so heavenly reward is bestowed only on those who have worked for the common good of the community.[10] Virtuous pagans who put the welfare of Rome before thoughts of personal gain are celebrated as paragons of the civic virtue that Remigio wishes to instill in his contemporaries.[11]

During the same period in which the first evidence of the influence of

Aristotelian thought can be observed in Italy in the writings of Remigio, the earliest major secular monument glorifying a like ideal of citizenship and the benefits to be derived from participation in the state was constructed: the Fontana Maggiore of 1278 in Perugia by Nicola Pisano and his circle (FIGURE 1).[12] There is in fact one positive link between Remigio and Perugia that suggests the existence of a cultural milieu receptive to Aristotelian ideas. Apart from Remigio's sojourns in Paris, the sermons and lectures given by him in Perugia between 1301 and 1304 represent the only documented instances of travels outside his native city of Florence.[13]

The city of Perugia, as it is described by the ambitious and multi-faceted program of sculptures, is placed at the center of the political and spiritual universe. There is a remarkable and unprecedented number of local references among the sculptures, including depictions of several of the heroic figures of the imagined past of the city and effigies of the leaders in the year in which the fountain was completed. Emphasis is placed on the ennoblement of the political life and on the history of the city. In this general fashion the fountain shares a primary thrust with the writings of Remigio: the enhancement of the state and the glorification of citizenship.

In a still more general yet related sense, one should observe that Nicola's structure has the shape of a baptismal fount, and that this is perhaps emblematic of the religious sanctity that the state had also assumed in the contemporary political literature. Inducting the citizen into the community through the sacramental rite of partaking of the waters furnished by the city is not unrelated to Remigio's view of the state as the object of religious deification.[14]

A more specific aspect of the total scheme describes the relation between Perugia and civic and ecclesiastical rule in Rome. The importance of Rome and of the church is affirmed only in so far as they grant a fiat to civic accomplishment. A mechanism of civic self-assertion is elucidated through a system of balances and proximate relations among the sculptures.[15] The city as an independent and free entity is at the core of the meaning of the decorations, just as a like ideal shaped the thought of Remigio.

As the natural necessity of the state and the economic and social benefits to be derived from participation in the human community were touchstones of Remigio's outlook, so too are these principles enunciated at Perugia.[16] The scenes of the labors of the months and of the food- and fish-producing regions of the Perugian *contado* illustrate the various forms of productive work for the common good and the rich and bountiful civic

FIGURE 1. Nicola Pisano: Fontana Maggiore, completed 1278. Perugia, Piazza IV Novembre (photo: Alinari Art-Reference Bureau).

resources from which the citizens draw as they work toward the welfare of the community. In simple, graphic terms, a symbiotic relationship between the citizen and the state is depicted. The personified figure of the city of Perugia holds a cornucopia as an attribute and thus distills the meaning of this relationship by symbolizing the prosperity that results from the fruitful utilization of the resources of the town. Here is the mechanism of work and reward that is the source of the common welfare, the goal of the Aristotelian state.

The importance attached to service to the state is given still greater force with the representation of the leaders and heroes of the city. The accomplishment of these individuals stands as an exemplum of the selfless devotion to the community that Remigio believed to be the sustaining element of the state. Civic service is conceived by both Remigio and Nicola as a positive action and is purged of any oppressive connotation as punishment for sin. Indeed, throughout the vast program of sculptures there is no overt reference to the problem of salvation nor is there any suggestion that the labors of the citizens have a theological source. As in the theory of Remigio, the state and good citizenship are defended as good and necessary in themselves.

A glance at two earlier monuments that include a political aspect indicates how the definition of secular power had changed under the impetus of the Aristotelian revival.

On the facade of San Zeno in Verona there is a sculptured relief from about 1150 depicting Bishop Zeno, patron saint of the city (FIGURE 2). He is shown with his hand raised in benediction and trampling a dragon, a popular symbol of evil used largely in ecclesiastical art. To either side of him are the armies of the commune. An inscription reads: "The Bishop gives to the people the standard [for] a worthy defense. Zeno gives the banner with serene heart" (Arthur K. Porter translation).[17] The bishop inspires the citizens with a Christian zeal in the defense of the city and stands, in the fashion of the ideal ruler of the time, as spiritual and secular leader of the city. The monumental scale of the bishop and the use of motifs derived from ecclesiastical imagery—the hand raised in benediction and the dragon—leave little doubt that the political community is portrayed as a theocracy. Contrast this religious gloss given to the actions and policies of the state with the intrinsic worth bestowed upon them on the Perugia Fountain. While at Verona service to the city assumes the aspect of a religious crusade, at Perugia the personified city, which stands as the symbol of the fruits of the labors of the citizens and is set among

FIGURE 2. Nicolao (attribution): The Veronese Commune receives a standard from St. Zeno, ca. 1150. Verona, tympanum of the west door of the monastery of St. Zeno (photo: Glenn Pryor).

a large range of civic emblems and attributes, has displaced the traditional religious locus found in the earlier work.

A second example of the treatment of a political subject before the Fontana Maggiore is the fresco cycle in the Palazzo della Ragione in Mantua, dated about 1200. Here, among other murals, are representations of the labors of the months and an enthroned figure of the Madonna and Child with saints and angels.[18] While at Perugia the scenes of the labors were related to the problem of how the common good is to be served, at Mantua the labors retain a conventional theological meaning as the good works needed to assure entry into heaven, depicted here as the Madonna, Queen of heaven, surrounded by her celestial court.[19]

A further extension of the influence of the Aristotelian revival on art is evident in another political monument in Perugia: the frescoes that decorate the spandrels of the arches of the Sala dei Notari in the Palazzo dei Priori, which are dated 1297 and are attributed to the circle of Pietro Cavallini (FIGURE 3).[20] The program of murals is only one of several commissioned during the 1290s to embellish the civic palaces constructed in the building boom of the time.[21] All the programs are, like that of the earlier Fontana Maggiore, unusually ambitious in scope—the Perugia decorations, for example, consist of 33 scenes—and this fact alone is testimony once more to the growth of civic art that accompanies the contemporary theoretical defense of the state. Public palaces had been constructed in earlier periods, but never before, if one is to judge correctly on the basis of the surviving evidence, had the palaces been seen as rivals to the ecclesiastical structures for the display of major decorative undertakings.[22]

The Perugia murals illustrate, among other Aristotelian ideas, the belief that the state was untainted by Original Sin. Among the first scenes are representations of the Creation of Adam and Eve and the First Labor, but the cycle is notable in that all reference to Temptation and Fall is excluded. More common in art was a fairly full exposition of the theme of the Fall; that is, Warning, Sin, Apprehension, and Expulsion.[23] As this expected theological course is bypassed at Perugia, the labor of Adam and Eve is not presented as a consequence of sin. The state is thus conceived in part as a stage for fruitful and virtuous labor; a setting, in other words, in which the political and social instincts of man are to be exercised. The preceding murals depict a paradigmatic form of government or rule and so the connection of the scenes of Adam and Eve with the political order

FIGURE 3. Circle of Pietro Cavallini (attribution): spandrel frescoes, 1297. Perugia, Palazzo dei Priori, Sala dei Notari (photo: Alinari Art-Reference Bureau).

is an explicit one. The following paraphrase of the words of Aristotle by Remigio's mentor, St. Thomas, express most succinctly the praise for useful work illustrated by the murals: "When we consider all that is necessary to human life . . . it becomes clear that man is naturally a social and political animal. . . . One man alone would not be able to furnish himself with all that is necessary, for no man's resources are adequate to the fullness of human life."[24]

The notion of good government outlined in the first murals at Perugia also reflects the theologically-based Aristotelianism of St. Thomas, and proclaims, as he did, that the rule of Moses be used as a model of ideal leadership.[25] That the illustrations from the life of Moses and from Genesis are not meant to express a biblical or theological purpose is even more manifest in the arrangement of the scenes: the Creation of Adam and Eve follows the scenes of Moses.

The belief in the importance of work for the community is represented in the same manner in the sculptural program of the Florence Campanile, attributed to Andrea Pisano and his circle and dated ca. 1337-1350. Here, as at Perugia, the Creation scenes lead directly to the representation of the First Labor. The redemptive spirit of work dedicated to the state is expanded at Florence with the addition of the Liberal Arts and the inventions of man, which, together with the scenes of the labors of the months, offer a seemingly exhaustive catalogue of the types of productive work that lead to the common good.[26] This encyclopedic scheme suggests a most specific point of contact with Aristotelian thought, for the flourishing of the crafts and professions was thought to be evidence of a well-run and purposefully ordered government.[27] Inclusion of one local reference among the sculptures, the figure of Hercules as emblem of the city, would appear to underscore the political implication of the reliefs.[28]

≈§ §≈

If the work of Remigio complements the initial stages of the artistic appropriation and transformation of Aristotelian theory, then the thought of another great Italian Aristotelian, Marsilius of Padua (1275/80-1342), best exemplifies the nature of the Aristotelian thought found in a second stage in the evolution of political art in Italy.[29]

Like Remigio, Marsilius was directly exposed to the fountainhead of the Aristotelian theory of the day, the University of Paris. This, together with the experience of the political life of his native city are the important formative influences on his thought. Like Remigio also, Marsilius was

FIGURE 4. Giotto: Justice, 1304-1312/13. Padua, Arena Chapel (photo: Alinari Art-Reference Bureau).

concerned with demonstrating how man is innately suited to political and social life—is in fact unable to survive without it—how the true purpose of government is to foster a self-sufficient and virtuous life in the community, and how the welfare of the community is of greater importance than the well-being of the individual. A primary distinction between the two theorists, only one among several, is that Marsilius was somewhat more concerned with the legalistic foundations of government and with the power of the state vis-à-vis the church. The pre-eminence of law, which is equated with and founded on a conception of justice, is a key to the thought of Marsilius.[30]

An interest in justice and a desire to establish the legalistic bases and obligations of the state are the most important and widely treated subjects in the political art of the first half of the trecento, as they are also the leading elements in the contemporary theory. Thus, while the political art and theory of the last years of the duecento deal principally with the general justification of the state and with the need for citizen participation in the community, the theory and art of the following period consider these same issues but generally in a more refined and legalistic way.

Developments in the art and thought of the time must not of course be viewed in isolation from the institutional evolution of the communes, for complementing the more pronounced concern with legalistic questions in the theory and art is a parallel institutional and constitutional development.[31] Neither should the interest in the meaning of justice be seen as something new, for a particular concept of justice is to be found at the core of any political theory. The continued infusion of Aristotelian influence, however, provided a shape and direction both for actual institutional change and for the modification of traditional meanings of justice.

A first major indication of the new cast given to the illustration of Aristotelian ideas in the art of the early trecento is found in Marsilius's native city in the fresco decorations by Giotto for the Arena Chapel (1304-1312/13). The program of the seven virtues and seven vices that runs along the bases of the two long walls has a distinctly political and social cast. The virtues illustrate the foundations of good government, which is embodied or personified by the figure of Justice, while the vices illustrate the foundations of bad government, which is identified as the figure of Injustice.[32]

The personification of Justice (FIGURE 4) is the formal and iconographic centerpiece among the virtues (FIGURE 5). She is the bridge between the earthly or cardinal virtues that precede her and the theological virtues that follow.[33] Giotto illustrates the Aristotelian belief that Justice is the mother

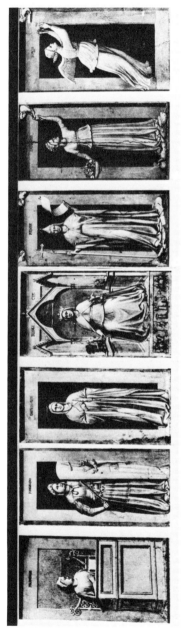

FIGURE 5. Giotto: Virtues, 1304-1312/13. Padua. Arena Chapel (author's illustration of scheme).

of the virtues.[34] The figure of Justice also represents the legalistic and theoretical conviction of the period, one derived ultimately from Aristotle, that Justice helps to bring about peace and order by punishing transgressors and rewarding law-abiding citizens.[35] Thus in the left pan of the scales of Giotto's Justice the transgressor is punished, and in the right pan the law-abiding citizen is rewarded.

The specific political meaning of the figure is developed in the scene found in the predella of the throne. Here are depicted scenes of the happy and flourishing countryside that results when the state is identified in Aristotelian fashion as Justice. And the scenes suggest as well the peace and common welfare that flow from the rule of Justice. This same convergence of the notions of peace, justice, and the common good is to be found in Aristotelian theory. St. Thomas, for example, wrote that "the common good means peace, which is the highest end of society, just as health is the greatest good of the body." And Marsilius would add later that the judge, as the dispenser of justice, is among the most important individuals in the state as his decisions affect most directly and immediately the "health" and peace of the community.[36]

While the figure of Justice would appear in the end to stand for a government devoted to serving the common good of the citizens, the opposing vice, Injustice (FIGURE 6), represents a political community dedicated to the pursuit of the self-interest of its leaders. Thus the ravaged and wasted landscape beneath the vice contrasts as strongly with the scene of peace beneath the throne of Justice as the menacing appearance of Injustice contrasts with the sense of control and reason radiating from the virtue.

Another sense in which the cycle of virtues and vices partakes of the Aristotelian erudition of the time is indicated by the fact that the cardinal or earthly virtues assume an importance equal to that of the theological virtues for the attainment of salvation. The figures of the virtues as a group lead directly to the region of the elect in the Last Judgment, an obvious expression of the belief that a life devoted to virtue is the necessary pre-condition for admittance into heaven. But as the pivotal figure of Justice is oriented at once toward the cardinal and the theological virtues, the program depicts the Aristotelian belief that earthly good works, as they are symbolized by the earthly virtues, are necessary for salvation. The exercise of the theological virtues was no longer sufficient, according to the Aristotelians, for avoiding eternal damnation.[37] Charity, one of the theological virtues, is in fact characterized by Giotto as having both a heavenly and an earthly aspect, so the social and political cast of the scheme is not limited to the earthly virtues.[38]

FIGURE 6. Giotto: Injustice, 1304-1312/13. Padua, Arena Chapel (photo: Alinari Art-Reference Bureau).

Further significance is lent to the importance of socially productive
work as a prerequisite for salvation in the formal and iconographic parallels between the figures of Christ the Judge in the Last Judgment (FIGURE
7) and Justice. Both are seated in the same position on their thrones, and

FIGURE 7. Giotto: Christ the Judge (detail from the Last Judgment), 1304-1312/13.
Padua, Arena Chapel (photo: Alinari Art-Reference Bureau).

both bodies are oriented in precisely the same fashion: the left arms be-
stow punishment, and the right arms, blessing. The fact that the processes
of earthly justice so closely follow the workings of heavenly justice suggests clearly how a life devoted to virtue is ultimately of transcendent
importance. And, following the theory of Remigio and Marsilius, the
similar appearance and meaning of the figures enforces the belief that a
fundamental role of the state was to establish conditions necessary for
achieving a good life and that perfection could not be attained outside
the conditions imposed by the state.[39]

A second set of works in Padua by Giotto that convey in a quite different sense the Aristotelianism of the first half of the trecento are the fresco
decorations of the 1320s devoted to astrological subjects that were once
found in the great hall of the Palazzo della Ragione. Little is certain about
the lost murals except that they were likely organized around depictions
of the months and the associated gods and planetary inclinations.[40] The
astrological works at Padua were by no means isolated ones in the art of
the age. Other important examples are elements in decorative programs
that have already been noted as bearing some relation to the Aristotelian
culture—the Perugia frescoes of 1297[41] and the Florence Campanile sculptures of 1337-1350[42]; and, let me add here, Ambrogio Lorenzetti's frescoes
of *Good and Bad Government* in Siena, dated 1337-1340, which will be
discussed shortly.[43]

A relation between astrological illustrations and Aristotelian ideology
might at first appear to be unlikely. The medieval astrological system,
which has been described by most historians as fatalistic, can hardly be
reconciled with a system stressing the relative freedom of action of man
within a political context.[44] But the astrology that developed in Padua
during the early fourteenth century was linked to the belief that the human community and the history of political man were subject to their own
natural laws which, if understood, would provide the means for a more
rational direction of the state in the name of the common welfare.[45] A
major principle of organization in the *Defensor Pacis* was, in fact, the
biological conception of a state believed to be subject to a fixed pattern
of laws.[46] Marsilius himself engaged in astrological researches and was
apparently a friend of Pietro d'Abano, the most influential astrological
thinker of the age and possibly the designer of the fresco program in the
Sala della Ragione.[47]

Far from locking man into a fatalistic universe, the astrology of the
period of the Aristotelian revival suggests a desire to comprehend those
laws to which the human community was subject. The contemporary his-

toriography, notably the writings of Giovanni Villani, points also to the special political meaning that must be assigned to astrological programs of the early fourteenth century in Italy.[48] For Villani, the study of the planets was intended to help transform statecraft into an exact science. Foremost among the axioms of this new science was the belief that when state policy is virtuous and when the citizens contribute in selfless fashion to the common good the stars will be favorable, or at least their evil influences can be anticipated and thus dealt with. Once more the connection between work for the state, virtue, and the general prosperity is propounded, the same pattern of ideas that we have observed in Aristotelian theory—and thus the concurrence in political art of subjects that glorify work for the community with planetary depictions.

In comparison to the preoccupation with the meaning of justice and the legalistic foundations of government, the issue of the organic functioning of the body politic is of secondary import. Several works from the second quarter of the fourteenth century in Florence are especially significant examples of the popular theme, seen first at Padua in the Arena Chapel, of how justice, conceived either as the ideal state or judge, is the fundamental concept of the political community.

The first of these works is a lost representation of an ideal judge by Giotto of about 1335 that decorated a meeting hall in the Palazzo del Podestà. A detailed description of the work by Giorgio Vasari allows a reasonably accurate reconstruction of its appearance.[49] A centrally placed figure of a judge, assisted by the four cardinal virtues, was depicted with a scepter in his hand and the scales of justice above his head. Such a conception is related to the theory of Marsilius, as the judge occupied an exalted position for him. According to Marsilius the principal end of the law is "civil justice and the common benefit," so that the judge, as interpreter of the law, is identified with the highest function of the state.[50] Thus, the inclusion of the cardinal virtues in Giotto's work is perhaps an illustration of Marsilius's opinion that in order for the judge to arrive fairly at his judgments he must himself be a paragon of virtue, notably of prudence and justice. Marsilius is also careful to acknowledge his debt to Aristotle for these views, which are found both in the *Politics* and the *Nichomachean Ethics*.[51] Such beliefs, it must be added, are not unique to Marsilius: a commonplace in the literature of the Aristotelian revival had long been the importance accorded to the judges of the state.[52] Giotto's work, and the others like it painted during the first half of the trecento, do reflect, however, the special refinement with which the problem of fair judgment is treated by Marsilius.

A similar work also in Florence, a mural dated about 1335 in the hall of the Arte della Lana, demonstrates in more thorough-going fashion the importance of the good judge and the requisite virtue of his personal life and decisions. Brutus, the first Roman consul, is enthroned as the just judge and is protected by the virtues from the attacks of the vices.[53] A passage from the *Defensor Pacis*, one which is acknowledged to derive from Aristotle, may serve perhaps as a description of the scene: ". . . if the judge has a perverted emotion, such as hate, love, or avarice, this perverts his desire . . . the law lacks all perverted emotion; for it is not made useful for friend or harmful for foe, but universally for all those who perform civil acts well or badly."[54] The judge's own virtue, the defeat of the vices in his own spirit, becomes a metaphor for the role of the judge as the expunger of vice in the state.

A final Florentine work, a mural from the Palazzo del Podestà dated about 1344 (FIGURE 8), shows how the Aristotelian perception of justice and of wise judgment came to influence the understanding of contem-

FIGURE 8. Maso di Banco (attribution): Expulsion of the Duke of Athens from Florence, ca. 1344. Florence, Palazzo del Podestà (photo: Glenn Pryor).

porary events. The fresco represents the expulsion of the Duke of Athens from Florence at the end of his short-lived tyrannical reign.[55] In the left side of the mural is the figure of St. Anne, on whose feast day the event took place. She is seated beside the Palazzo dei Priori in a defensive and militant posture, while the diminutive figure of the Duke is fleeing from his throne, clutching some hybrid beast, and, most importantly, trampling the sword and scales of justice. The moral is clear enough: Because the tyrant had not served the ideal of justice his subjects rose up against him. A prominent passage in the Aristotelian theory offered a defense of rebellion against a ruler for precisely such a reason.[56]

Absolute rulers, nevertheless, were also wont to announce that the purpose of their rule was to serve justice, personified, as in the republican representations, by the just and incorruptible judge, only now the ruler himself figures in the conception. Among the reliefs on the tomb monument of Bishop Tarlati in Arezzo, dated about 1323, are two scenes of special interest to us.[57] In the first of the sequential reliefs the citizens are seen in the pursuit of their separate interests, while the weak and dejected figure of the commune sits sadly at the center (FIGURE 9). In the following

FIGURE 9. Agostino di Giovanni and Agnolo di Ventura: Il commune pelato, 1330. Monument to Bishop Guido Tarlati, Arezzo, Duomo (photo: Glenn Pryor).

FIGURE 10. Agostino di Giovanni and Agnolo di Ventura: Commune in signoria, 1330. Monument to Bishop Guido Tarlati, Arezzo, Duomo (photo: Glenn Pryor).

scene a reign of justice and order has been established as the judge, with the help of the rulers of the city, and most importantly the Bishop himself, sentences to death the transgressors of the public peace (FIGURE 10).

A culmination and *summa* of the influence of Aristotelian thinking on the political art of the first half of the trecento is found in Ambrogio Lorenzetti's murals of *Good and Bad Government* (1337-1340) in the Sala della Pace in the Palazzo Pubblico in Siena (FIGURES 11 and 12). The frescoes deal with the common good and peace that follow from a government devoted to justice. The following words by St. Thomas have been cited as an appropriate elucidation of the program: "Since it belongs to the law to direct the common good . . . it follows that the justice which in this way is called general is called legal justice, because thereby man is in harmony with the law which directs the arts of all the virtues to the common good."[58]

Others have dealt thoroughly and persuasively with the way in which the theme is developed. A problem with the earlier examination of the cycle, however, is that the relevant artistic context and the specific precedents have not been investigated.[59] By viewing Ambrogio's works as sim-

FIGURE 11. Ambrogio Lorenzetti: Allegory of Good Government, 1336-1339. Siena, Palazzo Pubblico, Sala della Pace (photo: Alinari Art-Reference Bureau).

FIGURE 12. Ambrogio Lorenzetti: Allegory of Bad Government, 1336-1339. Siena, Palazzo Pubblico, Sala della Pace (photo: Alinari Art-Reference Bureau).

ply the most complete illustration of several motifs found in Italian art beginning soon after the translation of the *Politics*, as I believe we should, perhaps the murals lose some of their seemingly unique aspects.

Three walls of the Sala della Pace are covered with allegories of justice and injustice. On two of the walls the nature of ideal government and its effects are defined, and on the third wall the meaning and effects of tyrannical government are illustrated. Good government is conceived in a now familiar way in terms of justice, the common good, peace, and the cardinal and theological virtues. Here the ruler himself is presented as the incarnation of the common good, and the source of his judgments in justice and the resultant civic peace is represented as a cord held by the frieze of citizens that unites the figure of Justice with the figure of the Common Good. As in earlier representations of the Aristotelian state, the cardinal and theological virtues are the ends to which the citizens and their government dedicate themselves, and are the touchstones of good government and wise judgment. The inclusion of the *artes mecchanicae* illustrates, in familiar fashion, the various courses of socially and economically beneficial activity that contribute to the common welfare. Finally, similar to Giotto's representations of Justice and Injustice, we behold landscapes that depict the effects of good and bad government. Only through an iconographic evolution that began with the Fontana Maggiore in 1278 do we reach this comprehensive exposition of the ideal human community. The development is a direct consequence of the Aristotelian revival in Paris and of the Italian theorists who were directly involved in the Parisian movement. Doubtless the state would have assumed an enhanced position in the art of the late Middle Ages even without the *Politics*. But it is equally certain that this expanded role would have been expressed through a different set of iconographic formulas.

NOTES AND REFERENCES

1. The finest survey of political art during this period is Helene Wieruszowski, "Art and the Commune in the Time of Dante," *Speculum*, 19(1944): 14-33. See also, Jonathan B. Riess, "The Origin and Rise of Communal Art in Perugia: A Study of the Fresco Program of 1297 in the Palazzo del Popolo" (Ph. D. dissertation, Columbia University, 1977).

2. Nicolai Rubinstein, "Political Ideas in Sienese Art: The Frescoes by Ambrogio Lorenzetti and Taddeo di Bartolo in the Palazzo Pubblico," *Journal of the Warburg and Courtauld Institutes*, XXI(1958): 179-207, has shown how Aristotelian ideas are important for the Lorenzetti frescoes. I am greatly indebted to this and to other works by Rubinstein.

3. In general, see Martin Grabmann, *Studien über den Einfluss der Aristotelischen Philosophie auf die mittlealterlichen Theorien über das Verhältnis von Kirche und Staat* (Sitzungsberichte, Munich, 1934, No. 2).

4. On the translation of the *Politics*, see Martin Grabmann, *Guglielmo di Moerbeke O.P., il traduttore delle opere di Aristotle* (Rome, 1946). For the history of the study of Aristotle at Paris, see Hastings Rashdall, *The Universities of Europe in the Middle Ages*, I, eds. F. M. Powicke and A. B. Emden (Oxford: Clarendon Press, 1936, pp. 349-370).

5. Ernst H. Kantorowicz, *The King's Two Bodies; A Study in Mediaeval Political Theory* (Princeton: Princeton University, 1957, pp. 97-143).

6. For the medieval anticipations of the *Politics*, see Gaines Post, *Studies in Medieval Legal Thought; Public Law and the State, 1100-1322* (Princeton: Princeton University, 1964, pp. 496-561).

7. The basic source on Remigio is Charles T. Davis, "An Early Florentine Political Theorist: Fra Remigio de' Girolami," *Proceedings of the American Philosophical Society*, Vol. 104, no. 6(1960): 662-676.

8. Nicolai Rubinstein, "Marsilius of Padua and Italian Political Theory of His Time," *Europe in the Late Middle Ages*, eds. J. R. Hale, J. R. L. Highfield, B. Smalley (London: Faber and Faber, 1965, pp. 44-65).

9. Kantorowicz, *The King's Two Bodies*, pp. 479-483.

10. Kantorowicz, *The King's Two Bodies*, p. 479.

11. Davis, "Remigio de' Girolami," p. 666.

12. The most complete study of the fountain is Kathrin Hoffman-Curtius, *Das Programm der Fontana Maggiore in Perugia* (Dusseldorf: Rheinland-Verlag, 1968). More importance is granted to the communal context as a source for the iconography of the monument by John White, *Art and Architecture in Italy, 1250-1400* (Baltimore: Penguin, 1966, pp. 50-53). There has been some re-arrangement of the sculptures, and this has been taken into account in the above analysis: John White, "The Reconstruction of Nicola Pisano's Perugia Fountain," *Journal of the Warburg and Courtauld Institutes*, XXXIII(1970): 70-83.

13. Davis, "Remigio de' Girolami," p. 662.

14. Hoffmann-Curtius, pp. 60-63, has suggested the possible allusion to the sacramental rite of baptism into the state, but has stressed more the connection between the fountain and the meaning of the Fountain of Life.

15. White, *Art and Architecture in Italy*, pp. 50-51.

16. For a full exposition of this aspect of the Aristotelian revival, see Walter Ullmann, *Principles of Government and Politics in the Middle Ages* (New York: Barnes and Noble, 1961, pp. 231-280). The scenes of The Temptation and The Expulsion on the Fontana Maggiore should be grouped among the representations of the history of man and thus should not be interpreted as the theological foundation of the labors of man.

17. The precise date of the work is uncertain. The finest discussion of the relief and its place in the overall program of sculptures remains Arthur Kingsley Porter, *Lombard Architecture*, I (New Haven-London: Yale University, 1917, pp. 330-349).

18. Fragments of the work are found on the north wall of the council hall. A thorough description of the murals is found in Jürgen Paul, *Die Mittelalterlichen Kommunal-paläste in Italien* (Freiburg, Ph. D. dissertation, 1963).

19. For the traditional theological meaning of the labors of the months, see J. C. Webster, *The Labors of the Months in Antique and Medieval Art* (Princeton: Princeton University, 1938).

20. On the iconography of the murals, see Riess, "Communal Art in Perugia," pp. 62-199, and for the attribution, see Miklos Boskovits, *Pittura umbra e marchigiana fra medioevo e rinascimento* (Florence: Edam, 1973, p. 33).

21. Wieruszowski, "Art and the Commune," pp. 15-17.

22. On the movement from ecclesiastical structures for the display of political art to civic buildings, see Wolfgang Braunfels, *Mittelalterliche Stadtbaukunst in der Toskana* (Berlin, 1953, pp. 167-169).

23. Marvin Trachtenberg, *The Campanile of the Florence Cathedral* (New York: New York University, 1971, p. 92).

24. Saint Thomas Aquinas, *Selected Political Writings*, ed. A. P. D'Entrèves (Oxford: Oxford University, 1948, p. 5).

25. Riess, "Communal Art in Perugia," pp. 62-95.

26. Wolfgang Braunfels, "Giotto's Campanile," *Das Münster*, I(1948): 206-212.

27. Uta Felges-Henning, "The Picture Program of the Sala della Pace: A New Interpretation," *Journal of the Warburg and Courtauld Institutes*, XXXV(1972): 145-163.

28. L. D. Ettlinger, "Hercules Florentius," *Mitteilungen des Kunst-historischen Instituts in Florenz*, 16(1972): 119-142.

29. In general, see Alan Gewirth, *Marsilius of Padua*, 2 vols. (New York: Columbia University, 1951), and Michael Wilks, *The Problem of Sovereignty in the Later Middle Ages* (Cambridge: Cambridge University, 1963, pp. 84-117).

30. R. W. and A. J. Carlyle, *A History of Mediaeval Political Theory in the West*, VI (Edinburgh and London: Blackwood & Sons, 1930, pp. 3-12).

31. L. Zdekauer, " 'Iustitia,' immagine e idea," *Bollettino senese della storia patria*, 20(1913): 384-425.

32. The single derivative aspect of the discussion of the virtues and vices is the belief that the figures of Justice and Injustice stand for two opposed conceptions of government. See, Giulio Schlosser, "Giotto's Fresken in Padua und die Vorläufer der Stanza della Segnatura," *Jahrbuch der Kunsthistorischen Sammlungen des allerhochsten Kaiserhauses*, VI(1896): 92-106.

33. On the formal relations among the virtues and vices, see Milton L. Gendel, "Giotto's Representation of the Seven Virtues and the Seven Vices in the Arena Chapel at Padua" (M. A. thesis, Columbia University, 1940). I am presently preparing a study of the program of virtues and vices.

34. Kantorowicz, *King's Two Bodies*, p. 138.

35. Erwin Panofsky, *Renaissance and Renascences in Western Art* (New York: Harper & Row, 1962, p. 152 n. 2), discusses the precise Aristotelian source for the conception.

36. Saint Thomas, *Selected Political Writings*, p. 5. For Marsilius. see below, note 47.

37. Ullmann, *Principles of Government*, p. 243.

38. R. Freyhan, "The Evolution of the Caritas Figure in the Thirteenth and Fourteenth Centuries," *Journal of the Warburg and Courtauld Institutes*, XI(1955): 68-90.

39. For a discussion of this aspect of Italian Aristotelianism, see Rubinstein, "Political Ideas in Sienese Art," pp. 46-47.

40. Carlo Guido Mor, ed., *Il Palazzo della Ragione di Padova* (Venice: Neri Pozzi Editore, 1964, pp. 71-77).

41. Riess, "Communal Art in Perugia," pp. 167-199.

42. Trachtenberg, *Florence Campanile*, p. 85.

43. Felges-Henning, "Pictorial Program of the Sala della Pace," pp. 149-150.

44. See, for example, Fritz Saxl, "The Revival of Late Antique Astrology," *Lectures*, I (London: Warburg Institute, 1957, pp. 73-84).

45. For the astrological culture at Padua and its relation to Aristotelian philosophy, see J. K. Hyde, *Padua in the Age of Dante* (Manchester: Manchester University, 1966, pp. 306-307), and Nicolai Rubinstein, "Some Ideas on Municipal Progress and Decline in the Italy of the Communes," *Fritz Saxl: A Volume of Memorial Essays*, ed. D. J. Gordon (London: Warburg Institute, 1957, pp. 165-181).

46. Hyde, *Padua*, pp. 308-309.

47. Gewirth, *Marsilius of Padua*, I, pp. 20-23.

48. Louis Green, *Chronicle into History; An Essay on the Interpretation of History in Florentine Fourteenth-Century Chronicles* (Cambridge: Cambridge University, 1972, pp. 17, 29-35, 103-105, 149-150).

49. On this work, see S. Morpurgo, *Un affresco perduto di Giotto nel Palazzo del Podestà* (Florence, 1897).

50. Gewirth, *Marsilius of Padua*, II, p. 37.

51. Gewirth, *Marsilius of Padua*, II, p. 38.

52. Kantorowicz, *King's Two Bodies*, pp. 123-133.

53. S. Morpurgo, "Bruto, 'il buon Giudico,' nell' Udienza dell' Arte della Lana in Firenze," *Miscellanea di Storia dell'arte in onore di I. B. Supino* (Florence, 1930, pp. 45-57).

54. Gewirth, *Marsilius of Padua*, II, p. 37.

55. Frederick Antal, *Florentine Painting and its Social Background* (Boston: Boston Book & Art Shop, 1948, p. 262).

56. Carlyle, *Mediaeval Political Theory*, V, p. 93.

57. Morpurgo, "Bruto," pp. 155-158.

58. Quoted by Rubinstein, "Political Ideas in Sienese Art," p. 184, from the *Summa Theologica*, trans. Fathers of the English Dominican Province, X (London: Burns, Oates, and Washbourne, 1929, p. 122).

59. Rubinstein, "Political Ideas in Sienese Art," and Felges-Henning, "Pictorial Program of the Sala della Pace."

BARTOLUS DE SAXOFERRATO (Sassoferrato 1313-Perugia 1357) was to law what Dante was to poetry and Bach to music: one of the greatest of all times. Though little known today, for centuries he was famous and influential, as was *Bartolism*, the school he founded. His followers codified laws in France and Austria. What kind of man was he? A doctor of laws at 20, he became a judge, then professor at Perugia. As a Perugine ambassador, Bartolus met emperor Charles IV, who granted him arms and some privileges; his explanation of arms and similar marks of identification was the first treatise on the science of heraldry. He became the founder of international law; the first to discuss the problems of tyranny and of reprisals; to define property and eminent domain; to distinguish human from civil rights; to define sovereignty; to praise the rule of the people, by the people, and for the people as ultimately the best form of government.

MPC

Bartolus the Man

LEO MUCHA MLADEN

Institute for Medieval and Renaissance Studies
City College, City University of New York
New York, New York 10031

WHEN Alfred Rowse came out with his SHAKESPEARE, THE MAN in 1973, how many people, even among the less educated, wondered who Shakespeare might be? But Bartolus? Who is this man? Even many better educated people have never heard of him. Here at Columbia the well-known contemporary civilization course does not mention him and apparently his name does not appear anywhere in the college curriculum. THE COLUMBIA ENCYCLOPEDIA has no entry under Bartolus, not even the latest edition. The situation did not change even after the 1942 Columbia dissertation, BARTOLUS ON SOCIAL CONDITIONS IN THE FOURTEENTH CENTURY, by Anna Sheedy, was published by the Columbia University Press. Are there any civilization (current name for history) textbooks that mention him? In my undergraduate teaching I always added two persons not mentioned in the textbook, Bartolus and Comenius (Jan Amos Komenský, 1592-1670). Generally speaking, those who have heard his name seldom know who Bartolus really was, where and when he lived, what he achieved and why he deserves to be mentioned, let alone celebrated.

And yet, those who know about Bartolus de Saxoferrato (Sassoferrato 1313, p. Dec. 15—Perugia 1357, July 13), find him as important in his field as Dante Alighieri (1265-1321) and Johann Sebastian Bach (1685-1750) are in theirs. In 1954 Francesco Calasso (1904-1965) from Lecce, preside della facultà di jurisprudenza dell'università di Roma, hailed him as one of the great lawyers of all places and times, if not the greatest. (MEDIO EVO DEL DIRITTO, Milano, Giuffrè, 573: uno dei più grandi giuristi di ogni epoca e di ogni paese, se non forse il più grande). To him Bartolus was not only the greatest among the legal scholastics, a giant among the thinkers in general, but also along with Dante, (the Gothic painter) Giotto (di Bondone, 1266-1337), and the (activist nun, saint) Catarina (Benincasa) da Siena (1347-1380) one of the most exceptional human beings of the Italian trecento, this late Gothic, not yet a Renaissance century (574: la fama sua di massimo fra gli "scholastici" del diritto, la figura di Bartolo

[311]

0077-8923/78/0314-0311 $01.75/2 © 1978, NYAS

da Sassoferrato eccede, con la sua gigantesca opera di pensatore, la statura del giurista sia pure grandissimo, e prende il suo posto tra le figure umanamente più elevate del Trecento italiano: accanto a Dante, a Giotto, a Catarina da Siena). There were some famous contemporaries Calasso did not mention. He reminded us that for centuries it was repeated that only a Bartolist is a good jurist (nullus bonus jurista nisi sit bartolista). Lest we be tempted to suspect that this prominent Roman historian of Roman law was carried away by an exaggerated pride in a fellow national, we can easily find writers of different nations also speaking with admiration of Bartolus.

In 1883 a prominent law professor of Leipzig, Rudolf Sohm (1841-1917), came out with his INSTITUTIONEN DES ROEMISCHEN RECHTS (Leipzig, Duncker). In subsequent editions (definitely in the 8th, 1899, if not earlier) he stated:

> The central figure in the world history of the Middle Ages is neither Irnerius [actually Wernerius of Bologna who before 1100 started his glossator school of Roman law] nor any one of the glossators, but Bartolus. His commentaries determined the practice of the courts. In Spain and Portugal . . . they actually enjoyed statutory authority. The creation of the common law of Italy was due first and foremost to the labours of Bartolus. He is . . . the creator of the common law of Germany which sprang from the reception (THE INSTITUTES, translated by James Crawford Ledlie, Oxford, Clarendon press, 1st ed. 1892, based on 4th German, 1889, 2nd 1901, on 8th, 1899, 3rd, 1907, on 12th, 1905; a reprint New York, Kelley, 1970, witnesses its constant use).

In 1495, when Germany received the Roman law, it adopted not the CORPUS JURIS CIVILIS as such, but actually, in all truth, the commentaries of Bartolus (and Baldus) as the common law (11th, 1903, 151: Nicht das Corpus juris als solches wurde in Deutschland als geltendes Gesetz angenommen. In Wahrheit wurden die Kommentare des Bartolus (und Baldus) gemeinrechtlich rezipiert).

Yet, in 1896 in a similar textbook the Parisian law professor Paul Fréderic Girard (1852-1926), while admitting that Bartolus and the Bartolists (derogatorily called postglossators) created on the pretext of explaining Roman law a new law that gained influence over the whole of Europe, deplored that this development came out of the scholastic dialectic, with its diffused tracts encumbered by subtilities and useless divisions (meaning no doubt distinctions) that created an indigestible literature (MANUEL ELEMENTAIRE DE DROIT ROMAIN, Paris, Arthur Rousseau, 8 editions 1896-1919,

FIGURE 1. Portrait of Bartolus in ILLUSTRIUM IURECONSULTORUM IMAGINES QUAE INVENIRI POTUERUNT AD VIVAM EFFIGIEM EXPRESSAE. Ex musaeo Marci Mantuae Benavidii, patavini jureconsulti (1489-1582), Romae, 1566, 3. Engraved and printed by Antonius Lafrerius Sequanus from a portrait originally made at the time of Bartolus' death (vera effigies).

English translation 1906, German 1908, Italian 1909; 3rd ed., 1901, 83: L'étude directe des sources est absente des tracts diffus encombrés de subtilités et des divisions inutiles, dans lesquels les jurisconsultes posterieurs ont accomodé à l'exposition du droit procédé de la dialectique scholastique. . . . Les auteurs de cette litterature indigeste qui se multiplie du XIVe siècle au XVIe, les postglossateurs ou les Bartolists, comme on les appelle

du nom du plus célèbre d'entre eux l'Italien Bartolo de Sassoferrato (1314-
1357) . . . ont, sous prétexte du droit romain construit beaucoup de droit
nouveau et c'est l'explication á l'influence acquise par eux non seulement
en Italie, mais en France, en Allemagne et dans á peu prés toute l'Europe
savante). Thus what Sohm found so great, Girard found deplorable, referring to his colleague Jacques Flach (1846-1919), from Strasbourg, an
architect turned jurisconsult who in 1883 became the professor of legisla-
tion at the College de France and in the same year published in the NOU-
VELLE REVUE DU DROIT FRANÇAIS ET ETRANGER, Paris, he edited (7, 205-227),
an article on CUJAS, LES GLOSSATEURS ET LES BARTOLISTES. At variance with
Girard, Flach arrived at a peculiar conclusion: The influence of Bartolus
touched, but did not penetrate France. The French spirit is too logical to accept the bastardy of the Bartolists (L'influence de Bartole nous touche mais
elle ne nous pénétre pas: L'ésprit français est trop logique pour accepter
une creation aussi bâtarde que celle des Bartolistes). In 1883 another German law professor, Ferdinand Felix Hauptmann (1856-p. 1935), from
Bonn, but teaching at Fribourg in Switzerland, felt provoked to publish (if
my count is complete) the 74th edition and 1st German translation of a little work that made Bartolus founder of the science which was later and for
ages called heraldry, but which should better be called (borrowing the
term from Leo Frobenius, 1873-1938) cultural morphology (Do. Bartoli
a Saxoferrato . . . TRACTATUS DE INSIGNIIS ET ARMIS, mit Hinzufügung einer
Uebersetzung und der Citate neu herausgegeben von F. Hauptmann, Bonn,
Paul Hauptmann). Hauptmann sees the dominant influence of Bartolus
on all subsequent writers; in 1444, the canon of Zürich, Felix Hemmerlin
(Malleolus, 1388-ca. 1450, DIALOGUS DE NOBILITATE ET RUSTICATE, 29); in
1460, the canon of Colmar, decretorum doctor and canonist at Basel, Peter
von (=aus?) Andlau (in Alsace, Petrus de Andlo, ca. 1420-ca. 1480, DE
IMPERO ROMANO . . ., Argentoraci 1603, 1, 14); and in 1529, the eminent
jurist, avocat du roi at Autun, soon (1533) to become the first president du
parlement de Provence at Aix, Bartholomé Chasseneuz (Bartolomaeus
Cassaneus, 1480-1541, CATALOGUS GLORIAE MUNDI, Lugduni, consideratio
38) as lawyers generally lean on Bartolus, although the romantic lawyer
(Hauptmann) with a deep involvement in heraldry was not entirely happy
that a Roman jurist would apply Roman law to what to him has been a
German institution.

   If we wonder why Bartolus is so little known today, praised as he was
by some and criticized by others, we must realize that for more than 200
years after his death he was both famous and influential and that it was

only in the following centuries that his fame and influence slowly receded into near oblivion.

In 1936 Johannes van de Kamp listed in his exhaustive Amsterdam dissertation (BARTOLUS DE SAXOFERRATO 1313-1357, LEVEN, WERKEN, INVLO-ED, BETEEKENIS, door Dr. J. L. J. van de Kamp, Amsterdam, Paris, 53-126) hundred of manuscripts and editions containing the works of Bartolus; yet his listing was still not complete. Between 1481 and 1615 the Bartoli OPERA OMNIA alone appeared in 57 editions, 28 in Lyon, 17 in Venice, 4 in Basel, etc. The law professors for centuries recommended, sometimes demanded, that Bartolus be followed. Antonius de Mincucciis (1380-1468) in the proemium of his repertorium of Roman law that became simply known as REPERTORIUM BARTOLI (printed with Bartoli CONSILIA, . . . Lugduni 1511) regarded the authority of Bartolus such that his opinion substitutes for a defect of law (Bartoli . . . auctoritas tanta . . . ut fere deficiente lege . . . ejus opinio pro jure servetur). And in fact, the laws of Leon, Castile and Portugal (ORDENACOĒS ALFONSINAS), promulgated in 1427, 1433 and 1446 respectively, mandated exactly that. Even in 1603 the ORDENA-COĒS FILIPINAS, (fully abrogated in Brazil only in 1915) prescribed that the opinion of Bartolus be followed, as best conforming to reason (lib. 3, tit. 64, proemium: se guarde a opinião de Bartolo, porque sua opinião communamente he mais conforme á razão).

In 1467 Joannes Baptista de Caccialupis (ca. 1420–ca. 1485) calls Bartolus the mirror, father and lamp of civil law (speculum, pater et lucerna juris civilis); he asks, "Who would deny that Bartolus is the first [or prince if you prefer] in civil law?" (DE MODO STUDENDI . . . Venetiis 1472, Parisiis 1518, Lipsiae 1721: Bartolum in jure civili fuisse principem quis negabit?) Thus both the law students and the practicing lawyers needed Bartolus as their bible the same way as later the American lawyers of the first century of independence needed their bible, the 1765-1769 COMMEN-TARIES ON THE LAWS OF ENGLAND by Sir William Blackstone (1723-1780). Not only Italian law schools prescribed the study of Bartolus, and quite often to study law meant to study Bartolus, just as for Abraham Lincoln (1809-1865) to read the law meant to read Blackstone.

To England Bartolus came with civil law that found its place alongside the canon and common laws, including admiralty law, curia militaris, equity, and even the Star Chamber. Let us not forget that eminent domain is a term coined by Bartolus. Like Germany, also Bohemia, Hungary, Poland, etc., had their receptions of Roman law. Contrary to the derogatory remark by Flach, Bartolus did penetrate the French legal system as well.

On December 15, 1551 in an oratio François Le Douaren (Franciscus Duarenus, 1509-1559), the Breton chevalier and doyen docteur régent en
droict at Bourges, stressed that the French judges follow the opinion of
Bartolus as a law (OPERA OMNIA, Francofurti, 1592, 1114-1115: judices
nostri temporis Bartoli . . . opinionem judicando pro lege ac certo jure sequuntur). From whom other than the lawyers could simple folks have
learned the phrases "resolute as Bartolus" or "more resolute than Bartolus," that the lawyer Estienne Pasquier (1529-1615) overheard (LES RE-
CHERCHES DE LA FRANCE, Paris, lib. 8, 1564?, ch. 14: simple femmelettes et
autres idiots de la populace . . . le disent tantost plus resolu que Bartole
tantost resolu comme un Bartole)? In 1964 Leiden law professor Robert
Feenstra (1920-   ) pointed out in our Columbia Collegium for history
of legal and political thought that the definition of property in the CODE
NAPOLEON of 1804 (La propriété est le droit de jouir et disposer des choses
de la manière la plus absolu, pourru qu'on n'en fasse pas un usage prohibé
par les lois ou par des règlements) goes back to Bartolus' definition: Jus de
re corporali perfecte disponendi nisi lege prohibeatur. The Netherland
code of 1838 has the same in Dutch. Thus Bartolus for the first time in
the history of law provided a definition of property: A right to use and
dispose of things absolutely, as long as not prohibited by law. The CODE
does not mention Bartolus, yet his influence is there nevertheless, however submerged. One can say that this CODE and some others (1811 Aus-
trian, etc.) are the final products, the swan songs, of the dying Bartolism.
Bartolus himself had pointed the way to a synthesis of various laws, Roman, Germanic, statutory, customary, etc., and in this spirit the Bartolists
for centuries codified the laws of Burgundy, Savoy, etc. Unlike Dante who
sang the swan song for the past, Bartolus pointed to the future. At one
time he had superseded most of his predecessors. Yet, as time progresses,
even an Everest will disappear from view behind smaller but nearer mountains. This gradual loss of actuality may, of course, explain a growing
obsolescence and even oblivion, but hardly such a disdain as Girard and
contempt as Flach have shown or rather reported. Certainly it was not
their original insight. Rather it reflected the man Flach wrote about,
Jacques Cujaus or Cujas (1522-1590). Jacobus Cujacius was the first jurist
of importance who did not have a good word for Bartolus or room in his
library of 1300 volumes for his books. (His catalog is in the Bibliothèque
nationale, Paris, latin 4552, Flach 216.) Bartolus and other commentators
he implacably condemned en bloc: verbose in easy problems, mute in difficult, diffuse in narrow ones (verbosi in re facili, in difficili muti, in angusta

diffusi). He preferred the glossators, especially the author of their final GLOSSA, Accursius (1182-1260). The disagreements of Bartolus with the GLOSSA are to him idle fictions and dreams of a sick man (OBSERVATIONUM ET EMENDATIONUM liber 12, cap. 16: Accursium longe magis corona dona-verim, a quo quidquid aberrat Bartolus, vanae fictiones et aegri somnia videntur). But even Cujacius just followed the trend, a trend that origi-nated in Italy and was imported to France. It may go back to the contem-porary of Bartolus Francesco di ser Petracco (or Petracolo) di ser Parenzo di ser Garzo dall'Ancisa (now Incisa in Valdarno) alias Petrarca (1304-1374), the archdeacon of Parma, who deplored that the legists (civil law-yers) lacked an interest in the beginnings of law and in the ancient juris-prudents (EPISTULAE FAMILIARES lib. 20, ep. 4). After him the Camaldoli monk (and 1431 minister general) Ambrogio Traversari (1386-1439) ad-monished a friend to study and imitate the ancient jurisconsults, who are dignified and elegant rather than the new sluggish commentators who could not be (Ambrosii camaldulensis EPISTOLAE ed. P. Canneto, 1759, liber 5, n. 18 ad Marianum Porcinum: ut potius jurisconsultos veteres, quam commentatores ignavos tibi hauriendos atque imitandos moneam. Habent illi in se plurimum dignitatis veteremque elegantiam praeferunt, quam novi isti interpretes in tantum abest ut consequi potuerint. . . .)

As Traversarius' letter is undated, I am unable to tell at the present whether it preceded or followed the following two attacks, both in 1433. On 15 March of that year Maffeo di Bellorio Vegio da Lodi (Maphaeus Vegius laudensis, 1406-1458), then teaching at Pavia, came out with his insignificant DE VERBORUM SIGNIFICATIONE E PRISCORUM JURIS CONSULTORUM SCRIPTIS (ms. Paris BN latin 4599, ed. Vincentiae 1477). At the end of his dedication to the humanist from Cremona and Milanese archbishop, Bar-tolomeo di Francesco della Capra (1360/70-1433) he bemoans that Cinus (ser Quittoncino di ser Francesco de'Sighibuldi usually just Cino da Pis-toia, 1270-1336) or Bartolus are more believed than two ancient juriscon-sults, (most likely the founder of Roman legal Science Quintus Mucius) Scaevola (ca. 180-82, pontifex maximus 89) or (the last of the great an-cient jurists, the elegant Aemilius) Papinianus (ca. 140-212, praefectus praetorio 203, founder of the Papinianist school) blames the compiler and editor of the anthology of Roman legal texts (later known as CORPUS JURIS CIVILIS) the quaestor sacri palatii (530) Tribonianus (ca. 480-ca. 545) for changing some ancient texts and discarding the originals (the famous em-blemata Triboniani) with the result that he created (sic, not eliminated as we would expect) some contradictions and thus we now follow the latter day interpreters Bartolus and others as Apollon's oracles (Non possum sine dolore magno dicere eo deventum esse ut plus fidei adhibeatur Cino vel Bartolo quam Scaevolae aut Papiniano . . . quod non aliunde evenisse arbitror quam Tribuniani [sic] causa a quo absumptis jurisconsultorum libris necesse fuit oriri tot indifficiles quot in jure sunt contrarietates ac proinde posteros interpretes Bartolum at alios tamquam Apollinis oracula observamus).

But it was the other attack, also in 1433, that first scandalized practically everybody, but in the long run exerted a devastating influence on the reputation of Bartolus, at least among some elegant writers whom we now call humanists. The priest Laurentius Valla (1407-1457) can be called the real father of Renaissance humanism. At its base is the love affair with ancient Latin and the resulting hatred for the middle Latin of the scholastics. He demanded that all writers learn and exclusively use this ancient, to him elegant, language. A radical follower of the philologic or rhetoric tradition, he was in love with expression. Only elegant words can be pregnant with meaning. Armed with such fanaticism, he attacked all his life those who did not come up to his standards and dared to express inelegantly opinions that thus necessarily became inelegant, especially the scholastics, but also some fellow humanists. His dissertation (an attack too) showing that the Latin of Marcus Tullius Cicero (106-43) was not as elegant as that of Cicero's great admirer Marcus Fabius Quintilianus (ca. 35–ca. 100) got him in 1429 a teaching position at Pavia. A lawyer whom he urged to study Cicero told him that he preferred Bartolus and that all the writings of Cicero did not compare to the little treatise of Bartolus DE INSIGNIIS ET ARMIS. Valla exploded. The title alone was offensive to him: the word "insigniis" is to us middle Latin for the ancient "insignibus," but to Valla it was simply barbaric, simply wrong. In a letter to Pier Candido Decembrio (1392-1477), he first cursed the unjust (Flavius) Justinianus (483-565, Roman emperor 527) for abolishing the ornate jurisconsults (Laurentii Vallae OPERA, Basileae, 1518, 633: Candido Decembri. Dii itaque tibi male faciant, Justiniane injustissime, qui . . . ornatissimos illos jurisconsultos abolendos curasti) before he proceeded to demolish the geese Bartolus, (his pupil) Baldus (Francisci de Ubaldis, 1327-1400) and (their predecessor) Accursius, who speak a barbarous, not Roman language (anseres Bartolus, . . . qui non romana lingua loquantur sed barbara); and Bartolus who talks like an ass (635: libellum in quo, dii immortales, . . . asinum loqui, non hominem putes). He even quoted the lawyer who told him: We do not care about words, but about sentences (opinions), not about foliage but about fruits. You orators [today we mostly call them humanists] hunt words and miss the power and usefulness of sentences, preoccupied as you are with ridiculous inanities (634: Non est nobis cura de verbis, sed de sententiis, non de frondibus arborum, sed de pomis et fructibus: quemadmodum vobis oratoribus, que verba aucupamini, vim atque utilitatem sententiarum 635: omittitis et semper in ridiculis et rebus inanibus occupati estis). This stress on content rather than form did not enlighten Valla. He also entirely missed the subtle analysis of symbols of identification, distinguishing status symbols from group identifications (family arms and names, etc.), analysis still valid today. All he does is carp on the later part of the unfinished treatise where Bartolus described the usage of colors and flags of his days. As in most of his virulent attacks, Valla scandalized many, and he had to flee Pavia before the irate law students. Thus Bartolus became one of the many targets in Valla's shooting gallery, along with Aristoteles Nikomachu (384-322), the stoics, saint Eusebius Hieronymus (ca. 340-420), Anicius Manlius Severinus Boethius (480-524), saint Isidorus (episcopus) hispalensis (ca. 570-636), saint Thomas Aquinas OP (1224-1274), saint Bernardinus (degli Albizzeschi) senesis OFM (1380-1444) and even some humanists. Later Valla had his brushes with the inquisition, but he survived protected by the generous patron of humanists, Nicolaus V (Tommaso Parentucelli 1397-1455, pope 1447), cardinals Nicolaus Cusanus (1401-1464) and Ioannes Basileios Bessarion (1403-1472, abbot of Fonte Avellana), all eager to reconcile the new learning with the old religion. Valla even became a canon of the Roman cathedral, the Lateran archibasilica, and was buried in this first church of Christendom; yet on the whole he was a lonely man. It was his mother who had to erect his tomb with the epitaph: Laurentio Vallae harum aedium sacrarum canonico . . . qui sua aetate omnes in eloquentia superavit Caterina mater filio pientissimo posuit.

It took some generations before the venom of Valla's pen began to show. In 1444 his opus magnum, the ELEGANTIARUM LINGUAE LATINAE libri 6, came out. Between 1471 and 1536 it had 59 editions. Thus his ideas and his biases gradually spread into wider and wider circles. It helped that some of his attacks were hits, as the one on Bartolus certainly was not. But how many exercised discretion? Some of his works were a revelation to the capricious, witty and influential Augustinian canon and priest from Rotterdam with a theology degree from Torino, Desiderius Erasmus (1469-1536), who edited them for publication and privately or in letters kept casting aspersions on Bartolus while recommending Cicero. The funny and robust physician and former Franciscan François Rabelais

(1494-1553) got himself involved too: in 1532 he devoted a chapter of his first book to the significance of the colors white and blue, Gargantua's livery. To prove a point he could refer to the book of Valla against Bartolus (GARGANTUA 10: A quoy prover je vous pourrois renvoyer au livre de Laurens Valle contre Bartole). In his second book, 1534, he is more explicit: A controversy was patent and easy to judge, but was obscured with follies and unreasonable reasons and inept opinions of Accursius, Baldus, Bartolus, . . . and other old mastiffs, who never understood the least of the laws of PANDECTAE and were fat tithe calves ignoring everything needed to understand the laws. They know neither Latin nor Greek, only Gothic and Barbaric (PANTAGRUEL 10: Au cas que leurs controverse estoit patent et facile à juger, vous l'avez obscurcie par sottes et desraisonable raisons et ineptes opinions de Accurse, Balde, Bartole, . . . et ces autres vieux mastins qui jamais n'entendirent le moindre loy de Pandectes et n'estoint que gros veaulx de disme ignorans de tout ce qu'est necessaire à l'intelligence des loix. Car ils n'avoient cognoissance de langue ny grecque ny latine, mais seulement de gothique et barbare).

   It is amusing that from Petrarca on all the mentioned writers were ordained priests. They more than laymen started this rush into pagan antiquity, into Hellenism that is, mostly for esthetic reasons. They became enamored of a dead language, thus in another civilization, the parent civilization of our own Latin or Western. Mind you, this movement aimed to substitute the ancient, Roman, system of liberal arts, which balanced the cultural and natural sciences, with its own new studia humanitatis, where poetic, ethic, cosmography, and history replaced arithmetic, geometry, astronomy, and music. This movement gradually settled down into a system of secondary education of the elite. The Society of Jesus contributed more to its general acceptance than any one else. Thus many and gradually all students were exposed to these studia humanitatis (studies in humanity, note that they are several studies, but only one humanity meaning culture, thus properly translated cultural sciences but never humanities), before some of them embarked on the study of law. Thus they were indoctrinated (or should we say, had their consciousness raised) in humanism, before they were instructed in Bartolism. Most of them did not care. Some did. This they did either by entirely rejecting Bartolism like Cujas or by harmonizing Bartolism with humanism.

   The Bolognese canonist from Messina Andrea Barbazza (Andreas Barbatia siculus, p. 1400-1480) at least knew the humanist phrase when he wrote that he embraced the opinion of Bartolus as the truest oracle of

Apollon. (COMMENTARIA IN TERTIAM PARTEM DECRETALIUM, Venetiis, 1511, 10, 3, 26, 13: Ego autem amplector opinionem Bartoli tamquam Apollinis oraculum verissimum). The famous Jason de Maino (1435-1519) carefully collected all he could find about the life and works of Bartolus. He wanted posterity to know how devoted he was to Bartolus, whom he worshipped and whose footsteps he adored (Jasonis Maini OPERA, Venetiis, 1590, D.45, 1, 132, 42: praecipua meam erga Bartolum observantiam and summam devotionem posteris testatam relinquerem, quem semper in legibus ut terrestre numen colui et ejus vestigia semper quantum licuit adoravi). Yet, this Bartolist was also an elegant Latin orator, as his foremost pupil Andrea Alciato (1492-1550) testified (IN TRES LIBROS CODICIS, Bononiae 1513, Argentoraci 1515, praefatio: Jasonem . . . etiam in Latinis literis praestantem). Alciatus himself repeatedly calls Bartolus by far the first among the jurists (e.g., C.10, 32, 26: Bartolus qui inter juris interpretes longe primus est).

In 1531 Alciatus devoted a chapter of his popular EMBLEMATUM liber (Augustae, the first emblem book, immensely popular, with 90 editions before 1600) to the subject Bartolus discussed in his treatise DE INSIGNIBUS [sic] ET ARMIS. There is, he says, a book of accusations by Valla. On this book he does not need to pass judgment, thus avoiding stirring up a hornet's nest, irritating and offending either the lawyers or the grammarians (= humanists) (cap.13: Alium tractatum idem Bartolus composuit, quem DE INSIGNIBUS ET ARMIS inscripsit: in quo pleraque disputat ad hoc argumentum pertinentia: adversus quem exstat Laurentii Vallae accusatorius liber, de quo necesse nunc non habeo ferre judicium, ne forte crabrones excitem et vel jurisperitos vel grammaticos irritem atque offendam). Note that our elegant writer is here not sitting in judgment over the pertinent Bartolus, but over the impertinent Valla, whom he cannot praise, but finds it inexpedient to condemn. In his whole work Alciatus combines scholastic Bartolism with Renaissance humanism, extensively using both ancient authors and recent ones. This new methods of his will become known as the *mos gallicus*, the French style. The old Bartolism more or less undiluted with humanism will of course remain to be known as the Italian style, *mos italicus*. Alciatus was, to be sure, an Italian but he began to teach at Avignon (1518), later went to Bourges and afterwards taught in Italy. Thus this new trend he developed in France came to dominate some French schools, especially Toulouse, the place where Cujas was born.

There is an irony in all this: the *mos italicus* of Bartolus actually originated in France and the *mos gallicus* of Alciatus in Italy. Toulouse, now becoming a hotbed of legal humanism, was once a hotbed of the scholastic jurisprudence and even Bartolism. A sharp-minded French scholastic theologian Jacques de Révigny OSB (Jacobus de Raveniaco, usually but wrongly de Ravanis, ca. 1210-1296) went to Bologna to study law, which was taught there in the old Italian style of the glossators; in his teaching at Toulouse he applied the scholastic dialectic method, entirely French in its origin. His foremost pupil Pierre de Belleperche (Petrus de Bellapertica, ca. 1230-1308) continued the new trend. Both became bishops, Jacobus of Verdun; Petrus of Auxerre and also chancellor of France. The Italian poet and lawyer Cinus was so smitten with this novelty, that he promptly introduced it in Italy (LECTURA SUPER CODICE, ms. Paris BN latin 4547, ed. Papiae 1483 etc., opening words: Quia omnia nova placent, potissime quae sunt utilitate decora, bellissime visum est mihi Cino pistoriensi propter novitates modernorum doctorum SUPER CODICE breviter utilia scribere). Bartolus was his great pupil. It was the LECTURA of Cinus that formed his genius, as he will one day confide to his great pupil Baldus de Ubaldis (1327-1400) (IN LIBRUM FEUDORUM 2, 26, Vasallus: Dicebat autem mihi Bartolus quod illud quod suum fabricabat ingenium erat LECTURA Cini). Bartolus had thus not introduced a scholastic method into the discipline of law, but he greatly developed it. The new school of the commentators replaced the old exhausted school of the glossators. Bartolus was of course so much interested in speculation, the great joy of his life, that he simply did not have time or temperament to show disdain for his glossator teachers. Where he found them right, he used them; where wrong, he rejected their opinions, not them. He eagerly seized the opportunity to study Hebrew, a rarity in his time, but apparently did not have the opportunity to study Greek. Here Alciatus meant an improvement with his *mos gallicus*. The pupils of Alciatus often simply combined Bartolism with the new humanism. Some felt the need to sneer and reject, which irritated several prominent, elegant Bartolists.

In his already mentioned 1551 D. 15 oration, Duarenus exclaimed: Are you better informed that all the law teachers of Toulouse, Poitiers, Orléans who profess Roman law with such a perseverance and acclaim? Are you better teachers than Alciatus himself who was as good a Latin (ist = humanist) as you and yet, interpreted the GLOSSA of Accursius with the apparatus of Bartolus so magnificently? Are you wiser than the parlement de Paris which prescribed to teachers to teach this method (=Bartolism)? (OPERA 1592, 1114-1115: Ergo vos consultiores estis omnibus tolosanis, pictaviensibus, aurelianensibus legum doctoribus, qui tanta frequentia et celebritate profitentur jus civile? Meliores estis doctores Alciato ipso, qui cum non minus sit, quam vos, Latinus, glossas Accursii cum apparatu Bartoli magnifice interpretatur? Sapientiores denique senatu parisiensi, qui doctoribus eam formam rationemque docendi praescripsit?) In 1912 the native of the Breton Brest and Parisian Roman law professor Emile Jobbé-Duval (1851-p. 1918) just could not understand that this is Duarenus who speaks, the humanist Duarenus, pupil of the humanist who had introduced humanism to France and to legal studies as well, Guillaume Budé (Guillemus Budaeus, 1467-1540); he forgot that this erudite and eloquent Duarenus was not only a son of the sénéchal (seignorial judge) de la cour de Moncontour but also his successor (jurisdictioni in celtica nostra Britannia non omnino infeliciter praefui), a student not of Alciatus, as it was often and wrongly assumed, but of the Bartolists at Orléans and later also avocat at the parlement de Paris. Thus, just like Bartolus before him, Duarenus was imbued both with the scholastic juris-prudence (in his case Bartolism) and experience in legal practice, before he began to teach; and though he had his fights with the Bartolists, whose Latin he disdained as *scythica barbaries*, his main quarrels were with his fellow humanists, with his fellow Breton colleague and adversary of Bar-tolus, François Baron de Loc Eguiner (Franciscus Eguinarius Baro leonen-sis, 1495-1550) because of whom he had to leave Bourges and come back only after his colleague's death, and most importantly with Cujacius whom he prevented from coming to Bourges as long as he himself lived (MELANGES P.F. GIRARD, Paris, Rousseau, 1, 673-621, FRANÇOIS LE DOUAREN (DUARENUS) 1509-1559).

In 1576 Jean Bodin (1529-1596) published his magnum opus SIX LIVRES DE LA RÉPUBLIQUE. In the Latin preface he confesses that as a young teacher of Roman law at Toulouse (he studied there too) he appeared to himself very wise, surrounded as he was by his students, thinking that those princes of law Bartolus and Baldus did not understand much if any-thing (a typical humanist conceit). But later on when he was initiated in the forum of sacred jurisprudence and received training in daily court practice, he realized that the true and solid wisdom does not consist in churning up of scholastic (academic, thus humanist) dust but in grasping the court situation; not in stresses on syllables (philologic finesses) but in the pondering of equity and justice (Fuit enim tempus illud, cum populi romani pulice apud Tolosatos docerem ac valde sapiens mihi ipsi viderer in adolescentium corona. Illos autem juris scientiae principes, Bartolum, Baldum, ... nihil aut parum admodum sapere arbitrarer. Postea vero, quam in foro jurisprudentiae sacrae initiatus ac diurno rerum agendarum usu confirmatus sum, tandem aliquando intellexi non in scholastico pulvere, sed in acie forensi, non in syllabarum momentis, sed in aequitatis ac justitiae ponderibus veram ac solidam sapientiam positam esse). And for good measure Bodin turned the table on Cujas, who had preferred the African rhetor Lucius Apulejus (ca. 125–ca. 200) to Cicero, the Apulejus who first defiled the purity of Latin by foul barbarism (qui primus foeda barbarie latini sermonis puritatem conspurcavit).

   In 1582 Alberico Gentili (1551-1608) from San Ginesio, who studied law and learned to love Bartolus at Perugia before he became professor of civil law at Oxford (the only law taught in English universities before 1753, when Blackstone began his lectures in common law at Oxford), devoted a whole book to defending Bartolus against Cujas and other novices (=humanists) (DE JURIS INTERPRETIBUS dialogi (libri)6, Londini 1582, Lipsiae 1721). A humanist himself, he granted the superiority of the novices in fine letters, but wrote: The only thing they attack him [Bartolus] for is style; they attack him on account of etymology, hardly on account of analogy [perhaps content analysis would do]. They claim he is rude, as far as good letters are concerned. Granted, in etymological interpretation (philology) they are supreme, but in analogical interpretation, which is the proper method in law, they do amount to nothing (2, 23: Vix tamen est, ut in hoc interpretationis genere quod analogicum dicunt, novitii Bartolum lacessant, sed in etymologico: bonarum litterarum rudem clamant. 24: Modo etiam dicebamus, in etymologica interpretatione novitios interpretes valere plurimum, in analogica, quae propria nostra est, minimum.)

   Interestingly enough, at least the last of these defenders was (turned) a Protestant. With a dash of exaggeration one could say that Bartolus split the Renaissance but united the Reformation, as he had, quite innocently, split the Renaissance humanist writers into two camps: that of his detractors and that of his admirers. After the Reformation divided Latin Christendom into two camps, for quite different reasons, Bartolus becomes one of the unifying elements, bringing together the writers on both sides, from Spanish Jesuits to German Lutherans.

   Two eminent jurists were honored by his name: the Dutch Bartolus Nicolaus Everardi (1462-1519) and the Spanish Bartolus Diego de Covarrubias y Leyva (1512-1577). Both were Catholics.

   In 1905 a Protestant divine and well-known English historian John Neville Figgis (1866-1919) read a paper on BARTOLUS AND THE DEVELOPMENT OF EUROPEAN POLITICAL IDEAS (TRANSACTIONS OF THE ROYAL HISTORICAL

SOCIETY, London, n.s. 19, 147-168). Rector Figgis had behind him two books, THE DIVINE RIGHT OF KINGS (1896), and STUDIES OF POLITICAL THOUGHT FROM GERSON TO GROTIUS 1414-1625 (1900), and was then preparing POLITICAL THOUGHT IN THE 16TH CENTURY for the CAMBRIDGE MODERN HISTORY. He started by noting that "two names must be well known to anyone who has glanced at the margins of works on law and politics . . . in the 16th and 17th centuries . . . there stand out the great twin luminaries of Perugia, Bartolus and Baldus, his pupil, friend and adversary. Grotius and Gentilis and Bodin not merely quote Bartolus, but are what they are largely because of him. Pages might be filled with the epithets of laudation from time to time applied to him. He is the mirror and lamp of the law, his name is not so much that of a man as the very spirit of jurisprudence. Some say that he is the sole authority superior to the Roman Rota, while in Spain if there is a defect of law, the opinion of Bartolus is treated as itself decisive. It was said that he held the primacy in the schools, while his authority even in the courts was sacrosanct and neither professors nor judges dare to contradict his opinions. . . . How came it that it was sufficient to quote Bartolus and in many cases to ignore his predecessors? . . . (149) The reasons for his influence is twofold—the old explanation, the man and the milieu." Figgis now proceeds to give a brief outline of his life. He is wrong in calling Bartolus the son of a lawyer: Bartolus was the son of a farmer. He always signed his name Bartolus de Saxoferrato (if the occasional *a* Saxoferrato is an editorial emendation) and Figgis calls it pedantry to call him otherwise. In Italian records found more recently at Perugia he is listed as ser Bartolo da Sassoferrato. The Italian *da* clearly indicates that Sassoferrato is not his family name, let along a noble predicate, but simply a birthplace. In English it should thus be *from* (not *of*) Sassoferrato. The town is on the Adriatic side of the Apennine divide, close to Monte Catria on the river Sentino. In 1881 Cesare Bernabei stated that Bartolus was born in a nearby village Venatura (BARTOLO DA SASSO-FERRATO E LA SCIENZA DELLE LEGGI, Roma, Loescher, 15), located 9 km NW according to (1936) van de Kamp (3), who was told that this is a local tradition. In 1919, Guido Vitaletti further heard that Bartolus was born in the place (hamlet?) called La Raè (ARCHIVUM ROMANICUM, Genève, Olschki, 3, 1919, 409-510, TRADIZIONI CAROLINGHE E LEGGENDE ASCETICHE RACCOLTE PRESSO FONTE AVELLANA, 444: Bartolo, circolano a Venatura, paesello appolaiato sul versante orientale della Strega, 5km da Fonte Avellana. In questo villagio e precisamente nella località detta La Raè [. . . =canalo disrupato e sassoso] . . . sembra che egli veramente sortisse i natali). The local folklore made both Bartolo and Baldo natives of La Raè. While taking care of goats, they sat near a spring of water and wrote all the trees of laws. After they have written for a long time, they carried the books on a donkey to Perugia. The Perugines were alerted to the wisdom of the two strangers. People fêted them. . . . All said that the wisdom rose from their minds like water from the fountain (Bartolo e Baldo furono nativi della Raè, parrochia de Venatura. Andando con le capre si mettevano a sedere all'angolo di una fontana e scrissero tutti gli alberi delle leggi. Dopo di avere scritto molto tempo, presero i libri delle leggi li caricarono sopro un somarello e li portarono a Perugia . . . I Perugini accortisi della sapienza dei due forestieri . . . e il populo li festeggiò e ando loro incontro col concerto. E tutti dicevano che la sapienza sorgeva dalla loro mente come l'aqua dalla fontana). Of course, the local peasants lumped in Baldus who was in reality, a nobleman, a son of a Perugine physician and probably never even visited Sassoferrato. To these fantasies of local peasants we can add an imagination of our own: Imagine in 1318 a five-year-old Bartolo gazing on a stranger passing by! Dante, then 53 years old, stopped at the nearby monastery of Santa Croce di Fonte Avellana on the final stretch of his wanderings, on his way to Ravenna where he died in 1321 and was buried at San Francesco. All we can say is that these two men of genius were near each other in one moment of their lives.

In his diploma of 1334, in his citizenship 1348, and in his will of 1356 we find the full name of Bartolus: Bartolus Cecchi Bonacursii de Saxoferrato. (In 1576 the Tribonianus of Perugia, Joannes Paulus Lancelottus, 1511-1591, husband of a descendant of Bartolus, Marietta Alfani, appended these texts to his VITA BARTOLI JURECONSULTI . . . , Perusiae).

What was his family name? His life historians supplied him with several, all fruits of the imagination. The early ones listed Alfani or Severi: father Severi, mother Alfani; first Severi, later changed to Alfani and so on. In fact both names were adopted, but much much later (earliest evidence from 1478) by the descendants of his two grandsons Alfano and Severo. Still later, the Severi changed their name to Alfani, as the Alfani became more prominent. In 1827 Antonio Brandimarte OFM Conv. found in the archives of Sassoferrato two deeds of Cecco Bonacursii Bentevoli, the father of Bartolus: sales of lands and cows. The good friar leaped to the conclusion that the family name was Bentevoli rather than Alfani or Severi. In 1881 Bernabei (17) and in 1913 Leonard's older brother, Cecil Nathan Sidney Woolf (1887-1917), in his BARTOLO OF SASSOFERRATO, HIS POSITION IN THE HISTORY OF MEDIEVAL POLITICAL THOUGHT (Cambridge University Press, p. 395) followed the lead. Yet, as van de Kamp noticed in 1936, all the records convey is that Bartolus was a son of a Cecco (meaning Francesco, not Czech), who was a son of Bonaccorso, who was a son of Bentivoglio. No family name at all. As a matter of fact, only a few people in his days had a family name. Their pedigrees seemed to have identified them sufficently.

What is amazing is that this man, who did not have any family name, found just in family name the symbol of identification corresponding to family arms. The Roman *nomen* (family name) was to him a species of the same genus and thus the laws applying to *nomina* should apply to family arms as well. A less subtle mind would have fallen into the pitfall of identifying insignia (ensigns, flags) with the insignia mentioned in the ancient legal texts. The subtle Bartolus, on the other hand, dismissed these ancient insignia of proconsuls and legates as status symbols attached to a dignity (rank) or an office (function) and thus comparable to the insignia of bishops and doctors, reserved to those who have such a rank or function. Arms on the other hand are not status symbols, at least not *per se*, but rather symbols serving to identify certain groups of people, without ranking them. Both family names and arms belong to the whole family (agnatio or domus, thus clan, stirps) i.e. sum total of male descendants of a given man, not to any individual qua individual. Naturally, if the individual who adopted arms does not have any descendants, they die with him, yet this happens so to say by accident. Thus family names and family arms are *not* hereditary, i.e. one cannot leave them to a descendant and disinherit another from them. Rather they transit (*transeunt*) all new members at the moment of their becoming members. Some groups one can join only later in life (societies, universities, etc.), though one becomes member of some groups at the moment of birth: family, people (demos). Let us only remember, that all things hereditary either precede the birth (heredity) or follow (heritage, inheritance)— such is true even of groups one joins soon after birth, a nation (ethnos, Sprachnation) and cultural society (laos).

In 1387 an early popular writer doctor decretorum and prior Honoré Bonet OSB (b. 1340–d. 1409) used the TRACTATUS DE INSIGNIIS ET ARMIS for his L'ARBRE DES BATAILLES (part 4, ch. 124-129) without acknowledging Bartolus by name (*nos maistres*) and managing also to distort most of the subtlety of Bartolus: the *nomina* (family names) become *noms* (first names), the status symbols arms of dignity, and so on. Thus the fine analysis of Bartolus arrived at the gentlemen's libraries of France, Spain, or Britain, not only not identified with his name, but distorted as well. Especially in the British isles, the arms, in Europe collective, are individualized and proclaimed to be hereditary. Open a British book on heraldry and you face a statement, that the symbols become true heraldy only when they become hereditary. Poor Bartolus!

Cecco and Santa, the parents of Bartolus, were most likely more attracted to the San Francesco church and convent at Sassoferrato than to the nearer eremus of Fonte Avellana. At least there they found a teacher worthy of their precocius son. Later Bartolus will reminisce: I had a master who taught me first letters [=Latin]. His name was brother Peter of [=from] Assisi. Now in the Venetian city he is called Peter of Piety, thus called because he erected there a House of piety, where they take in and feed foundlings. He is an expert man, of no hypocrisy but of admirable holiness to me and to all who know him. I thank him . . . because while he taught me, he preserved me from a lapse and by God's grace trained me so well, that in my 14th year of age I began to study law at Perugia under lord Cinus of [=from] Pistoia and while His grace continued, I made such progress in a continuous study, that in my 20th year I publicly repeated and disputed [=became bachelor of laws] at Bologna and finally in my 21st year I made the doctorate. For the good friar Peter I bear so much love, that while the pen this writes, I cry in my heart. This venerable man told me himself that he gave many of these foundlings to others [=families] when they contracted [promised] that they will treat them as their own (D.45, 1, 132 =OPERA, 6, 1596, 49b-50a: Ego habui unum magistrum, qui me primas litteras docuit, qui vocabatur frater Petrus de Assisio, nunc vero in civitate Venetiarum vocatur frater Petrus Pietatis, sic dictus, quia locum ibi erexit, qui Domus pietatis vocatur, ubi infantes expositi nutriendi recipiuntur. Vir est expertus, nullius hypocrisis, mirae sanctitatis apud me et omnes, qui eum noscunt. Cui gratias ago . . . quia, dum me rexit, custodivit a lapsu. Et Dei gratia et sui doctrina me talem redidit, quod in 14. anno aetatis meae in civitate Perusii sub domino Cino de Pistorio jura civilia audire incepi. Et ejus perseverante gratia taliter continue studendo profeci, quod in 20. anno Bononiae repetendo et disputando publice de jure respondi. Et demum in 21. anno doctoratu fui. Et ex multo amore, quem ad illius fratris Petri bonitatem gero, cum calamus hoc scribit, cordis oculus lacrimatur. Ipseque vir venerabilis Petrus dixit mihi, quod plures ex illis pueris expositis aliis tradidisset hoc pacto; quod eos non aliter quam ut filios observarent). It took a long time before a reader of those words would publicly admit to be deeply moved. Some even misread the words to proclaim Bartolus himself a foundling. In 1550 doctor utriusque juris and canon of Vivien Bermond Choveron mentioned that some refer to Bartolus as a foundling and that before the gate of the Hospital Saint-Jacques at Toulouse at the Pont Saint-Subrac is a picture of Bartolus as a foundling, as one could read it in a French distich on the wall of this hospital (DE PUBLICIS CONCUBINARIIS, 97 = TRACTATUS ILLUS-TRIUM . . . , Venetiis, 11, 1588, 189a quamplures fuerunt alii virtuosi, qui ex spurca commixtione descenderunt, ut Bartolus quem . . . referunt qui-dam fuisse expositum et adhuc remanet effigies ante portam hospitalis sancti Jacobi juxta pontem sancti Subraci Tolosae, ibi apparet tamquam expositum et in hospitali fuisse nutritum, ut legitur in quodam disticho gallico adpicto parieti dicti hospitalis). The Spanish Bartolus Covarrubias evidently also reported this picture, but I think in 1964 Calasso was wrong when he saw in it a part of an anti-Bartolus propaganda (DIZIONARIO BIO-GRAFICO DEGLI ITALIANI, Roma, Istituto della Enciclopedia italiana, 6, 664-668, BARTOLO).

In 1905 finally came the long-expected exclamation: "One of those touches which help us to stretch hands across the ages. . . . I do not think anyone can read that for the first time in its place without loving the writer. How different from Wyclif, the most disagreeable beyond excep-tion, of all medieval thinkers!" (Figgis 154) How surprising a comment! The Protestant minister Figgis evidently did not like much the father of puritanism, John Wyclif (1324?-1384). In any case he was touched by a passage which to me is one of the most touching confessions there is.

Many Bartolist and many Franciscan writers kept mentioning this good frater Petrus de Assisio, but no one seems to have taken the trouble to find more about him in either Franciscan or Venetian archives, till in 1915 a Venetian Franciscan Leone Ranzato di Chioggia came out with the answer: Pietro or Petruccio de Guanchola, born ca. 1300 at Assissi, was the man who after a stay at Sassoferrato, went to the city on the lagoons, where he ca. 1335 established a home called Chasa de la Pietà (or Ospe-dale de la Pietade delli poveri fantolini) next to the Franciscan convent of San Francesco della Vigna; in 1346 he transferred the girls to another asylum at Santa Maria della Celestia and created schools in both institu-tions. As their prior he was in charge of both houses. As a good Lesser brother he begged along the canals crying Pietà! Pietà! Thus the Vene-tians began to call him Petruzzo de la Pietà. There he died in 1349. These homes are apparently still there. Ages later the abbate rosso Antonio Vi-valdi (1678-1741) was maestro di violino at the Ospedale della Pietà

(1703-1740), writing hot music for the orphans who sang and fiddled like angels. The terms "no hypocrit" and "man of admirable holiness" Bartolus did not just use because they flew into his pen. Like all great movements tend to, the Lesser brothers were then split into the rigorous Spirituals and the relaxed Conventuals. The two San Francesco convents at Sassoferrato and at Perugia were relaxed, but Bartolus seems to defend a Conventual and thus indirectly the Conventuals against the accusations of the Spirituals that they strayed from the path of holiness into hypocrisy. The lawyer Jacopo de' Benedetti, turned into the Spiritual poet Jacopone da Todi (ca. 1230-1306) said it emphatically in his LAUDI SPIRITUALE (lauda 19 and 31). Later Bartolus got himself involved in their disputes when he wrote his TRACTATUS MINORITARUM (or MINORICARUM DECISIONUM), making fine legal points on the vexed problems of Franciscan poverty. And the Lesser brothers have praised him ever since. Bartolus also wrote his last testament in 1356 at San Francesco at Perugia, asking to be buried in one of the two San Francesco churches and in fact was buried in the one at Perugia. Later his grave at San Francesco became famous for its laconic inscription: Ossa Bartoli. When he as a boy of 13 first came to do graduate studies in laws at Perugia, he might have come furnished with recommendations from his teacher and possibly even stayed at the convent. Years later Petruzzo must have stopped at Perugia to pay a visit to his now so-famous pupil, no doubt, on his way to the nearby Assisi. On this occasion he told him that he kept placing his orphans for adoption. (ARCHIVUM FRANCISCANUM HISTORICUM, ad Claras aquas, annus 8, p. 3-11, CENNI E DOCUMENTI SU FR. PIETRO D'ASSISI OFM (FR. PIETRUZZO DELLA PIETÀ) 1300-1349).

Whether it was this Franciscan connection that led him to go to Perugia rather than to the more famous school of Bologna, we just don't know. In any case, Cinus was making Perugia famous as the only place in Italy to soak in the new French style. It was Cinus who formed his genius, Bartolus later admitted to Baldus, whose genius he formed. And it must have been Cinus who also introduced him to the works of his friend Dante. Twice later Bartolus comes to speak of Dante. In one case he seems to agree with Dante on a rather complicated point of the relation of the Church and the empire (D.48, 17, 1, 1, 2=OPERA 6, 176b: tenemus illam opinionem quam tenuit Dantes, prout illam comperi in uno libro, quem fecit, qui vocatur MONARCHIA). In the second and more famous case he disputed with Dante the concept of nobility, though with a reverence for such a poet (C. 12, 1, 1-OPERA 8, 46rb-48va: salva reverentia tanti poetae). He called Dante an Italian poet worthy of praise and keeping his memory alive (Dantes Aligerius de Florentia, poeta vulgaris laudabilis recolendae memoriae). To my knowledge this is the only Italian work he ever quoted (CONVIVIO 4). Here it could only be said that the difference is between the idealistic, programmatic poet and the practical lawyer: is nobility an ennobling state of mind, a virtue or simply a privileged group in a given state? Is it something we want it to be or simply what we find it to be?

Bartolus stayed at Perugia as long as Cinus did. In 1333 both left: Cinus went to the Florentine studium, Bartolus turned to Bologna to get his degree. Here he made quick progress. On December 15, at 19, he publicly disputed a quaestio in the class of the leading glossator Jacobus Buttrigarius (1274-1348) and thus became a bachelor of laws. His sponsor (*pater meus*) presented him to Joannes Calderinus (ca. 1300-1365), adopted son of the leading canonist Joannes Andreae (1270-1348). Calderinus was the vicar of the archdeacon who (in lieu of the bishop) was in charge of the *studium generale* (=university to us). Bartolus was given two texts to dispute, one from the DIGESTA. the other from the CODEX. In 1334, Bartolus passed his doctoral examination before 10 doctors. On November 10 came the solemn rite in the cathedral of San Pietro. The vicar conferred on him the license and faculty to teach at Bologna and anywhere else and to promote his students to doctors of laws (legendi, docendi and doctorandi Bononiae et ubique de caetero plenam licentiam et liberam facultatem). His promotor Buttrigarius invested him with a book, a ring, and a biretta, those insignia of the doctoral dignity. A kiss of peace and a blessing concluded the ceremony. It was, naturally, a church rite, performed by a churchman in a church as only the Church could have underwritten the validity of degrees as well as their universality. What we call today a university was then known as a *studium generale*. It developed from a *studium*, that is cathedral or diocesan school, when the validity of its degrees was extended to the whole Latin Christendom. Only the Holy Roman church could have guaranteed that.

Thus at 20 Bartolus was a doctor, as he himself told us. In 1831 the prominent German legal scholar Friedrich Carl von Savigny (1779-1861) figured out that, if he was 20 when promoted, he must most likely have been born in 1314, as it leaves only 1 1/2 months in 1313, but 10 1/2 in 1314 (GESCHICHTE DES ROEMISCHEN RECHTS IM MITTELALTER, Heidelberg, Mohr, 6. Band, 1831, 125. Da nun von diesem sicheren Zeitraum nur 1 1/2 Monte in 1313 fallen, aber 10 1/2 Monate in 1314, so ist 1314 viel wahrscheinlicher als 1313). On his authority most writers and librarians rushed

to list 1314, even without a question mark. Thus it remained still in
Woolf (1913), Sheedy (1942), Calasso (1954), and as late as 1966 in
Guido Astuti (GRANDE DIZIONARIO ENCICLOPEDICO UTET 2, 744). The encyclopedias of different nations are about evenly divided, with BRITANNICA
repeatedly in the 1314 column. In 1936 Van de Kamp (18) noted that the
quaestio he defended at 19 is dated 1333 December 15. Thus Bartolus
must have been born after December 15. Some writers knowingly, most
of them quite unknowingly disregarded the rather conclusive evidence for
1313 in the form of a carta, a sort of obituary, composed not long after his
death and found ca. 1520 in the family papers by his descendant Teseo
di Severo Alfani (b. 1450?-1527) and about the same time reported by
one of his life historians Thomas Diplovataccius (1488-1541) in his VITA
BARTOLI (ARCHIVIO STORICO ITALIANO 16, 2, 1851, MEMORIE PERUGINE di
Teseo Alfani, 267; Bartoli OPERA, 1590, 8-9).

The five years following the doctorate are not fully documented. In
1338 Bartolus was present at Macerata and possibly instrumental as ad-
vocate general of the rector of marca anconitana in the absolution of the
commune and citizens of San Ginesio from an interdict they incurred when
they gave shelter to the outlawed Fraticelli de paupere vita, those irre-
concilable Franciscan flower children. Before that he was assessor or *offi-
cialis* (judge) at Todi and Cagli, two dependencies of Perugia, later at Pisa.
Thus he spent these years, most likely all of them in legal practice. It
marked him for the rest of life as a lawyer and teacher with a prime concern for court practice.

It looks as if all Italian *studia generalia* (universities) were founded
after an exodus of students from Bologna. In 1338 such an exodus gave
the opportunity to the Pisans to request from the pope a grant of a *studium
generale* of their own. In 1339 Bartolus was asked to become one of the
first teachers there. In 1343 he moved to Perugia, where he remained for
the rest of his short life. Here he began to excel (Baldus in CODICEM 1, 29;
. . . opinio Bartoli qui fuit homo multum inhaerens practicae. Et fuit assessor primo Tuderti, postea Pisis et ibi palam legere incepit. Et deinde venit
ad civitatem Perusii, unde legendo optimus factus est). And his foremost
pupil, Paulus de Castro or castrensis (1360-1441) overhead Baldus say that
at Perugia Bartolus became famous, began to be called Commentator, and
students from all over Italy flocked to Perugia to hear him (COMMENTARIA
IN JUS CIVILE, Lugduni, 1511, C.6, 26, 6: Et audivi a Baldo in voce . . . quod
Bartolus repetierit in studio perusino . . . et tunc Perusii acquisivit magnam
famam et incepit vocari Commentator et quod de tota Italia illic concurre-

FIGURE 2. Bartolus teaching. A miniature in a fourteenth-century manuscript (II.I. 217) in the Biblioteca nazionale centrale, Firenze. Bartolus is also shown in the initial Q.

bant omnes scholares). As Paulus both doctored and taught at Avignon, he spread this fame beyond the Alps.

The career of Bartolus and of so many others was interrupted by the great plague. It ravaged Perugia and environs (contado) in 1348, from April to August, taking 100,000 lives (says Graziani in his CRONACA, ARCHIVIO STORICO ITALIANO 16, 1, 1850, 158). Bartolus came several times to speak about the instant mortality of 1348. There was such a pestilence that laws were in abeyance in cities and countless people died. The hostility of God was stronger than the hostility of men (D.42, 4, 21 =OPERA 5, 1596, 92ra: tempore mortalitatis instantis de anno Domini 1348, prout scitis, erat tanta pestilentia, quod jura non reddebantur in civitatibus et moriebantur infiniti homies . . . Et fuit illa hostilitas Dei fortior, quam hostilitas hominum). Bartolus himself survived, though barely so. It was probably at this time, when he says, that he was so ill, that he ceased speculating about his teachings, something that normally gave him such a joy. Finally, he realized that his body will not regain its health unless the mind first recovered its joy of thinking (IN TRES POSTERIORES CODICIS LIB-

ROS, proemium =OPERA 8, 1596, B 2: Cum igitur diu fuerim gravi corporis infirmitate corporis gravatus et propter hoc a consueto gaudio speculationis doctrinae cessaverim, cognovi, quod in me vires corporis non poterant restaurari, nisi aliqualiter efficeretur animus meus gaudens). What an insight into the mind of Bartolus, where all the exciting things happen! How he loved to think! What a joy to ponder over the doctrines! And Bartolus seemed to have had an inkling about psychosomatics.

Bartolus might have known also something about diet. In 1433 Ludovicus de Ponte alias Pontanus (1409-1439) preserved us an interesting detail: To keep his mind always in a balance, he measured his food intake (SINGULARIA 712: Bartolus comedebat ad pondus, ut haberet intellectum pariter dispositum et numquam alteratum). One could wonder whether his survival was not helped by his disciplined way of living.

After the plague subsided, the trustees of the *studium* petitioned the Perugine government to grant Bartolus and his brother Bonaccursius Perugine citizenship, making an exception from the statute that forbade the citizens to teach and vice versa (1576 Lancellotti: Supplicatur vobis dominis prioribus artium populi Perusii pro parte sapientium studii perusini) both to reward him for past services (qui huic communi per plures annos servivit) and save him for Perugia in view of more lucrative offers from elsewhere (sit nuper ad alia studia vocatus cum majori salario). The convocation of citizens (convocato et coadunato publico et generali consilio majori civitatis) granted the request (sint . . . veri et originarii cives et populares istius civitatis et populi), obviously glad to keep this, no doubt their most famous, man for the local *studium*. Actually, Bartolus was from birth Perugine subject as his native commune Sassoferrato in 1297 submitted to Perugia. But from now on Bartolus proudly added to his signature *civis perusinus*. After all, the republic was at the peak of its power and he liked its working democracy. Better for a country to be born in the heart of a man than for a man to be born in a country!

As if expecting to die soon, Bartolus began to deposit some of his thoughts on various subjects in a novel form of short treatises or tractatus. (All are reproduced many times, thus in OPERA 10, 1596). In 1354 he published two: MINORITARUM, already mentioned, and REPRAESALIUM, on reprisals. These reprisals created a vexing problem. The CORPUS JURIS did not discuss them, as in the times of the flourishing Roman empire there was no place or need for them. Only after the empire disintegrated and new small but sovereign states (civitates superiorem non recognoscentes) replaced it, they began to apply them against each other, satisfying claims of their citizens against foreigners by seizing properties from their fellow citizens. Bartolus would allow reprisals in a very restricted sense only. Of course, the law exempted ambassadors, clerics, pilgrims and scholars.

In 1355 came another interruption, this time much shorter and much, much more enjoyable: Carolus IV (1316-1378), elected and twice crowned king of the Romans (as the German kings were then styled), a crowned king of Bohemia and an uncrowned king of Italy and Burgundy, descended on Italy to be crowned some more. With him was his third new queen (1353) Anna Piastowna (1339-1362). On Epiphany or Three Kings the archbishop Roberto Visconti crowned them king (with the iron crown of the Longobards) and queen of Italy at Milano; Petrarca, in the service of the Visconti, was present. It was said that Carolus was a virtual prisoner of the Visconti, his imperial vicars, who dined and wined him and his entourage, but asked his army (if you can call a troop of 300 an army) to wait outside. In 1331 his fasting on Good Friday saved his life, while the Visconti managed to poison his whole retinue. In 1332 at San Felice he scored a magnificent victory over them and their Gonzaga and della Scala allies Now he was in peace received by his vicars. At least in hindsight, I suspect, Carolus must have been glad that Petrarca refused to accompany him to the city where he in 1341 himself was crowned, though only a poet. Whatever his achievements as a *poeta laureatus*, and his great fame, the archdeacon of Parma was a great public relations man but not much of a statesman. Petrarca would have been disappointed at the imperial coronation because of his dreams of the restoration the ancient empire and to Carolus, with his much more modest attempts to pacify and consolidate Italy, Petrarca was simply a nuisance. On January 18, Carolus entered Pisa *con grande trionfo* and accepted Pisan lordship from the Bergolini party. Soon Siena and other cities made him their signore on March 20. Even the Florentines submitted paying off in cash the pleasure of his visit.

On Easter Sunday (April 15) Carolus was crowned in the Vatican basilica Roman emperor forever august (Romanus imperator semper augustus). The precedent established in 800 demanded that the Roman (not Holy Roman) pontifex himself crown the Roman (not Holy Roman) emperor, who only after this coronation is allowed to use this title. No one dreamt that centuries later German scholars would restyle him German emperor and the Americans Holy Roman emperor. Yet, as the popes dwelt mainly at Avignon, it had been more than 50 years since the last time a Roman bishop had set his foot in his cathedral and in his city. Were Clemens VI (Pierre Roger de Rosiers OSB, 1291-1352, pope 1342) still alive,

God only knows, he might have overcome all the innumerable obstacles
and crowned Carolus himself. After all, the Bohemian prince, baptized
Venceslav, became his pupil as a boy at Paris (1323-1331) and was later at
Avignon (1339) as a young man his choice for the Roman throne. The
king of the Franks, Charles IV le Bel (1294-1328, king 1322), not only
gave the beloved nephew of his queen and his godson at confirmation at
Saint Denis (1323) his own name. but also ordered the abbot Pierre to
teach him letters, though this handsome king himself was illiterate (FON-
TES RERUM BOHEMICARUM, Pragae, 3, 1882, VITA Caroli IV, 399: Dilexit me
praefectus rex valde et praecepit capellano meo, ut me aliquantulum in
litteris erudiret, quamvis rex praefatus ignarus esset littterarum). Both
the abbot and the prince greeted each other as the future pope and emperor
(VITA 362). Both Clemens and Carolus were builders, patronized the arts
and sciences, and protected Jews wrongly accused of spreading the plague.
But Clements was dead by now and his successor Innocentius VI (Étienne
Aubert, ca. 1300-1362, pope 1352), former law professor at Toulouse,
sent his prime cardinal, who succeeded him as bishop of Ostia and Vel-
letri, Petrus Maletonis Bertrandi de Columbario (ca. 1300-1361), to do
the honor. A German poet from Austria and herald in the service of the
Hohenzollerns and (later) Habsburgs, Peter (der) Suchenwirt (ca. 1325-ca.
1395), was present, but Bartolus apparently was not. Carolus again had
to leave his army outside, and, after knighting some Colonnas (who were
Ghibellins) in the Lateran archibasilica and banqueting in triclinium leoni-
anum, had to leave the city before midnight for San Lorenzo fuori le mura
where he next day knighted the Lateran count palatin Niccolo Orsini (a
Guelf). The new emperor was only too glad to have the first peaceful coronation in ages and most likely also to leave the city that gave its name
both to the Holy Roman empire (sacrum romanum imperium) and the
Holy Roman church (sancta romana ecclesia), but whose government was
a monstrosity (regimen monstruosum) to Bartolus and no doubt to Carolus too.

On their way back home by way of Pisa, both the *coronatus* and the
coronator made a stop at Perugia; so important was this republic to them.
The cardinal's chaplain did not take notice of Bartolus in his itinerary,
though he pays supreme encomium to the supposedly greatest poet of all
times on their visit to the Visconti (Johannis Porta de Annoniaco LIBER DE
CORONATIONE KAROLI IV IMPERATORIS, edidit Ricardus Salomon, Hanno-
verae et Lipsiae, Hahn 1913: Florentiae florem vero et toto in terrarum
orbe notabilem, imo verius unicum singularem poetam, quo nullus major

natus est umquam esse credatur Franciscum Petrarcam . . . per senatum et populum . . . laureatum in urbe, ubi solum poetas hujusmodi laureari fas est absque papae vel imperatoris praesentia). Petrarca no doubt supplied this information himself, as he fed them with the graces, prerogatives and conditions of Italy (de omnibus et singulis Italiae conditionibus, praerogativis et gratiis). The Perugines fêted the cardinal, in the city; but the emperor, who came first, was met outside the city walls, so afraid were the good Guelfs to let him in, even without his army. The fate of the Ghibelline cities that welcomed the king and made him their signore, only to go through a revolution, justified the precaution of the Perugines. Carolus, as much as he wanted, could not always have avoided some entanglement in the shifty and bloody city politics. Yet, on the whole, he managed to outwit them and contribute to the pacification of Italy, sometimes by legitimizing the illegitimate power, often by substituting legal solutions for military ones; he was always ready to issue golden bulls and grants of all kinds, thus making many Italians quite happy.

When the Perugines found out that he had something for them too, they sent Bartolus and four other ambassadors after him to Pisa to negotiate the details. The envoys were received on May 19. Next day the Pisans revolted. Bartolus witnessed the attempt by the Bergolini, who paid with their lives for it, to assassinate the emperor and take over the government. Carolus combined with the Raspanti and crushed the revolt. He also saw the emperor, after the Pentecost mass (May 24) sung in the duomo by the cardinal, crown with a laurel the Florentine poet Zanobi di Giovanni Mazzuoli da Strada OP (ca. 1310-1361) at the request of the Sicilian gran siniscalco and conte di Melfi, Niccolò from the Florentine family of Acciaiuoli (1310-1365). At Milano the jealous Petrarca took it rather badly and reacted with chagrin when Zenobius de Florentia (rather than he himself?) became the first poet (laureate) to become apostolic (papal) secretary. The next poet an emperor crowned was in 1449 his own secretary Aeneas Silvius Piccolomini (1405-1464) who became not pope's secretary but in 1458 pope himself as Pius II. From now on, not only the emperors but also the popes and some kings began to crown poets. In 1356 Petrarca was gradually pacified by a golden bull that created him count palatine. Bartolus never got such a title, though some writers credited him with it.

Carolus absolved Perugia from all past sins against the empire and confirmed it some privileges. The *studium* was confirmed in perpetuity, with the bishop of Perugia in charge and conferring the degrees. Carolus gave several *studia* such confirmations, but they all followed a papal bull creating such a *studium,* even in his own capital of Bohemia. The rectors of the Perugine university (=student union) should enjoy the same rights and privileges as elsewhere rectors do. The *scholares* (students) were exempt from reprisals and imposts.

Bartolus told us himself what he got: Carolus IV, the present emperor of the Romans, added him to the number of his counselors (rather than privy councillors) and domestics (members of his household) and decorated him with the privilege, extended to such of his descendants as happened to be law dactors, to legitimize bastards and make persons under age of age, and other privileges and graces (SUPER CONSTITUTIONE AD REPRIMENDUM, rubrica =OPERA 10, 1596, 94va: Caroli IV Romanorum imperatoris tunc regnantis . . . quia me suorum consiliariorum et domesticorum numero aggregavit et me meosque posteros, quod legum doctores esse continger-ent, legitimationis et concessionis veniae aetatis aliisque privilegiis et gratiis decoravit). Bartolus noticed that not only rulers and potentates but private persons as well, both nobles and commoners (shocking some later, even some recent writers who keep falsely insisting that arms are a sign of nobility) bear arms (heraldic arms that is), some of them granted by the emperor or some other lord (ruler). He saw the most serene prince, Roman emperor Carolus IV, king of Bohemia, grant such arms to many (!). To his counselor Bartolus and his male descendants he granted a red lion with two tails in the field of gold, or, to blazon it in heraldic English: or, a lion doublequeued rampant gules (DE INSIGNIIS ET ARMIS =OPERA 10, 124vb: Quaedam sunt insignia seu arma privatorum hominum seu nobili-um vel popularium. De istis quidam reperiuntur, qui habent arma et in-signia, quae portant ex concessione imperatoris vel alterius domini, ut vidi concedi multis a serenissimo principe Carolo IV, Romanorum imperatore necnon rege Bohemiae. Et mihi tunc consiliario ejus concessit inter caetera ut ego et caeteri de agnatione mea leonem rubrum cum caudis duabus in campo aureo portaremus). Bartolus answered questions he and others must have had about the meaning of this brand new custom of granting arms, of confirming new arms that is, and he was first to report it in his famous and later also unjustly maligned TRACTATUS DE INSIGNIIS ET ARMIS. This is not only the very first discussion or treatise on the subject of arms and flags, but also on hallmarks, watermarks, etc. Anyone can adopt such insignia and display them on a flag or shield, even without the benefit of a grant, in the same way as any one is free to adopt a family name without anyone's permission. Thus, the purpose of a grant is not to allow the use of something otherwise forbidden, but simply to register a licit and valid use, and thus to provide publicity, prestige, precedence and possibly even protection against contest. Per se, different people can adopt the same arms as they can adopt the same name for their families. There of course may be reasons for enjoining some people from doing so. A judge can forbid a duplicate use of arm or marks, thus copyright some arms (or marks), but his decision never can go against the arms (or marks) granted. i.e. recognized by a ruler.

When (as the carta records) in 1357 on July 13 Bartolus died, apparently suddenly, as he left this treatise and some other works unfinished, the task of publication was given to the husband of his daughter Paola, doctor Nicolaus Alexandri, who did publish it in January 1358. My guess is that this treatise— and we do not know whether he published it as Bartolus left it or added some lines at the end—was sent to the emperor either in Bohemia then or sometime thereafter or given to him in Italy when he came second time in 1368. My second guess is, that the emperor's eldest daughter, Anna (1366-1394), took a copy not only of a vernacular New Testament but also of the treatise of Bartolus with her to England, where she became in 1382 the beloved queen of Richard II (1367-1400), the second queen to be commonly known as the good queen (the first was also a Bohemian princess Margareta, 1186-1213, known as the good queen Dagmar of Denmark). In any case, about this time there developed in England the *curia militaris* (the court of chivalry), following not the common law of England, mute on the subject, but the common law of Europe, what is known as the civil law in England, i.e. Bartolus and his treatise. The good queen might have not inspired the translation of the Bible into English (maintained by some Czech writers, but in conversation doubted by Norman Davis of Oxford), but she did inspire the composition around 1395 of the second heraldic treatise by the mysterious Joannes de Bado Aureo. Only in 1943, did Evan John Jones print it with a Cymri text of the same treatise (and some other such works) identifying the author as legum doctor Ieuan ap Llywelyn (c1360-1410) (from) Trevor (in Powys Fadog), who was in 1389 involved in the *curia militaris* and who in 1389 and again in 1394 was elected a bishop of Saint Asaph, in 1399 deserted Richard II, in 1404 rebelled against Henry IV (1367-1413) and died in Paris (MEDIEVAL HERALDRY, Cardiff, Lewis, 95: Quoniam de armis multociens in clipeis depictis singula discernere inveniatur difficile, ad instantiam . . . specialiter dominae Annae quondam regnae Angliae hunc libellum compilavi.) He quotes all the time (and sometimes misquotes) dominus Bartolus in tractatu SUO DE ARMIS PINGEDIS [sic], mostly agreeing, sometimes disagreeing with him. Whereas to Bartolus arms, like names, serve to identify people, as a rule groups of people, and are thus collective symbols (Sicut enim nomina inventa sunt as congoscendum homines, ita etiam ista insignia seu arma ad hoc inventa sunt), Joannes distorts them into individual symbols, that a king, a prince, a king of arms or herald can grant, as Bartolus (supposedly) says (Quaero, quis potest dare arma? Dic, quod rex, princeps, rex armorum vel heraldus, ut dicit Bartolus. Dic ad cognoscendum unum hominem ab alio). Could this be the influence of Honoré Bonet? An excerpt of 1449, also included, returned to the Roman law and Bartolus (170: sicut nomina inventa sunt ad recognoscendum homines). To the present day arms are collective in Europe, but individual in the British isles, making the blessed cadency marks sprout all over them. Jones appended the treatise of Bartolus in the 75th (latest thus far) edition. Possibly already in 1842 and certainly in 1883, the most voluminous heraldic writer John Bernard Burke (1814-1892, Ulster king of arms 1853, knight 1854) blazoned Trane family arms *tempore* Henry IV (1399-1413) as: Or, a lion rampant doublequeued gules (THE GENERAL ARMORY OF ENGLAND, SCOTLAND, IRELAND AND WALES comprising registry of armorial bearings from the earliest to the present time, London, Harrison, 1025 a). No doubt, only a reader of Bartolus' treatise would have adopted for his own family arms identical with those of the family of Bartolus. Bartolus would not have objected, as he himself established the heraldic rule still observed today that the copyright can be enforced only within one country: A German went to Rome during the indulgence (anno santo 1350) where he finds an Italian bearing the same family arms. He could not complain, because the distance between their domiciles is such that no one could feel offended. (Unus Teutonicus tempore indulgentiae ivit Romam, ubi reperrit quendam Italicum portantem arma antiquorum suorum et insignia et de hoc volebat conqueri. Certe non poterat tanta enim distantia est inter utriusque domicilium quod et hoc ille primus laedi non possit).

The Trane family member either did not know or disregarded that Joannes disapproved of the arms Bartolus had received as it is against nature for an animal to have two tails (quia contra naturam est, it unum animal haberet duas caudas). In 1902 baronet George Reresby Sitwell (1860-1943) found this foul finding more cruel than the violent diatribe in the vituperative pamphlet of Valla (THE ANCESTOR 1, 58-103, THE ENGLISH GENTLEMAN, 84).

Writers not versed in heraldry had difficulties in blazoning (describing) the Bartolus' family arms correctly. In 1578 Bodin mentioned that the emperor ennobled Bartolus granting him a lion gules in the field of argent and also the power to octroy the benefice of age (grant fullness of age to persons not having it yet) . . . (138: qui annoblist Bartol, et lui donna le lyon de Guelles en champ d'argent, et puissance d'ottroyer benefice d'aage, pour lui et pour les siens qui feroient profession d'enseigner le droit. .) and that Bartolus in recognition for this favor declared heretics all those who would not believe that the emperor is lord of the world (*dominus mundi*). In 1913 Woolf (25, n. 2) quoted this passage trying to prove that the *mundus* (world) is not as absolute as Bodin would understand. In process he entirely missed that Bodin not only wrongly blazoned the arms granted to Bartolus but also wrongly assumed that this grant was a grant of nobility. Bartolus as all *legum doctores* was styled *dominus*. Whether we translate it sir (in Italian, ser) or lord, it implies a noble status. So do the titles the emperor gave him. Yet, Bartolus never styled himself *nobilis*. As matter of fact, the Roman law granted him after 20 years of legal profession an automatic title of a count (*comes*). Yet, just about this time Bartolus called himself *civis*. A noble citizen, a noble commoner!

The Bartolus family arms were, of course, adapted from and inspired by the royal arms of Bohemia: gules, a lion double-queued rampant argent, crowned or, but only inspired. Technically, the arms are different, and the arms of Bartolus is not an augmentation, as some earlier and many later arms are and any statement that the emperor granted Bartolus family his imperial or royal Bohemian arms is simply a misstatement.

The summer of 1355 Bartolus spent in a landhouse near Perugia on the river Tevere. Here he began to contemplate the winding and flow of the river, its changes, islands formed in it and got quite excited by its legal implications (Cum igitur a lectura vacarem et recreationis causa in quandam villam prope Perusium super Tiberim constitutam accederem, incepi Tiberis circuitus, alluviones, insulas in flumine natas mutationesque alvei contemplari et circa multa dubia, quae de factor occurerant at alia, quae ego ipse ex aspectu fluminis excitabam, quod juris esset.) In a dream he was encouraged to overcome his reluctance to use figures (diagrams) that were perhaps usual in a treatise on geometry but highly unusual in a legal text. Thus he used them in his famous TRACTATUS TIBERIADIS sive DE FLUMINIBUS. Dreams always played a role in human lives, but in his age they seem to have reached the peak of popularity. Granted, some must be dismissed as literary ploys. Yet, it would be foolish to dismiss them all as ploys. After all many people solved their personal and even their scientific problems in real dreams, not pretended dreams. I have no doubt that this is what happened to Bartolus. Once Doris Hellmann invited me to discuss this treatise before an international congress of the history of science. Unable to accept then, I am reserving it for a similar occasion in the near future. Here I must only say that this treatise became a textbook of geometry both in original and translation. In 1964 Guido Astuti introduced a facsimile of the 1576 edition by a Bolognese erudite Ercole Bottrigari (1531-1612), containing in the frontispiece a reproduction of the oldest known portrait of Bartolus, quite likely a *vera effigies* (Bononiae apud Roscium; Torino, Bottega d'Erasmo). In 1936 van de Kamp collected and reproduced practically all known images of Bartolus. It would perhaps have been wiser for the Columbia Law School to have commissioned a statue of Bartolus rather than that of Bellerofon taming Pegasos.

Following the flow of the river in his mind, Bartolus arrived at Todi, where he once was a judge. In a TRACTATUS DE GUELPHIS ET GEBELLINIS, first known discussion of political parties, he recommended a two-party system, practiced at Todi, but unknown elsewhere in Latin society. The Ghibellins and Guelfs, those old imperial and papal parties in the times of the investiture controversy, became just the Republicans and Democrats of trecento Italy. As Perugia only the parte guelfa was functioning, though some Guelfs wished to be Ghibellins for old times sake.

The Tevere finally flows through the *civitas romana*, the city that once went through all three (good) forms of government; ruled by the *populus* when small, by the few (*senatus*) when larger and at the end by one imperator when she became *caput mundi*. Now she lacks any government worthy of its name, capable of responsibility and ensuring stability and peace. Her "government" is a monster with several heads, all feeble (the barons or bosses of the regions unable to impose their rule), all violent (opposing anybody else's rule), fighting each other. Bartolus must have visited the eternal city, at least during the *anno santo* of 1350, if not oftener. He might have also met the Roman poet adventurer Cola di Rienzo (1313-1354), who once proclaimed himself tribune, later fled, summoned Carolus, went after him to Bohemia, was imprisoned, shipped to Avignon, freed by the pope (Petrarca had his fingers in it), sent to Perugia in 1353 and after months of waiting there back to his city as papal governor (styled senator), only to be murdered on Campidoglio soon after. He might have even discussed the Roman government with him. Adding this *regimen monstruosum* to the Aristotelian six, three good ones and three bad, he analyzed all seven in his TRACTATUS DE REGIMINE

CIVITATIS, giving them his own names to make it easier for lawyers to understand. To him the rule of the people, by the people and for the people (regimen populi, per populum, ad populum) is the overall best and most divine (God inspired and God pleasing) and especially so in small countries (to him civitas magna). Under circumstances a large country (civitas major) can do well under the rule of a good (rich) few and a very large country (civitas maxima) prospers under monarchy. If one has to have a monarchy, an elective monarchy is preferable to a hereditary (regimen quod est per electionem est magis divinum quam illud quod est per successionem) and the king should not be a foreigner (periculosum est habere regem alterium nationis). Here Bartolus anticipated some French ideas of a relative value of a form of government, such as in 1748 Charles-Louis de Secondat, baron de la Brède et de Montesquieu (1689-1755) proposed in his DE L'ESPRIT DES LOI and after him Jean-Jacques Rousseau (1712-1778) and Alexis-Charles-Henri-Maurice Clérel de Tocqueville (1805-1859). Bartolus was no doubt unaware of earlier statements in that direction by two Parisian doctors. Around 1303 Joannes Quidort parisiensis OP (ca. 1240-1306), a Parisian by birth as well, stated that various conditions produce various governments and what is good for one people is not necessarily good for another (DE POTESTATE REGIA ET PAPALI, ed Melchior Goldast, Francofurti 2, 1614, 112: quod est virtuosum in una gente, non est virtuosum in alia; this is not a plea for situation ethics). In 1324 Marsilius Mainardini (1270-1343), a Padovan by birth, Parisian rector in 1313, made a similar statement on relativity of forms in his famous DEFENSOR PACIS (edited repeatedly 1517-1933, English translation by Alan Gewirth, published by Columbia University Press 1956; dictio 1, cap. 9).

Both Marsilius and Bartolus were in fact democrats, though neither called himself so. The term democracy still had the Aristotelian stigma of a mob rule, what Bartolus called *regimen populi perversi* and we may sometime call ochlocracy; only since Rousseau the word has begun to receive an acceptable connotation. Bartolus groped for the Aristotelian *politeia*, though he comes out with *policratia*. But when he adopts the term, *regimen populi* (rule of the people) he no doubt unwittingly, uses the Latin translation of the Hellenic *demokratia*. Thus he is a democrat to us, but not of the same kind as Marsilius. Marsilius seems to allow the majority (pars principans) rule, without any safeguards for the minority. He was thus a totalitarian democrat, as was Rousseau after him and many others to the present day. Bartolus is rather of the sort of liberal democrat. The majority (multitudo) rules, but to be a good government it must rule for the benefit of all (bonum commune), not against any group, at the expense of any group or to the detriment of any group, be it the very poor, be it the very rich. On the other hand, the loafers and ne'er-do-wells (people not ready to participate) and some unruly magnates (nobles not willing to cooperate) may be excluded from an active participation (Dico per multitudinem intelligo exceptis vilissimis . . . item ab isto regimine possunt excludi aliqui magnates). Bartolus of course thinks in terms of a direct democracy, as existed at Perugia and still does in some Swiss cantons and American town meetings, not in terms of universal suffrage and elected representatives, so essential to a democracy today. Unlike Bartolus, most thinkers of his time were for monarchy. Bartolus examines the arguments for monarchy that after 1280, another Parisian doctor, Aegidius romanus (Giglio Colonna, 1247-1316, general of the Eremites of Saint Augustinus 1292, archbishop of Bourges 1295) developed in his popular DE REGIMINE PRINCIPUM, declaring the rule of the people good, the rule of the few better, and the rule of one best. Though he called him not only *in theologia magister* but also *magnus philosophus* (influenced perhaps by Cinus, who listened to him at Paris), he refuted his arguments in favor of the rule of the people, using the Bible for support. When the people of Jisrael demanded a king just like everybody else, God told the complaining Šemuel: They have not rejected you, but me. They do not want me to be their king (1 Sam. 8, 7). Thus the rule of the people means a free people under God, not a bunch of serfs, slaves or subjects. When he presented this exquisite argument at Pisa to the emperor, Carolus commended him highly (Carolus IV cum apud eum essem, maxime commendavit). After all what choice he had but to agree with God himself? Bartolus noted that at Pisa and Siena the arrival of Carolus (as signore) meant an arrival of the rule of the people. After what was said, it is curious that in 1945 he can be blamed for producing a class of learned lawyers devoted to an absolute monarchy and alienated from the legal life of the people (SCHWEIZER LEXIKON, Zürich, Encyclos, 1, 902: Er beherrschte in seinen Schriften trotz scholastischer Überspitzung die Kunst das für seine Zeit Praktische herauszuarbeiten. Damit erhielten seine Erläuterungen grössten Einfluss auf die Rezeption, ja auf die Anfänge der modernen Rechtswissenschaft überhaupt. Sie trugen aber auch bei zur Entfremdung eines vornehmen, absolut-monarchisch denkenden Juristenstandes gegenüber einem volksnahen Rechtsleben).

The three bad modes of rule can all be called tyranny. The tyrants not being exactly sharing types, the tyranny of the people and that of the few do not last long; they both slide soon into a tyranny of one. He ends with a final salvo: Today Italy is full of tyrants (Hodie Italia est tota plena tyrannis), echoing Dante: che le terre d'Italia tutte piene son di tiranni (PURGATORIO 6, 124).

Bartolus wrestled with the vexing problem of tyranny in another treatise, specially devoted to the subject, the first to be so. In his TRACTATUS TYRANNIDIS sive DE TYRANNIS he made his famous distinctions between veiled and manifest tyrants and among the manifest between an illegitimate usurper (ex defectu tituli) and the legitimate ruler who becomes a tyrant by abusing his power (ex parte exercitii). In 1925 Ephraim Emerton (1851-1935), Leipzig doctor, Unitarian minister and Harvard professor of divinity translated two of the Bartolus' treatises, both first in their fields, the one on tyranny and the one on the political parties in his HUMANISM AND TYRANNY, STUDIES IN ITALIAN TRECENTO (Harvard University Press; recently reprinted). Bartolus to him reflected the most pressing problems of trecento Italy, and one could get its complete history just by filling in the names and dates.

The only other published English translation of a Bartolus text was in 1914 by another Harvard professor, this time the Royall professor of law, Joseph Henry Beale (1890-1945) in his BARTOLUS ON THE CONFLICT OF LAWS (Harvard University Press), in turn translated into Portuguese. The professors of international private law never forgot their Bartolus. In 1834 the associate judge of the United States and first Harvard law professor Joseph Story (1779-1845) whose texts helped to form American jurisprudence started his CONFLICT OF LAWS with Bartolus. Thanks to him, the Library of Congress has an impressive collection of works by Bartolus and the Bartolists, a librarian once told me.

In 1894 the Swiss professor of international law Friedrich Meili (1848-1914) began a set of publications, pointing to Bartolus as the head of the first school of international private and criminal law. The role of Bartolus as the founder of the international public law was for long more controversial. For many lawyers, sometimes even today, international law starts with Hugo Grotius (Huigh de Groote, 1583-1645), for a few with Gentilis, and the concept of sovereignty of course with Bodin. Granted, the comity of nations was an idea of Grotius, and Bodin came out for the absolute sovereignty of a ruler. But if we heed the 1905 words of Figgis that Bodin, Gentilis, and Grotius not only quote Bartolus, but are what they are because of Bartolus, we realize that all the arguments against Bartolus and his school as too lost in Roman law and universality of the

Roman empire to be able to develop such modern concepts, may not be valid. In 1933 the Palermo law professor Placido Zancla (1902-p. 1969) presented convincingly the case for Bartolus as founder of international law,
public as well as private, in his LA DOTTRINA DELLA SOVRANITÀ DELLO STATO
E IL PROBLEMA DELL'AUTORITÀ INTERNAZIONALE IN BARTOLO (Palermo, Ci-
uni). There is no space here to document the whole controversy, but only to
sum up the case. Bartolus took the French formula that the king is emperor
in his kingdom, i.e. does not recognize any superior power and univer-
salized it: all kingdoms and republics claiming sovereignty and able to
defend it are free to handle their affairs, not subject to any empire, but
still bound to pursue the happiness and welfare of their people. This did
not abolish the empire entirely. The Roman empire remains a symbol of
the unity of the Roman people. By *populus romanus* Bartolus meant
what other writers would call *res publica christiana*, Western Christendom,
Western civilization, Latin society and so on. The empire remains also the
guardian of peace between states, the enemy of tyranny within and imperialism without. Thus Bartolus advocated the rule of law, where other
writers called for tyrannicide and other such remedies. Thus the Roman
empire, not the Roman church, has the role of a superstate organization.
Grotius, in whose time the Holy Roman empire was no more Holy, no more
Roman, no more an empire, suggested a comity of nations to regulate international relations. But when the states formed the League of Nations
and later the United Nations, they abandoned the be-nice-to-each-other of
Grotius and returned to Bartolus and his suprastate organization, re-
specting the independence of member states, but limiting it in cases of
tyranny and imperialism, which were to be dealt with, if possible, by rule
of law, by peaceful intervention rather than by killing, rebellion, war. How
modern!

It was repeatedly pointed out that Bartolus, quoting and slightly
stretching the text of the CODEX 1, 1, 1, distinguished between the personal
statute (actually status if not integrity) by which every person has the
right to be respected by others and the real statutes regulating his pos-
sessions. From there is a logical step to the distinction between the human
rights any human being carries with him wherever he goes and the civil
rights that link him to the community where he is a citizen. With Bartolus
began the line that goes toward a declaration of the rights to life, liberty
and pursuit of happiness and finally to the rights of confession, nation,
profession and region, the rights that is, to one's own religion, mother
tongue, choice of occupation and choice of place to live. I plan to elaborate

this statement further and hope to present it before the Columbia Collegium on Human Rights, now being formed.

True, not everything Bartolus said can possibly be valid today. Like
everyone else he was a child of his times. What is amazing is how many
of his ideas are seminal of later and even modern thought. Let me finish
with the words of my dear teacher Garrett Mattingly (1900-1962): "Bartolus of Sassoferrato who gave his name to the leading school, no more
thought of the law he taught at Perugia as something fixed since the age
of Justinian than he thought of the terse, serviceable Latin in which he
wrote and lectured as a language fixed in the age of Augustus. Though
he tried to connect his precepts with the great tradition of the Roman past,
he shared and encouraged the interest of his practical-minded students in
the actual law of their contemporary world. His followers, the Bartolists,
continued . . . his effort to assimilate into the civil law the teaching of the
Church and the customs of the Italian cities and of the transalpine peoples,
so as to provide a single rational doctrine for the legal relations of the
Western world" (RENAISSANCE DIPLOMACY, Boston, Houghton Mifflin,
1955, 21-22).

Another Columbia professor once told me: Translate Bartolus and political science will not be the same. No doubt, a better knowledge of Bartolus, his life and thought, will make the difference.

To achieve this better knowledge, I am working on a new life of Bartolus. After a critical history of the life and work of Bartolus, I would like to
see a popular life story. I also have in mind a set of critical editions and
translations of at least some of his works, starting with his tractatus. An
increased knowledge of the life and work would then lead to some works
of art, a play or even an opera, oratorio or cantata. Bartolus has his statues
in Italy, but he deserves one in front of the United Nations, and another
perhaps on this campus of Columbia. Can you imagine statues of those
figure umanamente più elevate, Dante, Giotto, ser Bartolo and santa Catarina, looking down on Amsterdam avenue and Casa italiana with the Law
school in the background? Dreams? Perhaps, but I hope fruitful dreams,
pregnant with future reality.

ACKNOWLEDGMENTS

I wish to dedicate this little contribution to the memory of my Columbia
teacher Garrett Mattingly and to the memory of Francesco Calasso whom
I met during a historical congress at Stockholm in 1960. Professor Calasso

sent me his MEDIO EVO DEL DIRITTO, and this work was an eye opener to me.

◆◎ ◎◆

Professor Paul Oskar Kristeller very kindly took time to read my paper and offer some very helpful suggestions.

Also thanks to Gabriella Oldham for her careful and patient typing and retyping of the manuscript.

Last but not least the editors at the Academy, Bill Boland and Joyce Hitchcock were very helpful and understanding with what to them must have been my foibles. One day I will issue my own manual of style, presenting a unified field theory of bibliographing. Even a casual reader will notice that I capitalize titles and avoid footnotes: notes first at the foot of a page, later of a chapter and now at the end of a book. In this I follow Bartolus.

[NOTE ADDED IN PROOF: Belatedly, I acquired the information illuminating the controversy in Pavia in 1433. Valla got the treatise of Bartolus from a law professor, Catone Sacco, and to him he originally addressed the letter berating Bartolus. It created such a controversy that both Valla and Sacco lost their jobs. This all happened between the 4th and the 18th of March. Sacco later returned to his job, and my guess is that to facilitate that Valla readdressed the letter to Candido Decembrio (Girolamo Mancini's VITA DI LORENZO VALLA, 1891).]